In Honor of
Philip M. Morse

Editors Herman Feshbach and K. Uno Ingard

In Honor of Philip M. Morse

THE M.I.T. PRESS
Cambridge, Massachusetts, and London, England

Set in Monotype Times New Roman.

ISBN 978-0-262-06028-8 (hc : alk. paper)
978-0-262-56159-4 (pb : alk. paper)

Library of Congress catalog card number: 77-87297

Contents

Preface

Ours frankly is a sentimental project. Those of us who have been closely associated with Phil want to take this opportunity to express in a public fashion on the occasion of his sixty-fifth birthday our deep affection and high regard for him. Some of us have been Phil's students; some have collaborated with him; some have been his colleagues on the M.I.T. faculty. But all of us feel a debt of gratitude to the good fortune which made us one of his circle of friends. We all have responded to Phil's extraordinarily wide range of interests and his deep commitment to the understanding and improvement of the human condition.

The papers presented in this volume form an image of Phil's own wide ranging interests. There are papers on astrophysics, on general relativity, on atomic physics and the solid state, on nuclear and elementary particle physics, on accoustics, on operational research and on mathematical methods. Unfortunately it was not possible for many of Phil's friends to contribute to this volume. We know that they share our feelings.

This preface would be incomplete if we did not also express our thanks to Annabelle Morse for her warm hospitality and understanding which we have all enjoyed on so many occasions.

HERMAN FESHBACH
UNO INGARD

Cambridge, Mass.
April 1969

Foreword

Philip McCord Morse retires in June 1969 as Professor of Physics at M.I.T. It is conceivable that he may then take advantage of the privilege of an emeritus to relinquish a few of his many commitments and responsibilities. But those of us who know him well suspect that there will be hardly a perceptible change in the pace of his work, his teaching, and his writing.

I first met Phil and Annabelle Morse early in the summer of 1931. They and my friend Will Allis had spent the winter in Munich, and before returning home, we joined forces in Cambridge, England, to study a little, to talk a lot, and to roam on our bicycles about the lovely English countryside. Together we found lodgings in a pleasant boarding house on Brookside. The cuisine was deplorable, highlighted regularly, I recall, by a disaster in pastry which we described as a "Burberry pie." But the company was excellent. That was an exciting time for physics, and in the presence of such men as Rutherford and Dirac, of Fowler and Kapitza, of Blackett and Mott, even a very brief visit to Cambridge was an exhilarating experience. In the contemporary idiom, it was for me an experience of learning by a kind of total immersion, with Phil Morse and Will Allis, fresh from their stay with Sommerfeld, guiding me along the way. And happily, too, this was England, for at tea time we put aside quantum theory to talk of other things. Whatever the subject of conversation, Phil would display, as he does to this day, the amazing span of his reading, his knowledge of history and literature, his familiarity with art, his love of music and the theater. We have heard a good deal lately about science and the humanities — about the ideal schooling for the truly educated man. But in reality how much more goes into the making of that truly educated man than simply a course in college! — a quality of mind, an insatiable curiosity about man and his world, and the intellectual energy and resolve to pursue the quest despite every obstacle and diversion.

From time to time that summer we spoke, too, about M.I.T., for the Morses were to come here in the fall. What sort of place was it, really? With Karl Compton as the new president and John Slater as head of the Department of Physics, would pure science indeed be taken seriously? And how was this to be brought about?

The special strengths and biases of an institute of technology would be no new experience to Phil. As an undergraduate at the Case School of Applied Science he had been one among only three in his class majoring in physics. Resources were limited, but in Professor Dayton Miller he found at Case a great teacher whose personal attention and generous help played a major part in the launching of his career. He owes, I believe, to Dayton Miller his continuing interest in acoustics. And it was Professor Miller who recommended him as his most brilliant student in twenty years to Dr. Compton — then at Princeton — for a graduate fellowship.

At Princeton his accomplishments were prodigious. With Karl Compton he published a paper on the theory of glow discharges. Then followed two papers in the *Physical Review* on diatomic molecules and the publication, with E. U. Condon, of a basic text on quantum mechanics — all over a three-year period on the way to his doctorate in 1929. Next came a brief stay with Davison at the Bell Telephone Laboratories, where he delved into the theory of electron motions in crystals, and in 1930 an International Research Fellowship provided the opportunity for his year abroad.

These, then, were the early achievements, the manifest ability, and the experience that Compton and Slater clearly recognized and that Philip Morse brought with him to M.I.T. He came at a time when the Department of Physics was deep in the throes of reformation, and his contributions to the building of an essentially new department marked by the highest standards of teaching and scholarship were enormous.

Looking back over the years, it is fascinating to observe how consistently Phil has remained true to the principal themes of his early work and how logically they have led him into diverse other fields. The three main strands, of course, were acoustics, quantum theory, and the reduction of theory to numerically usable results. The first course listed under his name at the Institute was "Sound, Speech, and Audition." The following year, 1932, he offered "Quantum Mechanics," the "Theory of Spectra," and a "Theoretical Seminar." To these he then added "Boundary Value Problems in Theoretical Physics."

Out of his teaching of acoustics came his text on *Vibration and Sound*, published in 1936, and recognition as an authority in the field. With the onset of the war he was inevitably drawn in, first as Chairman of a National Research Council Committee on Sound Control and then as director of a project concerned with the exceedingly critical problems of submarine detection and the design of counter measures to acoustical mines and torpedoes. As our losses from submarine warfare grew increasingly serious, Morse was one of the first to recognize the potential of a new method of operations analysis developed in Britain by A. V. Hill, Blackett, Tizard, and others. In 1942 he was instrumental in the formation of the Operations Research Group (O.R.G.) attached to the Chief of Naval Operations. Under Phil's direction the O.R.G. made a profound impression upon our naval high command, and its influence upon a widening range of operational problems steadily expanded throughout the remainder of the war. Indeed, historically we must look upon this original effort as a milestone in the development of mathematical analysis and logic for the design of operational strategies, not only in the military domain but equally in the management of industry, of government, of our cities — in short, an approach to the bafflingly complex problems of contemporary society. Over the past twenty years, as we all know, operations research has developed into a major discipline of its own. And in this development, both at home and abroad, Phil Morse has been a prime mover and a pre-eminent contributor.

From his Princeton days onward Phil has followed closely the progress of virtually every branch of theoretical physics and contributed to many, as his published books and papers amply reveal. But without any doubt, the most striking characteristic of all his scientific work has been an unwillingness to allow theory to remain in a vaguely suggestive form and an ability to press solutions to the point of numerically quantitative results. He has a most powerful command of the mathematical methods of theoretical physics — a subject which he has developed and taught at M.I.T. almost from the day of his arrival. But his feeling for physical reality, for the importance of usable results, inevitably led him to devote much time and thought to the task of practical computation. During the depression years of the mid-thirties, with the help of W.P.A. stipends, he planned and carried through the preparation of numerical tables of various functions commonly occurring in problems of scattering and diffraction. No one who had observed at firsthand the appalling labor and the monotony of such calculations could fail to be looking for a better way, and very shortly M.I.T. proved to be precisely the place to find it. The analog computer, developed by Vannevar Bush and his students, was rapidly becoming a most useful device for the numerical solution of all manner of physical problems. And then, out of work begun during the war, Jay Forrester took the basic steps which made possible the modern high-speed digital computer. As one might readily have anticipated, Phil Morse entered into this new field with his customary vigor. It related directly to so many problems that had occupied his attention in earlier years, and he followed closely the development of the technology as well as the theory of the new generation of electronic computers. By the middle of the nineteen fifties, many were beginning to recognize the tremendous import of this miraculous device not for science alone but as an indispensable aid to teaching and research in every field of study. We began to move from the stage of computer development to that of computer service and to grapple with the problems of assuring access to such service for the university community in its entirety. In 1955 the decision was made at M.I.T. to consolidate our institutional resources and to create a major Computation Center. It was only natural that we should then turn to Phil Morse, with his commanding knowledge of the use of mathematics in science, to plan and organize that Center and to set it on its course.

* * *

In this brief and personal foreword I have endeavored to touch only upon a few highlights in the brilliant career of a friend and colleague of nearly forty years. During that time we have counted Phil Morse among the most creative, the most stimulating, and the most distinguished members of our faculty. His interests have ranged far and wide, and his contributions to research have won him recognition throughout the world of science. Yet his scientific work over the years tells only a part of the story. The rest lies in the quiet, constructive force he has been for all of us who have enjoyed the good fortune to work with him — guidance and inspiration to his students, sound advice and support to his colleagues.

The pages which follow provide a penetrating insight into the character of his own achievements and a measure also of his profound influence upon his students, who in their turn have become leaders in science and engineering. I am delighted to join with these former students, colleagues, and friends who on the occasion of his retirement have prepared this volume as a tribute to Philip Morse — in appreciation and as an expression of good wishes for the many years to come.

J. A. Stratton
President Emeritus
Massachusetts Institute of Technology

Cambridge, Massachusetts
October, 1968

List of Publications

Published Articles

"A Study of Solar Motion by Harmonic Analysis," (with J. J. Nassau), *Astrophys. J.* **65**, 73–85, No. 2, March 1927.

"Theory of Normal Cathode Fall in Glow Discharges," (with K. T. Compton), *Phys. Rev.* **30**, 305–317, No. 3, September 1927.

"Ionization in Positive Ion Sheaths," (with W. Uyterhoeven), *Phys. Rev.* **31**, 827–832, No. 5, May 1928.

"A Theory of the Electric Discharge Through Gases," *Phys. Rev.* **31**, 1003–1017, No. 6, June 1928.

"Space Charge Sheaths in Positive Ray Analysis," (with R. W. Gurney), *Phys. Rev.* **33**, 789–799, No. 5, May 1929.

"Diatomic Molecules According to the Wave Mechanics I: Electronic Levels of the Hydrogen Molecular Ion," (with E. C. G. Stueckelberg), *Phys. Rev.* **33**, 932–947, No. 6, June 1929.

"Diatomic Molecules According to the Wave Mechanics II: Vibrational Levels," *Phys. Rev.* **34**, 57–64, No. 1, July 1929.

"Storungsrechnung des Wasserstoffmolkulions und des Wasserstoffmolekuls," (with E. C. G. Stueckelberg), *Helv. Phys. Acta* **2**, 304–306, No. 5, 1929.

"Recombination of Electron and Alpha-Particle," (with E. C. G. Stueckelberg), *Phys. Rev.* **35**, 116–117, No. 1, January 1, 1930.

"Quantum Mechanics of Electrons in Crystals," *Phys. Rev.* **35**, 1310–1324, No. 11, June 1930.

"Computation of the Effective Cross Section for the Recombination of Electrons with Hydrogen Ions," (with E. C. G. Stueckelberg), *Phys. Rev.* **36**, 16–23, No. 1, July 1930.

"Strahlungslose Stossprozesse bei kleinen Geschwindigkeiten," (with E. C. G. Stueckelberg.), *Ann. Physik* **9**, 579–606, No. 5, 1931.

"Die spezifische Warme von quasifreien Elektronen," (with E. C. G. Stueckelberg), *Z. Physik* **69**, 666–677, Nos. 9 and 10, 1931.

"Theorie der Streuung Langsamer Elektronen an Atomen," (with W. P. Allis), *Z. Physik* **70**, 567–582, Nos. 9–10, 1931.

"Unelastische Stösse zwischen Molekülen," (with E. C. G. Stueckelberg), *Helv. Phys. Acta* **4**, 136–137, Nos. 3–4, 1931.

"Lösung des Eigenwertproblems eines Potentialfeldes mit zwei Minima," (with E. C. G. Stueckelberg), *Helv. Phys. Acta* **4**, 337–354, No. 5, 1931.

"Unelastische Streuung von Kathodenstrahlen," *Z. Physik* **33**, 443–445, No. 2, January 1932.

"Quantum Mechanics of Collision Processes," *Rev. Mod. Phys.* **4**, 577–634, No. 3, July 1932.

"On the Vibrations of Polyatomic Molecules," (with N. Rosen), *Phys. Rev* **42**, 210–217, No. 2, October 1932.

"Variable Scale Atomic Wave Functions," (with J. P. Vinti), *Phys. Rev.* **43**, 337–340, March 1, 1933.

"The Effect of Exchange on the Scattering of Slow Electrons from Atoms," (with W. P. Allis), *Phys. Rev.* **44**, 269–276, August 15, 1933.

"Electrons, Photons and Waves," *School Sci. Math.* **34**, 200–206, No. 2, February 1934.

"Addition Formulae for Spheroidal Functions," *Proc. Natl. Acad. Sci. U. S.* **21**, 56–62, No. 1, January 1935.

"Velocity Distributions for Elastically Colliding Electrons," (with W. P. Allis and E. S. Lamar), *Phys. Rev.* **48**, 412–419, September 1, 1935.

"Tables for Determining Atomic Wave Functions and Energies," (with L. A. Young and Eva S. Haurwitz), *Phys. Rev.* **48**, 948–954, December 1935.

"Collision of Neutron and Proton," (with J. B. Fisk and L. I. Schiff), *Phys. Rev.* **50**, 748–754, October 15, 1936.

"The Elastic Scattering of Neutrons by Protons," (with J. B. Fisk), *Phys. Rev.* **51**, 54–55, No. 1, January 1, 1937.

"Collision of Neutron and Proton II," (with J. B. Fisk and L. I. Schiff), *Phys. Rev.* **51**, 706–710, May 1, 1937.

"Acoustic Vibrations and Internal Combustion Engine Performance I. Standing Waves in the Intake Pipe System," (with R. H. Boden and Harry Schecter), *J. Appl. Phys.* **9**, 16–23, No. 1, January 1938.

"Acoustical Analysis of the Pressure Waves Accompanying Detonation in the Internal-Combustion Engine," (with Charles S. Draper), *Proc. Fifth Intl. Cong. Appl. Mech.*, 727–732, 1938.

"The Diffraction of Waves by Ribbons and by Slits," (with Pearl J. Rubenstein), *Phys. Rev.* **54**, 895–898, December 1, 1938.

"Some Aspects of the Theory of Room Acoustics," *J. Acoust. Soc. Am.* **11**, 56–66, July 1939.

"The Transmission of Sound Inside Pipes," *J. Acoust. Soc. Am.* **11**, 205–210, October 1939.

"The Opacity of Gas Mixtures in Stellar Interiors," *Astrophys. J.* **92**, 27–49, No. 1, July 1940.

"Acoustic Impedance and Sound Absorption," (with Richard H. Bolt and Richard L. Brown), *J. Acoust. Soc. Am.* **12**, 217–227, No. 2, October 1940.

"Sound Waves in Rooms," (with Richard H. Bolt), *Rev. Mod. Phys.* **16**, 69–150, No. 2, April 1944.

"Of Men and Machines," *Technology Review* **49**, 29–31, No. 1, November 1946.

"Mathematical Problems in Operations Research," *Am. Math. Soc.* **54**, 602–621, No. 7, July 1948.

"Pure and Applied Research," *Am. Scientist* **38**, 253–259, No. 2, Spring Issue 1950.

"The Thermal Rayleigh Disk in Liquid He II," (with John R. Pellam), *Phys. Rev.* **78**, 474–475, No. 4, May 15, 1950.

"Physics and Radiation," *Am. Scientist* **38**, No. 3, July 1950.

"Must We Always Be Gadgeteers?," *Phys. Today* **3**, 4–5, No. 12, December 1950.

"Operations Research," *Technology Review* **53**, 1–6, No. 4, February 1951.

"Operations Research, What Is It?," *J. Appl. Phys.* **23**, 165–172, No. 2, February 1952.

"Universities or Project Centers?," *Phys. Today* **5**, 4–5, No. 4, April 1952.

"Excitation of Molecular Rotation-Vibrations by Electron Impact," *Phys. Rev.* **90**, 51–55, No. 1, April 1, 1953.

"Operations Research—An Application of Scientific Method," *Technology Review* **55**, 2–8, No. 7, May 1953.

"Trends in Operations Research," *J. Opns. Res. Soc. Am.* **1**, 159–165, No. 4, August 1953.

"Report on the First Summer Program on Operations Research at the Massachusetts Institute of Technology, June 16-July 2, 1953," *J. Opns. Res. Soc. Am.* **1**, 303–305, No. 5, November 1953.

"Operations Research," *Mech. Eng.*, 231–235, March 1954.

"Operations Research," *Appl. Mech. Rev.*, 89–93, March 1954.

"A Family of Queuing Problems," (with H. N. Garber and M. L. Ernst). *J. Opns. Res. Soc. Am.* **2**, 444–445, No. 4, September 1954.

"Operations Research: Past, Present and Future," *Advanced Management* **19**, 10–15, No. 11, November 1954.

"Operations Research," *Commun. Pure Appl. Math.* **8**, 1-12, No. 1, February 1955.

"Acoustics and Basic Physics," *J. Acoust. Soc. Am.* **27**, 213–216, No. 2, March 1955.

"Stochastic Properties of Waiting Lines," *J. Opns. Res. Soc. Am.* **3**, 255–261, No. 3, August 1955.

"Where Is The New Blood?," *J. Opns. Res. Soc. Am.* **3**, 383–387, No. 4, November 1955.

"Statistics and Operations Research," *J. Opns. Res. Soc. Am.* **4**, 2–18, No. 1, February 1956.

"Attendance and Use of the Science Library at M. I. T.," (with G. C. Bush and H. P. Galliher), *Am. Documentation* **7**, 87–109, No. 2, 1956.

"Waves in a Lattice of Spherical Scatters," *Proc. Natl. Acad. Sci.* **42**, 276–286, No. 5, May 1956.

"On the Use of Digital Computers," *Phys. Today* **9**, 19–23, No. 10, October 1956.

"Uber den Feldbegriff in den Theoretischen Physik," (with H. Feshbach), *Physikalische Blatter* **12**, 439–441, No. 10, 1956.

"Training in Operations Research at the Massachusetts Institute of Technology," *Opns. Res.* **4**, 733–735, No. 6, December 1956.

"Operations Research Is Also Research," *Proc. First Intl. Conf. O/R*, Oxford, September 1957.

"Teaching Machines to Reckon," *Technology Review* **60**, 1–6, No. 6, April 1958.

"Solutions of a Class of Discrete-Time Inventory Problems," *Opns. Res.* **7**, 67–78, No. 1, January-February 1959.

"Dynamics of Two Classes of Continuous Review Inventory Systems," (with H. P. Galliher and M. Simond), *Opns. Res.* **7**, 362–384, No. 3, June 1959.

"MIT Physics Graduate Alumni, Graduate Record vs. Achievements," (with G. F. Koster), *Phys. Today* **14**, 20–29, No. 8, August 1961.

"Report on International OR Activities," *Opns. Res.* **9**, 910–912, No. 6, November–December 1961.

"Concerning Bottlenecks," (review), *Science* **137**, 742–743, No. 3532, September 7, 1962.

"The Prospects for Mechanization," *College and Research Libraries* **25**, 115–119, No. 2, March 1964.

"Computers & e.d.p.," *Ind. Res.* **6**, 62–70, No. 6, 1964.

"Computing Machines as Research Assistants," *Proc. Am. Phil. Soc.* **108**, 291–297, No. 4, August 1964.

"Acoustic Scattering from an Inhomogeneity of the Medium," *Helv. Phys. Acta* **38**, 53–55, No. 1, 1965.

"Transmission of Sound Through a Circular Membrane in a Plane Wall," *J. Acoust. Soc. Am.* **40**, 354–366, No. 2, August 1966.

"Lines at the Airport," (review), *Science* **155**, 993, No. 3765, February 24, 1967.

"George E. Kimball," *Opns. Res.* **16**, 871–874, No. 4, July–August, 1968.

"A Probabilistic Model for Obsolescence," (with Caroline Elston), *Opns. Res.* **17**, 36–47, No. 1, January–February 1969.

Contributed Chapters

"Operations Research," chapter in *Frontiers of Numerical Mathematics*, ed. by R. E. Langer, The University of Wisconsin Press, Madison, Wisconsin, 1960.

"Linear Acoustic Theory," (with K. Uno Ingard), chapter in *Handbuch der Physik*, Vol. **XI**/1, ed. by S. Flugge, Springer-Verlag, Berlin, 1961.

"Dynamics of Operational Systems: Markov and Queuing Processes," chapter in *Progress in Operations Research*, Vol. **I**, ed. by R. Ackoff, John Wiley & Sons, Inc., New York, 1961.

"Computers and Operations Research," chapter in *Applications of Digital Computers*, ed. by W. E. Freiberger and W. Prager, Ginn & Co., Boston, 1963.

"Queues and Markov Processes—The Response of Operating Systems to Fluctuating Demand and Supply," chapter in *System Engineering Handbook*, ed. by Robert E. Machol, McGraw-Hill Book Co., New York, 1965.

"On the Prediction of Library Use," (short form) Appendix N in "*INTREX, Report of a Planning Conference on Information Transfer Experiments*," ed. by Carl F. J. Overhage and R. Joyce Harman, The M. I. T. Press, 1965.

"Design for a Brain," chapter in *Science and the Modern World*, ed. by J. Steinhardt, Plenum Press, New York, 1966.

"John Clarke Slater, a Biographical Note of Appreciation," introduction to *Quantum Theory of Atoms, Molecules, and the Solid State*, ed. by P. Lowdin, Academic Press, 1966.

Books

Quantum Mechanics, (with Edward U. Condon), McGraw-Hill Book Co., New York, 1929.

Vibration and Sound, McGraw-Hill Book Co., New York, 1936; second ed. 1948.

Elliptic Cylinder and Spheroidal Wave Functions, (with J. A. Stratton, L. J. Chu, R. A. Hunter), The Technology Press of M. I. T. and John Wiley and Sons, Inc., New York, 1941.

Tables for the Variational Determination of Atomic Wave Functions, (with H. Yilmaz), The Technology Press, 1956.

Methods of Operations Research, (with George E. Kimball), The Technology Press of M. I. T. and John Wiley and Sons, Inc., New York, first ed. revised 1951; Japanese ed. 1958.

Methods of Theoretical Physics, (with Herman Feshbach), Parts I and II, McGraw-Hill Book Co., New York, 1953; Russian ed. 1958.

Queues, Inventories and Maintenance, John Wiley & Sons, Inc., New York, 1958; French ed. 1960.

Notes on Operations Research, (Editor) The Technology Press, Cambridge, Mass., 1959.

Thermal Physics, W. A. Benjamin, New York, 1962.

Operations Research for Public Systems, (Editor with Laura W. Bacon), The M. I. T. Press, Cambridge, Mass., 1967.

Theoretical Acoustics, (with K. Uno Ingard), McGraw-Hill Book Co., Inc., New York, 1968.

Library Effectiveness: *A Systems Approach*, The M. I. T. Press, Cambridge, Mass., 1968.

In Honor of
Philip M. Morse

J. C. Slater
Massachusetts Institute of Technology

1 Morse's Contributions to Atomic, Molecular, and Solid-State Physics

Philip Morse has made so many contributions to so many fields — mathematical physics, acoustics, operations research, computer technology, science administration, to name a few — that one may lose sight of the fundamental nature of his work in the field of the quantum theory of matter. In this note I should like to call attention to some of his papers, and to the main ideas which have underlain his work.

First of course was his collaboration on the Condon and Morse text on Quantum Mechanics, one of the very first texts in English on the subject and written in 1929 when he was still a very young man at Princeton. This established him at once as an authority in this new field, and his first applications were to the theory of diatomic molecules. His joint paper with Stueckelberg, in 1929, was one of the first successful attempts to apply Schrödinger's equation to the discussion of the electronic motion of the hydrogen molecular ion H_2^+, which has proved to be the problem more than any other that has pointed the way to an understanding of the nature of electronic energy levels of diatomic molecules. This was carried out in the spirit of the Born-Oppenheimer approximation, which divides the discussion of a molecular problem into two parts, one the study of the electronic motion and the resulting energy levels as functions of the internuclear separation, the other the motion of the nuclei under the influence of the resulting potential.

The second paper in the series on diatomic molecules, by Morse alone, and also in 1929, dealt with the other half of the problem, the vibrational levels. It was in this paper that the famous Morse curve was introduced, a simple analytic approximation to the energy of a diatomic molecule as function of internuclear distance, which allowed an exact analytic solution of the vibrational problem. The importance that this paper has had in the development of molecular theory can hardly be overestimated. Morse in this pioneering paper not only carried through the whole mathematical discussion of the problem but he correlated the experimental data then available, and gave the constants describing the potential curves for the molecular states of diatomic molecules for which numerical values existed.

This book, and two fundamental papers, in 1929 would have been enough to satisfy most new Ph.D.'s for several years, but with Morse it was different. He spent a summer working with Davisson at the Bell Telephone Laboratories and became interested in the theory of low-energy electron diffraction by crystals. He realized that neither the tight-binding nor the almost free-electron approximation for electrons in crystals was satisfactory and in 1930 he wrote a paper on the quantum mechanics of electrons in crystals, which formed one of the first proper treatments of the motion of electrons in a periodic potential. He analyzed the potential into a three-dimensional Fourier series, as Bethe had done shortly before, but went much farther than Bethe in analyzing the problem mathematically, showing its relation to Hill's equation, and working out in detail the soluble problem of the three-dimensional sinusoidal potential, which results in Mathieu's equation. This formed one of the very early contributions to the study of energy bands in crystals.

Next he went to Germany to work with Sommerfeld, who suggested that he work on the theory of the Ramsauer effect, the experimental phenomenon which had been discovered some years before, according to which the collision cross section of some atoms, such as the rare gases, for electrons, became very small for very slow electrons. Morse, and Allis, who was also working with Sommerfeld, realized the relation of this problem to the diffraction of light by spherical objects, which had been treated by Mie and Debye. They formulated the problem by wave mechanics, much along the lines that had been used shortly before by Faxen and Holtsmark, but realized that to treat the problem realistically, they would have to approximate the self-consistent potential fairly accurately and carry out the problem by detailed analytic study of the actual cases. In their joint paper in the *Zeitschrift für Physik* in 1931, and in Morse's later review article in 1932, the whole theory of scattering of electrons by atoms is treated in a realistic manner, adapted to slow electrons rather than using the Born approximation which is only suitable for fast electrons. This work forms a real landmark in scattering theory.

Here within a four-year period were papers on diatomic molecules, on the motion of electrons in crystals, and on the scattering of electrons by atoms, all of fundamental importance, and at the same time all showing some of the characteristics which have been striking in all of Morse's work. In all of them, he showed his mastery of mathematical methods — Laguerre functions, Hill's equation, Mathieu functions, and so on. But at the same time he showed his desire to carry out his calculations on the actual case of physical importance, getting physically applicable results, rather than to leave results in a vague state of mathematical generalities.

These same characteristics were shown in some of Morse's next pieces of work, on atomic wave functions. In collaboration with Vinti in 1933, he explored analytic approximations to atomic wave functions, and in the work with Young and Haurwitz in 1935, the authors set up an analytic approximation to atomic wave functions that had been shown to be useful and proceeded to determine the parameters in these wave functions by variational methods. so as to lead to numerically accurate and practically usable functions. This same type of work was carried further in Morse's work with Yilmaz, after the war in 1956. These calculations were the precursors of the more recent and more accurate analytical wave functions that have been worked out since 1956 by Roothaan, Clementi, and others. This work of Morse and Yilmaz, as well as the more recent work, was made possible by the use of the digital computer, and of course it was partly Morse's interest in computers that led him to continue with this atomic problem.

The last paper that I shall comment on was also written in 1956, dealing with waves in a lattice of spherical scatterers. This very interesting piece of work touched on many of Morse's areas of concern. It deals directly with the motion of electrons in crystals, and in particular with the approximation in which the potential is assumed to be spherically symmetrical in spheres representing the various atoms and constant between the spheres. The present author had proposed one method of solving this problem in 1937, a method which we now know as the augmented plane wave method. Other methods of handling the problem had been suggested in 1947 by Korringa and in 1954 by Kohn and Rostoker. Morse's work, however, was largely independent of this other work, and in fact he became aware of it only after he had arrived himself at a good many of the results of his 1956 paper.

He realized that one could build up the wave function of a scattered electron in such a potential as a superposition of scattering problems such as he and Allis had treated in 1931 and 1932. He realized also that the problem was mathematically equivalent to one which he was meeting in acoustics, the scattering of sound waves by a collection of regularly spaced spherical scatterers. All of these facts, plus his thorough knowledge of Green's functions, and his recollection of his 1930 work on the quantum mechanics of electrons in crystals, combined to suggest the method which he did not actually publish until 1956, but which he had been turning over in his mind for a considerable period before. This paper has now taken its place as one of the important pieces of work which has led to our present understanding of methods of finding energy bands in crystals.

Here, then, we see contributions which Morse has made over a period of nearly 30 years to many different aspects of the theory of atoms, molecules, and solids, contributions which would have given him a secure place in the history of mathematical physics even if it had not been for his many other contributions in different, though related, fields.

General References

E. U. Condon and P. M. Morse, *Quantum Mechanics*, McGraw-Hill Book Co., Inc., New York (1929).

P. M. Morse and E. G. Stueckelberg, "Diatomic Molecules According to the Wave Mechanics. I. Electronic Levels of the Hydrogen Molecule Ion," *Phys. Rev.*, **33**, 932 (1929).

P. M. Morse, "Diatomic Molecules According to the Wave Mechanics. II. Vibrational Levels," *Phys. Rev.* **34**, 57 (1929).

P. M. Morse, "Quantum Mechanics of Electrons in Crystals," *Phys. Rev.*, **35**, 1310 (1930).

W. P. Allis and P. M. Morse, "Theorie der Streuung langsamer Elektronen an Atomen," *Z. Physik*, **70**, 567 (1931).

P. M. Morse, "Quantum Mechanics of Collision Processes. II.", *Rev. Mod. Phys.*, **4**, 577 (1932).

J. P. Vinti and P. M. Morse, "Variable Scale Atomic Wave Functions," *Phys. Rev.*, **43**, 337 (1933).

P. M. Morse, L. A. Young, and E. S. Haurwitz, "Tables for Determining Atomic Wave Functions and Energies," *Phys. Rev.*, **48**, 948 (1935).

P. M. Morse and H. Yilmaz, *Tables for the Variational Determination of Atomic Wave Functions*, Technology Press, Cambridge, Mass., 1956.

P. M. Morse, "Waves in a Lattice of Spherical Scatterers," *Proc. Nat. Acad. Sci.*, **42**, 276 (1956).

E. U. Condon and H. Odabasi
*Joint Institute for Laboratory Astrophysics**

2 Self-Consistent Field Calculations of the $1s^2 2s^2 2p3p$ Configuration in the Carbon Isoelectronic Sequence

It is a pleasure to dedicate this paper to Professor P. M. Morse, recalling a forty-year association with him of the senior author, which started when we worked together in Palmer Physical Laboratory of Princeton University in the fall of 1928

We have recently completed some self-consistent field calculations of several low configurations in the carbon, nitrogen, and oxygen isoelectronic sequences which are presented in detail in a recent JILA (Joint Institute for Laboratory Astrophysics) report.[1]

In this work, we found an interesting case of strong mixing of two Russell-Saunders levels, because of the magnetic interaction terms, of a type which has not occurred before, so far as we are aware. We will report on this briefly here, referring to the report for fuller details.

The self-consistent field treatment is based on the nonrelativistic Hamiltonian with omission of magnetic terms

$$H = \sum_{i=1} \left[-\frac{\hbar^2}{2m} \nabla_i^2 + \frac{Ze^2}{r_i} \right] + \sum_{\text{pairs}} \frac{e^2}{r_{ij}} \tag{1}$$

We write

$$u_a(i) = r_i^{-1} P(n_a, \ell_a) Y(\ell_a, m_a; \theta_i, \varphi_i)(m_{sa}, \sigma) \tag{2}$$

* The National Bureau of Standards and the University of Colorado.

for the one-electron wave function of the ith electron whose coordinates are $r_i \theta_i \varphi_i$ and σ_i and whose quantum numbers are a which stands for $(n_a, \ell_a, m_{\ell a}, m_{sa})$.

In the Hartree method the N-electron wave function is a simple product over N one-electron wave functions

$$\Psi_H = u_1(1)\, u_2(2) \ldots u_N(N) \tag{3}$$

When a wave function of this type is used, the mean energy of a configuration is readily computed to be

$$E_H(\text{config}) = \sum_a [q_a I(a) + \tfrac{1}{2} q_a(q_a - 1) F^0(a, a)] + \sum_{a,b} q_a q_b F^0(a, b) \tag{4}$$

where the sums on a and b are over distinct shells, specified by (n_a, ℓ_a) and q_a is the number of electrons in each shell. The double sum is on each pair of shells counted once. Here

$$I(a) = \int_0^\infty P(a; r)\left[-\frac{\hbar^2}{2m}\frac{d^2}{dr^2} + \frac{\ell_a(\ell_a + 1)}{r^2} - \frac{Ze^2}{r}\right] P(a; r)\, dr \tag{5}$$

$$F^0(a, b) = \int_0^\infty P^2(a; r) \frac{e^2}{r}\, Y_0(b, b; r)\, dr \tag{6}$$

and

$$Y_0(b, b; r) = \int_0^r P^2(b; s)\, ds + r \int_r^\infty P^2(b; s) \frac{ds}{s} \tag{7}$$

The Hartree equations[2] for the occupied ($q_a \neq 0$) radial functions are found by varying these functions in the expression for E_H(config) subject to the condition that each one of them is normalized, to find the radial functions which give E_H a stationary value.

In the Hartree-Fock method,[3] the determinantal form of N-electron wave function is used. This gives for the configuration average energy,

$$E(\text{config}) = \sum_a [q_a I(a) + \tfrac{1}{2} q_a(q_a - 1) J(a, a)] + \sum_{a,b} q_a q_b J(a, b) \tag{8}$$

which is formally the same as Equation 4 but with J in place of the F. These are defined as

$$J(a, b) = F^0(a, b) - \frac{1}{2}\sum_k \begin{pmatrix} \ell_b & k & \ell_a \\ 0 & 0 & 0 \end{pmatrix}^2 G^k(a, b) \tag{9}$$

and

$$J(a, a) = \frac{2}{4\ell_a + 1}\left[(2\ell_a + 1) F^0(a, a) - \frac{2\ell_a + 1}{2} \sum_k \begin{pmatrix} \ell_a & k & \ell_a \\ 0 & 0 & 0 \end{pmatrix}^2 F^k(a, a)\right] \tag{10}$$

Here the $3j$ symbol or Wigner coefficient is related to the reduced element of the Racah[4] tensor $C^{(k)}$ as

$$\begin{pmatrix} \ell & k & \ell' \\ 0 & 0 & 0 \end{pmatrix} = (-1)^{\ell'} \frac{\langle \ell' || C^{(k)} || \ell \rangle}{\sqrt{(2\ell + 1)(2\ell' + 1)}} \tag{11}$$

and

$$F^k(a, b) = \int_0^\infty \int_0^\infty P^2(a; r_1) P^2(b; r_2) e^2 \frac{r_<^k}{r_>^{k+1}} \, dr_1 \, dr_2 \tag{12}$$

and

$$G^k(a, b) = \int_0^\infty \int_0^\infty P(a; r_1) P(b; r_1) P(b; r_2) P(a; r_2) e^2 \frac{r_<^k}{r_>^{k+1}} \, dr_1 \, dr_2 \tag{13}$$

where $r_<$ and $r_>$ are respectively the lesser and the greater of r_1 and r_2.

The Hartree-Fock equations for the radial wave functions are those that express the condition that E(config) have a stationary value for variations of the $P(a; r)$ subject to the requirement that the radial functions be normalized. The two methods differ because of the difference between Equations 4 and 8. The presence of the exchange integrals of Equations 13 in 8 greatly complicates the Hartree-Fock equations.

In the Hartree-Fock-Slater[5] method, the effect of the exchange integrals is approximately taken into account by a method which preserves the simpler form of the Hartree equations. In this method, the equations for the radial function are

$$\frac{\hbar^2}{2m} \left[-\frac{d^2}{dr^2} + \frac{\ell_a(\ell_a + 1)}{r^2} \right] P(a; r) + V(r) P(a; r) = E_a P(a; r) \tag{14}$$

in which

$$\begin{aligned} V(r) &= V_0(r) && r < r_0 \\ &= -\frac{(Z - N + 1)e^2}{r} && r > r_0 \end{aligned} \tag{15}$$

in which r_0 is the value of r for which these two expressions are equal and the expression for $V_0(r)$ is

$$V_0(r) = -\frac{Ze^2}{r} + \frac{e^2}{r} \sum_a q_a Y_0(a, a; r) - 3e^2 \left[\frac{3}{8\pi} \rho(r) \right]^{1/3} \tag{16}$$

in which

$$\rho(r) = \frac{1}{4\pi r^2} \sum_a q_a P^2(a; r)$$

The term involving $\rho(r)$ is the Slater approximation to the exchange terms. The device of making the change expressed in Equation 15 to give $V(r)$ the correct behavior was introduced by Latter[6] and adopted by Herman and Skillman.[7]

The electrostatic interaction in Equation 1 has the effect of splitting the configuration energy into *terms*, accurately labelled by S and L, the quantum numbers which give resultant spin and resultant orbital angular momentum. Terms so labelled are said to be in Russell-Saunders coupling.

We write $E(Z, N, \text{config}, \text{term})$ for the energy of a term measured from the state of complete N-fold ionization, and $W(Z, N, \text{config}, \text{term})$ for the departure of the term energy from the configuration average, that is,

$$E(Z, N, \text{config}, \text{term}) = E(Z, N, \text{config}) + W(Z, N, \text{config}, \text{term}) \tag{17}$$

When magnetic interaction within states of the same configuration is taken into account, each term of $(2S + 1)(2L + 1)$ states is split into *levels*, labelled by J values which run from $|L - S|$ to $L + S$ by unit steps. Here $J(J + 1)$ is the value of the squared vector resultant of spin and orbital angular momentum, $(\mathbf{L} + \mathbf{S})^2$.

Nondiagonal matrix elements of the magnetic interaction connect levels of the same J but differing S and L values. In consequence S and L are no longer accurate quantum numbers, the wave function for a particular level becoming a linear combination of those for the interacting Russell-Saunders levels. This situation is called intermediate coupling[8]. Usually one particular Russell-Saunders level is predominant in the linear combination and spectroscopists attach that name to the level, while recognizing that the wave function contains components of other levels as well.

The magnetic terms in the Hamiltonian were first studied by Breit.[9] A modern account is given by Bethe and Salpeter[10] and by Slater[3]. They are all formally proportional to α^2 where $\alpha = e^2/\hbar c$ is the fine structure constant. Written in atomic units, they are conveniently expressed as the sum of three parts,

$$H_a = \tfrac{1}{2}\alpha^2 Z \sum_i \frac{1}{r_i^3} \mathbf{L}_i \cdot \mathbf{S}_i \tag{18a}$$

$$H_b = -\tfrac{1}{2}\alpha^2 \sum_{i,j} \frac{1}{r_{ij}^3} (\mathbf{r}_{ij} \times \mathbf{P}_i) \cdot (\mathbf{S}_i + 2\mathbf{S}_j) \tag{18b}$$

$$H_c = \tfrac{1}{2}\alpha^2 \sum_{\text{pairs}} \frac{3}{r_{ij}^3} \left[\mathbf{S}_i \cdot \mathbf{S}_j - \frac{3}{r_{ij}^2} (\mathbf{S}_i \cdot \mathbf{r}_{ij})(\mathbf{S}_j \cdot \mathbf{r}_{ij})\right] \tag{18c}$$

The H_a represents the interaction of each electron's magnetic moment with the magnetic field arising from the electron's motion in the Coulomb field of the nucleus. In H_b each pair occurs once as (i, j) and again as (j, i) where $i \neq j$. The terms in $\mathbf{S}_i$ represent the interaction of the ith electron's magnetic moment with the magnetic field due to its own motion in the electric field of the other $(N - 1)$ electrons. Its sign is opposite to that of H_a so the effect of these terms is to reduce the effect of H_a in a manner that is analogous to electrostatic screening.

The terms proportional to $\mathbf{S}_j$ in H_b arise from the interaction of the jth electron's magnetic moment with the magnetic field because of the orbital motion of the other $(N - 1)$ electrons.

Finally H_c, commonly called spin-spin interaction, represents the magnetic dipole interaction of all pairs of electrons.

Since the parts H_b and H_c involve sums over all electron pairs of two-electron symmetric operators, their matrix elements with determinantal wave functions will involve exchange integrals, as well as direct integrals, as first pointed out by Marvin.[11] The treatment of these terms was studied later by Horie[12] and Blume and Watson.[13]

The older treatment (Reference 8, p. 195), of magnetic interaction was based on the approximation that the magnetic part could be put in the form

$$H^I = \sum_i \xi(r_i)\mathbf{L}_i \cdot \mathbf{S}_i \tag{19}$$

in which, in atomic units,

$$\xi(r_i) = \tfrac{1}{2}\alpha^2 \frac{1}{r_i}\frac{\partial U_a(r_i)}{\partial r_i} \tag{20}$$

and $U_a(r_i)$ is the effective central field for the electron with quantum numbers (n_a, ℓ_a), namely,

$$U_a(r) = -\frac{Ze^2}{r} + \frac{e^2}{r}\sum_b (q_b - \delta_{ab})Y_0(b, b; r) \tag{21}$$

in which b is summed over the occupied shells.

In some presentations $\xi(r)$ is defined as in Equation 20 but using $V(r)$ from Equation 15 rather than $U_a(r)$ from Equation 21. The difference is quite small but Equation 21 is preferable as having a firmer theoretical foundation.

Horie has shown that the major part of H_b can be put in the form of Equation 19. The sum of H_a and this part of H_b is called *spin-orbit* interaction. The residual and the spin-spin interactions are often quite small so that in most cases the magnetic interaction is quite accurately given by the spin-orbit part alone.

The spin-orbit interaction in the Russell-Saunders limit (neglect of nondiagonal matrix elements) gives rise to an energy term (Reference 8, p. 194),

$$\Gamma(\alpha, S, L) = \tfrac{1}{2}\zeta(\alpha, S, L)[J(J+1) - L(L+1) - S(S+1)] \quad (22)$$

Here the ζ for levels are related to the one-electron $\zeta(n_a, \ell_a)$ of Reference 8, where

$$\zeta(n_a \ell_a) = \int_0^\infty \xi(r) P^2(n_a \ell_a; r)\, dr \quad (23)$$

The formulas for the $\zeta(n_a, \ell_a)$ are considerably more complicated when the full spin-orbit part of H_b is included with its exchange integrals. For $\zeta(2p)$ the formula is

$$\begin{aligned}\zeta(2p) = \tfrac{1}{2}\alpha^2 Z \int_0^\infty \frac{1}{r^3} P^2(2p, r)\, dr &- 4M^0(2p, 1s) \\ &- 4M^0(2p, 2s) + 2V^0(2p, 1s) \\ &+ 2V^0(2p, 2s) + 2N^1(2p, 1s) \\ &- 4N^{-1}(2p, 1s) - 4N^{-1}(2p, 2s) \\ &+ mM^0(2p, 2p) \end{aligned} \quad (24)$$

in which $m = 1$ for $N = 6$; $m = 3$ for $N = 7$ and $m = 5$ for $N = 8$ and the various terms in Equation 24 are

$$M^k(a, b) = \tfrac{1}{4}\alpha^2 \int_0^\infty \int_0^\infty \frac{r_2^{\,k}}{r_1^{k+3}} \varepsilon(r_1 - r_2) P_1(a) P_2^{\,2}(b)\, dr_1\, dr_2$$

$$N^k(a, b) = \tfrac{1}{4}\alpha^2 \int_0^\infty \int_0^\infty \frac{r_2^{\,k}}{r_i^{k+3}} \varepsilon(r_1 - r_2) P_1(a) P_1(b) P_2(a) P_2(b)\, dr_1\, dr_2$$

$$V^k(a) = \tfrac{1}{4}\alpha^2 \int_0^\infty \int_0^\infty \frac{r_<^{\,k}}{r_>^{k+3}} r_1 r_2 P_1(a) P_2(b) \left[r_2 \frac{\partial}{\partial r_1} - r_1 \frac{\partial}{\partial r_2}\right] \times \frac{P_2(a) P_1(b)}{r_1 r_2}\, dr_1\, dr_2$$

where

$$\begin{aligned} \varepsilon(x) &= 1 \quad \text{for} \quad x > 0 \\ \varepsilon(x) &= 0 \quad \text{for} \quad x < 0 \end{aligned}$$

These complications over the older theory have a quite small effect in the examples so far calculated. The calculated ζ parameters depend both on the formulas used and the specific choice of radial wave functions. Froese[14] has made the best calculations of any known to us. She used radial functions from a full Hartree-Fock solution and the full theoretical formula as given by Blume and Watson. This gives somewhat smaller values of the $\zeta(n\ell)$ for the cases tested than the ones calculated by us[15] using Equations 20, 21, and 23 with radial wave functions from the Hartree-Fock-Slater solution.

For the study of intermediate coupling, the nondiagonal elements of magnetic interaction must also be included. This gives a matrix to be diagonalized for each J, and results in producing wave functions which are mixed in L and S.

An interesting example of this occurs in $1s^2 2s^2 2p3p$ of the $N = 6$ carbon isoelectronic sequence. This configuration is known from analysis of spectra for $Z = 6, 7, 8$ but only a few levels are known for $Z > 8$.

For this configuration the matrices of electrostatic interaction are given by Slater[3] (Volume 2, pp. 288–290). The matrices of spin-orbit interaction are given by Condon and Shortley[8] (p. 268). The radial functions were obtained by an adaptation to local computer facilities of the program of Herman and Skillman[7] for the Hartree–Fock–Slater methods. Table 1 presents the observed and theoretical term energies relative to configuration averages. The observed values are calculated from the values given in the compilation by C. Moore.[16] Table 2 gives the theoretical values of the energy levels relative to the configuration average. Figure 1 presents the same material graphically.

Of the four levels having $J = 1$, the one which we denote by "a" is almost purely $^3D_1$ and the one we denote by "b" is almost purely $^1P_1$ for low Z. The coefficients for the expansion of the $\Psi(a)$ and $\Psi(b)$ in terms of Russell-Saunders states are given in Table 3.

With increasing Z, the admixtures of $^3P_1$ and $^3S_1$ in "a" and "b" remain quite small, but the mixing of $^3D_1$ and $^1P_1$ increases rapidly. Figure 2 shows the projections of $\Psi(a)$ and $\Psi(b)$ on the plane of $\Psi(^3D_1)$ and $\Psi(^1P_1)$. Because the $^3P_1$ and $^3S_1$ components remain small, $\Psi(a)$ and $\Psi(b)$ are nearly orthogonal, nearly unit vectors as Z increases. Only for $Z > 14$, does $\Psi(a)$ become quite appreciably less than a unit vector in the plane of Figure 2, because of the increase of its components out of this plane.

As Figure 2 shows, $\Psi(a)$ and $\Psi(b)$ have a smooth dependence on Z. However, for $Z > 17$, the $\Psi(a)$ contains less of $\Psi(^3D_1)$ than of $\Psi(^1P_1)$. Hence, if the usual rule were to be followed of giving a Russell-Saunders

Table 1. Observed and Theoretical Term Energies Relative to the Configuration Average for the $1s^22s^22p3p$ Configuration of the $N = 6$ Sequence (Unit: Rydbergs. First entry is observed and second is theoretical)

	^3D	^3P	^3S	^1D	^1P	^1S
$Z = 6$, CI	−0.00862 −0.01103	0.00644 0.00771	0.00069 −0.00079	0.01771 0.02179	−0.01649 −0.01785	0.03015 0.04292
7, NII	−0.02193 −0.02826	0.01471 0.02101	−0.00118 −0.00602	0.04730 0.05516	−0.04020 −0.04309	0.08430 0.10637
8, OIII	−0.03544 −0.04469	0.02194 0.03341	−0.00372 −0.01099	0.07854 0.08724	−0.06387 −0.06716	0.14431 0.16797
9, FIV	−0.06039	0.04504	−0.01554	0.11808	−0.09030	0.22765
10, NeV	−0.07561	0.05617	−0.01978	0.14810	−0.11283	0.28601
11, NaVI	−0.09051	0.06696	−0.02382	0.17757	−0.13497	0.34349
12, MgVII	−0.10519	0.07754	−0.02773	0.20666	−0.15683	0.40034
13, AlVIII	−0.11971	0.08796	−0.03153	0.23546	−0.17849	0.45672
14, SiIX	−0.13411	0.09825	−0.3525	0.26406	−0.20001	0.51278
15, PX	−0.14841	0.10846	−0.03891	0.29250	−0.22141	0.56855
16, SXI	−0.16265	0.11859	−0.04252	0.32081	−0.24273	0.62413
17, CXII	−0.17683	0.12867	−0.04610	0.34902	−0.26398	0.67954
18, ArXIII	−0.19097	0.13871	−0.04966	0.37716	−0.28518	0.73483
19, KXIV	−0.20507	0.14872	−0.05319	0.40523	−0.30633	0.79001
20, CaXV	−0.21914	0.15869	−0.05670	0.43325	−0.32744	0.84510

Table 2. Theoretical Energy Levels for $N = 6$ $1s^2 2s^2 2p3p$ Relative to Configuration Average (Unit: Rydberg)

Z	$^3D_3$	$^3D_2$	$^1D_2$	$^3P_2$	a	$^3P_1$	b	$^3S_1$	$^3P_0$	$^1S_0$
6	−0.010773	−0.011155	0.021798	0.007837	−0.011396	0.007594	−0.017860	−0.000793	0.007459	0.042920
7	−0.027556	−0.028627	0.055171	0.021365	−0.029297	0.020680	−0.043133	−0.006027	0.020295	0.106386
8	−0.043118	−0.045510	0.087274	0.034190	−0.046953	0.032679	−0.067302	−0.101999	0.031800	0.168008
9	−0.057349	−0.061988	0.118185	0.046532	−0.064684	0.043654	−0.090699	−0.015551	0.041894	0.227749
10	−0.070273	−0.078444	0.148345	0.058756	−0.082893	0.053809	−0.113855	−0.019800	0.050584	0.286255
11	−0.081796	−0.095212	0.178099	0.071141	−0.101851	0.063272	−0.137355	−0.023835	0.057719	0.344020
12	−0.091725	−0.112598	0.207715	0.083895	−0.121592	0.072123	−0.161973	−0.027691	0.063003	0.401410
13	−0.099791	−0.130916	0.237435	0.097174	−0.141912	0.080439	−0.188776	−0.031349	0.066011	0.458754
14	−0.105679	−0.150529	0.267626	0.111114	−0.162511	0.088352	−0.219078	−0.034731	0.066207	0.516398
15	−0.109014	−0.171833	0.298569	0.125803	−0.183261	0.096035	−0.254190	−0.037654	0.062920	0.574691
16	−0.109394	−0.195317	0.330748	0.141319	−0.204488	0.103760	−0.295288	−0.039806	0.055365	0.634101
17	−0.106367	−0.221538	0.364701	0.157701	−0.226978	0.111868	−0.343381	−0.040678	0.042619	0.695134
18	−0.099443	−0.251151	0.401077	0.174975	−0.251907	0.120786	−0.399450	−0.039568	0.023629	0.758383
19	−0.088096	−0.284906	0.440617	0.193164	−0.280706	0.131006	−0.464511	−0.035605	0.002779	0.824524
20	−0.071769	−0.323652	0.484141	0.212310	−0.314921	0.143081	−0.539643	−0.027860	−0.037882	0.894297

Table 3. Mixing Coefficients for $(^3D_1)$ and $(^1P_1)$ in $N = 6\ 1s^2 2s^2 2p3p$

$(a) = a_1\,(^3D_1) + a_2\,(^3P_1) + a_3\,(^1P_1) + a_4\,(^3S)$

$(b) = b_1\,(^3D_1) + b_2\,(^3P_1) + b_3\,(^1P_1) + b_4\,(^3S)$

Z	a_1	a_2	a_3	a_4	b_1	b_2	b_3	b_4
6	0.999	−0.008	−0.035	0.001	0.035	0.006	0.999	−0.012
7	0.999	−0.008	−0.046	0.001	0.047	0.006	0.999	−0.016
8	0.997	−0.010	−0.070	0.003	0·071	0.008	0.997	−0.023
9	0.994	−0.014	−0.106	0.006	0.107	0.011	0.994	−0.033
10	0.987	−0.020	−0.157	0.012	0.158	0.014	0.986	−0.046
11	0.974	−0.029	−0.225	0.023	0.227	0.018	0.972	−0.062
12	0.949	−0.040	−0.311	0.041	0.314	0.021	0.946	−0.079
13	0.909	−0.055	−0.408	0.067	0.415	0.023	0.905	−0.096
14	0.855	−0.075	−0.502	0.104	0.515	0.023	0.850	−0.110
15	0.795	−0.099	−0.580	0.149	0.601	0.022	0.790	−0.121
16	0.737	−0.128	−0.633	0.200	0.667	0.020	0.733	−0.129
17	0.686	−0.162	−0.662	0.255	0.716	0.018	0.685	−0.135
18	0.641	−0.201	−0.672	0.312	0.750	0.015	0.646	−0.141
19	0.602	−0.244	−0.666	0.367	0.775	0.013	0.615	−0.145
20	0.567	−0.288	−0.650	0.417	0.793	0.011	0.591	−0.150

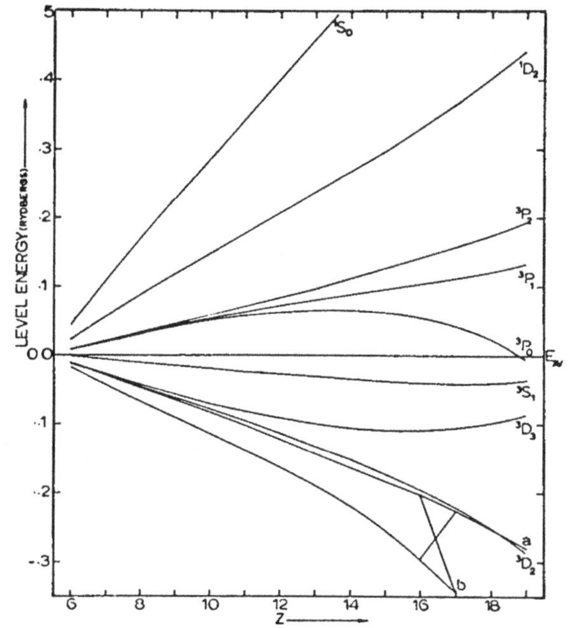

Figure 1. Theoretical level energies of the $1s^22s^22p3p$ configuration in the $N=6$ sequence.

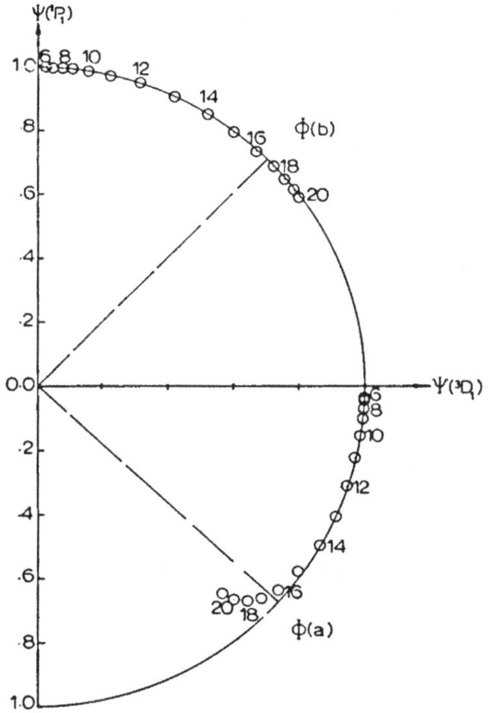

Figure 2. Projection of $\Psi(a)$ and $\Psi(b)$ on the $\Psi(^3D_1)$, $\Psi(^1P_1)$ plane.

name according to the term of largest coefficient in the state of intermediate coupling, then the names should be interchanged between $Z = 16$ and 17 as indicated by the crossover lines in Figure 1.

Table 4. Intervals of the ${}^3D_{3,2,1}$ and ${}^3P_{2,1,0}$ Russell-Saunders Levels for the $1s^22s^22p3p$ Configuration of the $Z = 6$ Sequence (Unit: 10^{-4} Rydberg. First entry is observed, second is theoretical)

	${}^3D_3-a$	${}^3D_2-a$	${}^3P_2-{}^3P_0$	${}^3P_1-{}^3P_0$
Z = 6, CI	4.98	1.93	3.00	1.13
	6.23	2.41	3.78	1.35
7, NII	14.30	5.54	8.53	3.21
	17.41	6.70	10,70	3.85
8, OIII	32.48	12.42	19.38	7.48
	38.35	14.43	23.90	8.79
9, FIV	63.56	23.83	38.77	15.55
	73.35	26.96	46.38	17.60
10, NeV				
	126.20	44.49	81.72	32.25
11, NaVI				
	200.55	66.39	134.22	55.53
12, MgVII			191.82	108.62
	298.67	89.94	208.92	91.20
13, A1VIII				
	421.21	109.96	311.63	144.28
14, SiIX				
	568.32	119.82	449.07	221.45
15, PX				
	742.47	114.28	628.83	331.15
16, SXI				
	950.94	91.71	859.54	483.95
17, ClXII				
	1206.11	54.40	1150.82	692.49
18, ArXIII				
	1524.64	7.56	1513.46	971.57
19, KXIV				
	1926.10	−42.00	1903.85	1282.27
20, CaXV				
	2431.52	−87.31	2501.92	1809.63

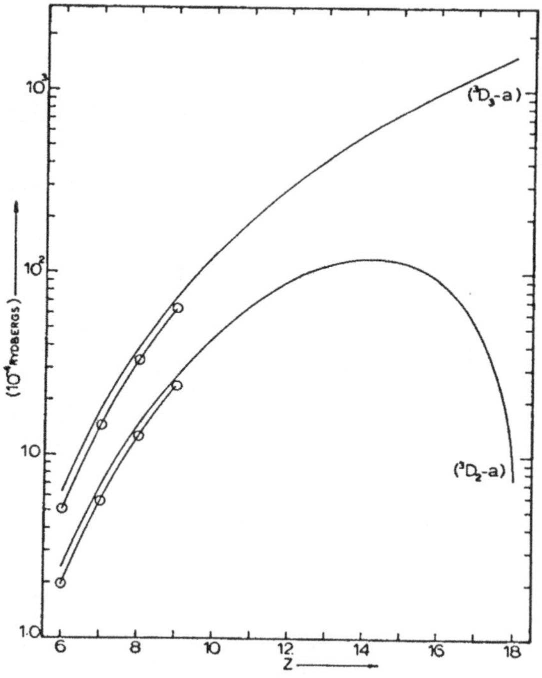

Figure 3. Variation with Z of the theoretical and observed $(^3D_3 - a)$ and $(^3D_2 - a)$ intervals in the $N = 6$ sequence $1s^22s^22p3p$. (Observed points are circled.)

The calculated values of $\zeta(2p)$ and $\zeta(3p)$ lead to good agreement with observed term intervals as shown in Table 4. Figure 3 shows a plot of the Z dependence of the $(^3D_3 - a)$ and $(^3D_2 - a)$ intervals, including the few known observed intervals.

The large degree of mixing involved in this example also shows up in the Zeeman effect. (Condon and Shortley,[8] Section 3[16]). In addition to the strong mixing in $2p3p$ of $^3D_1$ and $^1P_1$ there is also a strong mixing in $2p3d$ of $^1D_2$ and $^3F_2$. Table 5 gives the calculated Landé g-factors of interest in these cases. Figure 4 shows the Z-dependence of the g-factors for $J = 1$ in the $2p3p$, showing the crossover between $Z = 16$ and 17. Figure 5 is a similar graph of the g-factors for the $J = 2$ levels in the $2p3d$ configuration showing the crossover interaction of $^1D_2$ and $^3F_2$ and the lack of it for $^3D_2$ and $^3P_2$ owing to the vanishing of the non-diagonal matrix element of the magnetic interaction connecting these two levels.

Table 5. Landé g-Factors for the $2p3s$, $2p3p$ and $2p3d$ Excited Configurations of the N = 6 Sequence. Intermediate Coupling

	$L-S$	$Z=6$	7	8	9	10	11	12	13	14	15	16	17	18	19	20
$2p3s\ {}^3P_1$	1.5	1.500	1.500	1.499	1.498	1.497	1.494	1.491	1.485	1.479	1.471	1.462	1.452	1.442	1.433	1.423
$2p3s\ {}^1P_1$	1.0	1.000	1.000	1.001	1.002	1.003	1.006	1.010	1.015	1.021	1.029	1.038	1.048	1.058	1.067	1.077
$2p3p\ {}^3D_2$	1.167	1.167	1.167	1.167	1.167	1.167	1.167	1.167	1.167	1.168	1.168	1.168	1.169	1.170	1.171	1.171
$2p3p\ {}^3P_2$	1.5	1.500	1.500	1.500	1.500	1.499	1.498	1.497	1.494	1.491	1.486	1.480	1.471	1.460	1.446	1.431
$2p3p\ {}^1D_2$	1.0	1.000	1.000	1.000	1.000	1.001	1.002	1.003	1.005	1.008	1.012	1.019	1.027	1.037	1.050	1.065
$2p3p\ {}^3D_1$	0.5	0.501	0.501	0.503	0.506	0.513	0.527	0.553	0.593	0.648	0.711	0.776	0.762	0.739	0.721	0.708
$2p3p\ {}^3P_1$	1.5	1.500	1.500	1.500	1.501	1.501	1.502	1.504	1.506	1.510	1.515	1.523	1.534	1.547	1.564	1.585
$2p3p\ {}^3S_1$	2.0	2.000	2.000	1.999	1.998	1.996	1.993	1.987	1.978	1.962	1.940	1.907	1.861	1.802	1.731	1.652
$2p3p\ {}^1P_1$	1.0	1.000	0.999	0.998	0.996	0.990	0.978	0.957	0.923	0.880	0.834	0.794	0.843	0.912	0.983	1.054
$2p3d\ {}^3F_3$	1.083	1.100	1.085	1.084	1.084	1.084	1.085	1.085	1.086	1.087	1.088	1.090	1.092	1.094	1.097	1.100
$2p3d\ {}^3D_3$	1.333	1.316	1.331	1.333	1.333	1.332	1.332	1.331	1.330	1.328	1.326	1.323	1.320	1.316	1.312	1.307
$2p3d\ {}^1F_3$	1.0	1.001	1.000	1.000	1.000	1.000	1.001	1.001	1.001	1.002	1.002	1.003	1.004	1.006	1.008	1.010
$2p3d\ {}^3F_2$	0.667	0.700	0.672	0.675	0.689	0.727	0.789	0.829	0.796	0.777	0.766	0.760	0.756	0.753	0.752	0.752
$2p3d\ {}^3D_2$	1.167	1.137	1.165	1.168	1.170	1.175	1.183	1.196	1.210	1.224	1.235	1.239	1.234	1.221	1.200	1.174
$2p3d\ {}^3P_2$	1.5	1.498	1.499	1.498	1.495	1.489	1.480	1.465	1.447	1.427	1.409	1.392	1.379	1.367	1.358	1.351
$2p3d\ {}^1D_2$	1.0	0.999	0.998	0.994	0.979	0.942	0.882	0.844	0.880	0.905	0.924	0.943	0.965	0.992	1.023	1.057
$2p3d\ {}^3D_1$	0.5	0.503	0.502	0.503	0.507	0.514	0.525	0.541	0.560	0.582	0.605	0.628	0.649	0.668	0.685	0.700
$2p3d\ {}^3P_1$	1.5	1.496	1.498	1.497	1.483	1.487	1.476	1.460	1.441	1.420	1.398	1.376	1.356	1.339	1.323	1.309
$2p3d\ {}^1P_1$	1.0	1.002	1.000	1.000	1.000	1.000	1.000	0.999	0.999	0.998	0.997	0.996	0.995	0.994	0.992	0.990

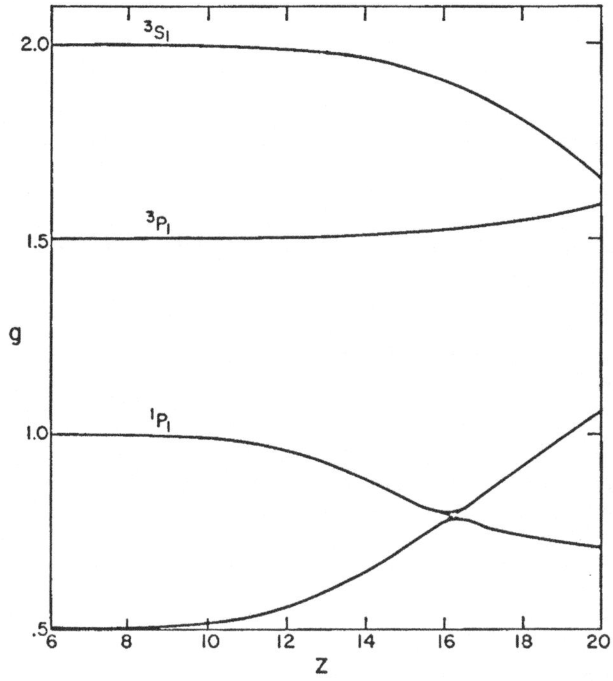

Figure 4. Landé g-factors of the $J = 1$ levels of $1s^2 2s^2 2p3p$.

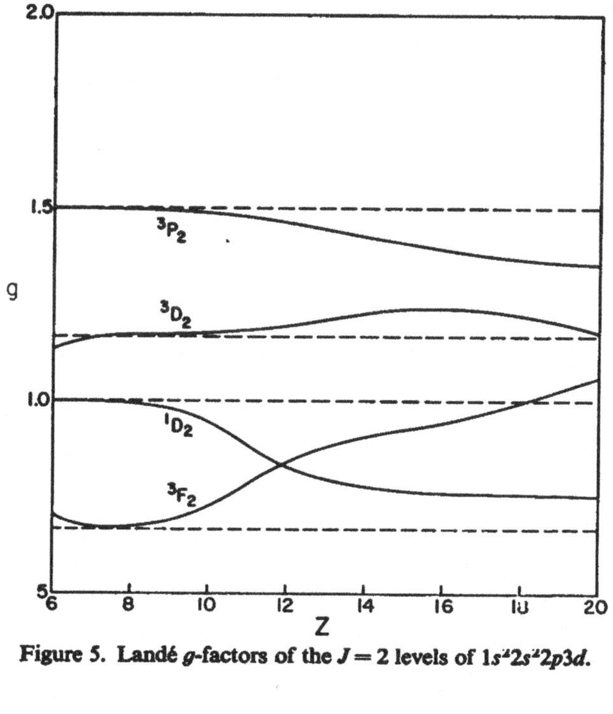

Figure 5. Landé g-factors of the $J = 2$ levels of $1s^2 2s^2 2p3d$.

References

1. E. U. Condon and H. Odabasi, " Self-consistent Field Calculations for Energy Levels of 6, 7 and 8 Electron Isoelectronic Sequences." *JILA* (Joint Institute for Laboratory Astrophysics, Boulder, Colo.) *Report No. 95* (August 6, 1968).
2. D. R. Hartree, *The Calculation of Atomic Structures.* John Wiley & Sons, Inc., New York (1957).
3. J. C. Slater, *Quantum Theory of Atomic Structure*, 2 Vols. McGraw-Hill Book Co. Inc., New York. (1960). (See Vol. 2, Chap. 24.)
4. G. Racah, *Phys. Rev.*, **62**, 438 (1942); U. Fano and G. Racah, *Irreducible Tensorial Sets*, Academic Press Inc., New York (1959).
5. J. C. Slater, *Phys. Rev.*, **81**, 385 (1951). Also Reference 3 above, Vol. 2, Chap. 17.
6. R. Latter, *Phys. Rev.*, **99**, 510 (1955).
7. F. Herman and S. Skillman, *Atomic Structure Calculations*, Prentice-Hall Englewood Cliffs, N.J. (1963).
8. E. U. Condon and G. H. Shortley, *The Theory of Atomic Spectra*, Cambridge University Press, New York (1935). (See Chap. 11.)
9. G. Breit, *Phys. Rev.*, **34**, 553 (1929); **36**, 383 (1931); **39**, 616 (1932).
10. H. A. Bethe and E. E. Salpeter, *Encyclopedia of Physics*, Vol. 35, p. 267. Springer- Verlag, Berlin (1957).
11. H. H. Marvin, *Phys. Rev.*, **71**, 102 (1947).
12. H. Horie, *Progr. Theoret. Phys.* (*Kyoto*), **10**, 296 (1953).
13. M. Blume and R. E. Watson, *Proc. Roy. Soc.* (*London*), **A270**, 127 (1962).
14. C. Froese, *Can. J. Phys.*, **45**, 1501 (1967).
15. E. U. Condon and H. Odabasi, *JILA Report* No. 61, also pp. 185–201 in *Quantum Theory of Atoms, Molecules and the Solid State. A Tribute to John C. Slater*, Per-Olov Löwdin, ed., Academic Press Inc., New York (1966).
16. C. Moore, *Atomic Energy Levels.* Circular 467 of National Bureau of Standards.

William P. Allis
Massachusetts Institute of Technology

3 Nonlinear, Collisionless, Plasma Waves*†

Introduction

The author learned from Morse[1] that the complete Boltzmann equation determines the steady-state distribution function. The collisionless Boltzmann equation does not do so but only determines the evolution in time of an arbitrary distribution. The reflection of particles at a potential barrier appears as a branch line of the distribution. Linearization of the Vlasov equation reduces this branch line to a pole, the Landau pole, and does not explicitly consider reflected particles, the electrons trapped in the potential wells.

We postulate a distribution, Equation 21, which is no more arbitrary than that (Equation 5) postulated in linear theory and reduces to the latter for small amplitudes. Its integral, Equation 26, is real and Poisson's equation is solved, as in the theory of space charge sheaths, as an integral over the potential, Equation 41. Potential profiles (Equations 42, 48, 55, 65) and undamped dispersion relations (Equations 43, 51, 58, 66) follow immediately. The two constants of integration determine the amplitude of the wave and the current, Equation 40, "induced" by it. The current is proportional to the arbitrary constant in van Kampen's dispersion relation, Equation 13.

* This work was supported principally by the U.S. Atomic Energy Commission (Contract AT(30-1)-1842).

† This work was done partly at The Los Alamos Scientific Laboratory under U.S. Atomic Energy Commission Contract W-7405-Eng-36 with the University of California, and also at St. Catherine's College, Oxford University, Oxford, England.

The above dispersions follow from a distribution, Equation 24, of trapped electrons which is called "normal." It is shown: (a) that an antisymmetric distribution of trapped electrons leads to damping and to oscillations at the pendular period, Equation 14, (b) that abnormal symmetric distributions introduce the subharmonic and do not tend towards the linear theory for small amplitudes.

1. Linearized Theory

The motions of electrons in a plasma may be described by the collisionless Boltzmann equation

$$\frac{\partial f}{\partial t} + \mathbf{v} \cdot \nabla f = \frac{e\mathbf{E}}{m} \cdot \nabla_v f \tag{1}$$

where, for self-consistency, the electric field must satisfy Poisson's equation

$$\nabla \cdot \mathbf{E} = -\frac{e(n_- - n_+)}{\varepsilon_0} \tag{2}$$

and, of course

$$n_- = \int f \, d^3v \tag{3}$$

We shall consider the ions to be infinitely heavy so that they contribute a uniform positive space charge $n_+ e$, and we have already neglected collisions and short-range electric fields.

These three constitute the Vlasov set of equations, and we shall seek plane wave solutions of this set; that is solutions in which f has already been integrated over v_y and v_z and is a function only of $v_x = v$ and of $x - ut$.

$$\begin{aligned} (v-u)\frac{\partial f}{\partial x} &= \frac{eE}{m}\frac{\partial f}{\partial v} \\ \frac{\partial E}{\partial x} &= \frac{-e(n_- - n_+)}{\varepsilon_0} \\ n_- &= \int_{-\infty}^{+\infty} f \, dv \end{aligned} \tag{4}$$

If the distribution function f can be expanded in terms of the normalized functions $f_0(v^2)$, and $f_1(v)$ and the densities n_0 and $n_1(x - ut)$, then

$$f = n_+ f_0 + n_1 f_1 \qquad \text{with} \quad n_1 \ll n_0 \tag{5}$$

the term in En_1 on the right of Equation 4 can be neglected. Equation 4 can then be integrated to give

$$(v - u)n_1 f_1 = -\frac{n_+ eV}{m} \frac{\partial f_0}{\partial v} \tag{6}$$

This equation is now linear in V, and we know that it can be solved by trigonometric functions. If we set

$$V = V_1 e^{ik(x-ut)} \tag{7}$$

we find

$$f_1 = \frac{\omega_p{}^2}{k^2} \frac{\partial f_0/\partial v}{v - u} \tag{8}$$

where

$$\omega_p{}^2 = \frac{n_+ e^2}{\varepsilon_0 m}$$

and normalization yields the dispersion relation

$$\frac{k^2}{\omega_p{}^2} = \int_{-\infty}^{+\infty} \frac{\partial f_0/\partial v}{v - u}\, dv \tag{9}$$

where the integral is, of course, taken along the real axis.

We do not propose to discuss this integral as innumerable papers have been written about it. Vlasov[2] takes the principal part and finds, for $f_0 = e^{v^2/a^2}$

$$\left(\frac{ka}{\omega_p}\right)^2 = Z_r'(u/a) \tag{10}$$

where $Z(u)$ is the plasma dispersion function[3] and Vlasov takes only the real part. This leads to undamped waves. Landau[4] uses Fourier–Laplace transform theory, and the pole at $v = u$ in the integrand of Equation 9 yields the complex dispersion relation

$$\left(\frac{ka}{\omega_p}\right)^2 = Z'(u/a) \tag{11}$$

and this leads to "Landau damping" of the waves. Van Kampen[5, 6] returns to real variables but argues that the pole in Equation 8 has been artificially introduced by division and may be compensated by adding an arbitrary constant λ times a delta function

$$f_1 = \frac{\omega_p^2}{k^2}\left(\frac{\partial f_0/\partial v}{v-u} + \lambda\delta(v-u)\right) \tag{12}$$

This leads to the dispersion relation

$$\left(\frac{ka}{\omega_p}\right)^2 = Z_r'(u/a) + \lambda a^2 \tag{13}$$

which gives undamped waves with "arbitrary" dispersion. The particular Van Kampen solution with $\lambda = 0$ agrees with Vlasov. Landau damping is retrieved out of this only by forming a wave packet.[7]

This multiplicity is impossible to understand because linearization has reduced finite objects to delta functions and amplitudes to zero. Furthermore, there is a clear indication of being on the wrong track: the third term of the series in Equation 5 which is proportional to V^2 diverges!

Two papers have used a less miniaturized approach: O'Neil[8] has calculated the phase mixing of the electrons in their pendular motion in the potential troughs of a wave and shown that Landau damping occurs only during times less than one pendular period

$$t_L < \tau_{el} = \frac{2\pi}{k}\sqrt{\frac{m}{2eV_1}} \tag{14}$$

Phase mixing is substantially completed and the Vlasov wave continues undamped for times $t_V > N\tau_{el}$ where N is a large number. Bernstein, Greene, and Kruskal[9] have shown that the lowest significant power of V to enter is not the first but is $\sqrt{V}$ and that arbitrary dispersion relations hold if arbitrary wave forms are allowed.

All of these results fall into their proper place in the following nonlinear analysis.

2. *Nonlinear Distribution Function*

We shall use the method of Bernstein, Greene, and Kruskal but in the reverse direction, starting with the distribution function, and the clue to the correct distribution is obtained from Equations 5 and 8. Let the relative velocity be

$$w = v - u \tag{15}$$

Then Equations 5 with 8 can be written

$$f = n_+\left[f_0 - \frac{2eV}{m}\frac{\partial f_0}{\partial w^2}\right] \tag{16}$$

and these are the first two terms in the Taylor expansion of

$$f(w^2 - 2eV/m).$$

This suggests that

$$f\left(u + \sqrt{w^2 - \frac{2eV}{m}}\right) \tag{17}$$

might be the appropriate nonlinear distribution function for a wave in a plasma whose distribution in the absence of the wave is $f(v)$.

Let us introduce coordinates in a frame moving at the phase velocity of the wave and scaled to thermal quantities

$$\begin{aligned} x' &= (x - ut)\sqrt{\frac{m}{2\kappa T}} \\ w' &= (v - u)\sqrt{\frac{m}{2\kappa T}} \\ \eta &= \frac{eV}{\kappa T} \end{aligned} \tag{18}$$

and let us drop the primes as unnecessary. In this frame let us assume that $\eta(x)$ is independent of the time. The energy constant (Figure 1) is

$$c^2 = w^2 - \eta \tag{19}$$

and we shall take

$$c = w\sqrt{1 - \frac{\eta}{w^2}} \tag{20}$$

to have the sign of w. This implies a cut in the complex w-plane between the branch points at $w = \pm\sqrt{\eta}$.

The rule 17 suggests that v be replaced by $u + c$ so that a Maxwell distribution becomes

$$f = \frac{n_0}{\sqrt{\pi}}\, e^{-(u+c)^2} = \frac{n_0}{\sqrt{\pi}}\, e^{-u^2-c^2}(\cosh 2uc - \sinh 2uc) \tag{21}$$

This distribution is appropriate for a wave in a Maxwellian plasma because it tends to a Maxwell-Boltzmann distribution

$$f \to \frac{n_0}{\sqrt{\pi}}\, e^{\eta - v^2} \quad \text{for} \quad \begin{cases} \text{either} & \eta \ll w^2 \\ \text{or} & u^2 \ll v^2 \end{cases} \tag{22}$$

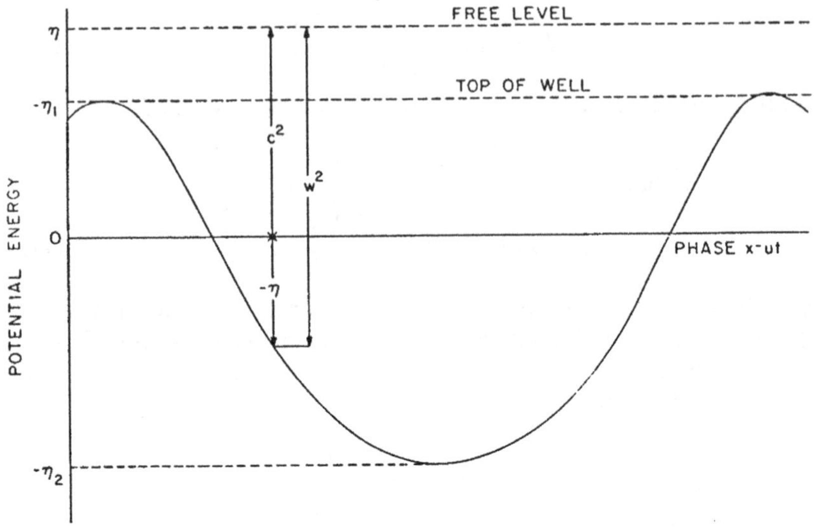

Figure 1. Wave form and energy diagram.

The relation of this function to linear theory is best seen by expanding the square root in Equation 20 in powers of η/w^2

$$-(u+c)^2 \to -v^2 + \frac{\eta v}{w} - \frac{\eta^2 u}{4w^2} \cdots \tag{23}$$

The first term yields f_0 in Equation 5, the second yields f_1 in Equation 6 with its pole, the third yields the divergence in f_2. The pole of linear theory is artificially introduced by the linearization. The true singularities are branch points at $c = \pm\sqrt{-\eta}$ representing reflections by the potential walls. By assiduously avoiding expanding the square root all integrals of the function 21 will converge and yield power series in η which converge for all η. The integration of Poisson's equation will then yield a two-parameter family of wave forms which may look like Figure 1. We shall limit our solutions to those for which η_1 is negative and η_2 positive so that c is real for untrapped electrons.

Because we are considering a collisionless plasma the phase orbits of the electrons have independent existences. The phase orbits for a potential such as in Figure 1 are sketched in Figure 2. There are open orbits corresponding to free electrons, separated by a separatrix $c^2 = -\eta_1$ from the closed orbits which correspond to trapped particles. The open orbits are indexed by values of c, and by Liouville's theorem *any* function of c and u will satisfy the collisionless, time independent, Boltzmann

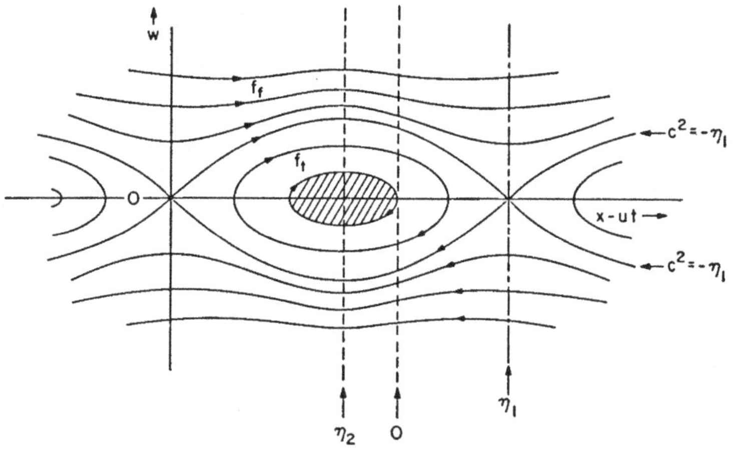

Figure 2. Phase orbits.

Equation 4 in the wave frame. This arbitrariness is inherent in the collisionless assumption, but Equation 21 is also single-valued, continuous with infinitely many derivatives, and satisfies boundary conditions 22.

The closed orbits are indexed by c^2, *not* c, which has a different sign on the upper and lower halves of each orbit. Accordingly, Equation 21 is not a time-independent solution of the collisionless Boltzmann Equation for $c^2 < -\eta_1$. The discontinuity at $w = 0$ will flow around the orbits, the period for the inner orbits being similar to that given in Equation 14 and the outer orbits having progressively longer periods so that the line of discontinuity winds itself up and eventually disappears in the grain size of the distribution function. This is the process studied by O'Neil for a slightly different distribution. The final distribution must be a function of c^2, that is the term in cosh 2 uc is left unchanged but sinh 2 uc is washed out, and the final steady-state solution for the trapped electrons is

$$f_t = \frac{n_0}{\sqrt{\pi}} e^{\eta - u^2 - w^2} \cosh 2uc \tag{24}$$

Phase mixing transfers the discontinuity at $w^2 = 0$ to one at $c^2 = -\eta_1$. If the wave is generated adiabatically, the trapped distribution will certainly be symmetric in w and similar to Equation 24. Figure 3 shows a distribution given by Equations 21 and 24. We shall call this a "normal" distribution.

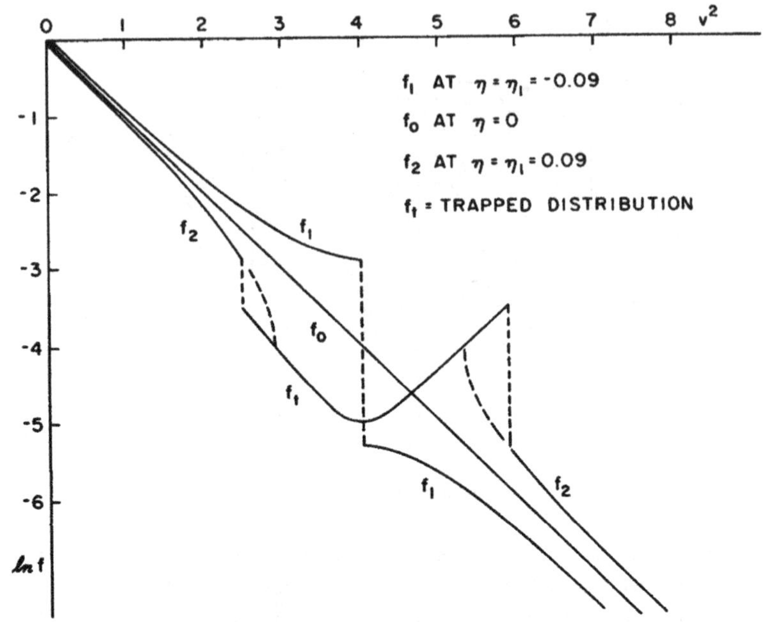

Figure 3. The distribution function.

The distribution 21 becomes exactly Maxwellian each time the potential goes through zero, that is along lines similar to that marked 0 in Figure 2. One can imagine the potential of Figure 1 applied to a plane grid in a Maxwellian plasma, starting at potential zero. If the amplitude and wave form are correct, this potential will propagate in both directions at the velocity u, and the electrons will follow those orbits of Figure 2 which are intercepted by the line $v = 0$, leaving the orbits at the bottoms of the wells, shaded in Figure 2, empty. This picture has two errors because the electrons that cross the grid come from a wave going in the wrong direction, and electrons faster than the phase velocity of the wave will disturb the plasma ahead of the wave. It is a useful picture nevertheless as it shows how empty wells and asymmetric distributions originate in impulsively applied potentials, in contrast to adiabatically applied potentials. We shall consider first the distribution given by Equations 21 and 24 extending to the bottoms of the wells, however this may have been generated. Then we shall consider the effect of adding or subtracting proportional amounts to the trapped distribution.

3. Density, Flow and Energy.

The electron density n_- is found by integrating Equation 24 over all values of w as the term in sinh $2uc$ is odd and gives nothing:

$$n_- = \int_{-\infty}^{+\infty} f\, dv = \frac{n_0}{\sqrt{\pi}} e^{n-u^2} \int_{-\infty}^{+\infty} e^{-w^2} \cosh 2uc\, dw \tag{25}$$

If we expand cosh $2uc$ in powers of c^2, the negative values give no difficulty. All integrals converge absolutely and yield

$$n_- = n_0 \sum_0^{\infty} \frac{M_a \eta^a}{a!} \tag{26}$$

where M_a is the confluent hypergeometric function, $F_1(a, \frac{1}{2}, -u^2)$. It has the convergent and asymptotic series

$$M_a = 1 - a2u^2 + \frac{a(a+1)}{1 \cdot 3}\frac{(2u^2)^2}{2!} \cdots = \frac{3 \cdot 5 \cdots (2a-1)}{(-2u^2)^a}$$
$$\times \left[1 + \frac{a(2a+1)}{2u^2} + \frac{a(a+1)}{2!}\frac{(2a+1)(2a+3)}{(2u^2)^2} \cdots\right] \tag{27}$$

In particular,

$$M_0 = 1, \qquad 2M_1 = -Z_r'(u) \tag{28}$$

and M_1 is negative for $0.857 < u^2$. Higher order functions can be calculated by the recursion formula:

$$2aM_{a+1} = (4a - 1 - 2u^2)M_a + (1 - 2a)M_{a-1} \tag{29}$$

and

$$(u^{2a}M_a)' = au^{2a-2}M_{a+1}$$

The functions $2u^2M_1$ and $2u^2M_2$ are shown in Figure 4. M_a has a zeroes.

Three limits are of particular interest and will be studied in detail in the next section. For slow waves,

$$u^2 \to 0 \qquad n_- \to n_0 e^{\eta} \tag{30}$$

the electrons have a Boltzmann distribution. For cool plasmas,

$$T \to 0 \qquad n_- \to \frac{n_0}{\sqrt{1 + \eta/u^2}} \tag{31}$$

and for weak waves the first two or three terms of Equation 26 suffice.

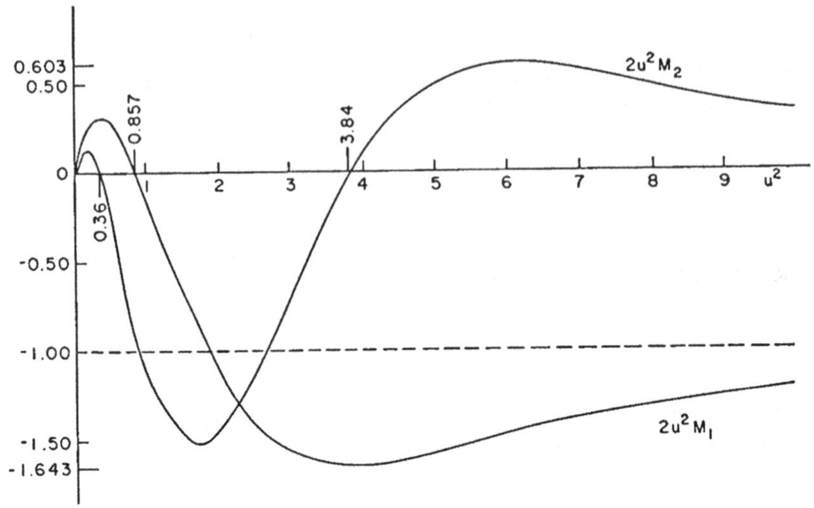

Figure 4. First and second dispersion functions.

The electron flow is given by

$$\Gamma = \int_{-\infty}^{+\infty} fv\, dv = \int_{-\infty}^{+\infty} f(u+w)\, dw = n_- u + \int_{-\infty}^{+\infty} fc\, dc - \int_{-\sqrt{-\eta_1}}^{\sqrt{-\eta_1}} fc\, dc$$

$$= (n_- - n_0)u + \Gamma_s = \Gamma_0 + \Gamma_s \tag{32}$$

The first integral makes use only of the cosh $2uc$ term in Equation 27, the last two only of the term in sinh $2uc$. The last term is the correction introduced by dropping the term in sinh $2uc$ in the trapped distribution. It is the current which is driven by the wave during phase mixing and produces the initial damping. The flow Γ_0 is also caused by the wave but is unrelated to phase mixing or to damping.

The kinetic energy is obtained by noting that

$$v^2 = w^2 + 2uv - u^2 \tag{33}$$

and evaluating $\int fw^2\, dw$ by power series. The result is

$$T = \int fv^2\, dv = \frac{n_0}{2}\left[1 + 2u^2\eta(1 - e^\eta) + \sum_1^\infty \frac{M_{a-1}\eta^a}{a!}\right] + u\Gamma_0 \tag{34}$$

The first term is the kinetic energy of the undisturbed plasma, and the last term the extra energy caused by a current.

4. *Space Charge Integral*

The average electron density $\langle n_- \rangle$ need not be equal to n_0. Therefore we shall write

$$n_+ = n_0(1 + C_2) \tag{35}$$

whence

$$n_- - n_+ = n_0 \left[\sum_1^\infty \frac{M_a \eta^a}{a!} - C_2 \right] = n_0 \frac{dF}{d\eta} \tag{36}$$

In our units, Poisson's equation is

$$\frac{d^2\eta}{dx^2} = 2\omega_0{}^2 \frac{dF}{d\eta} = \frac{1}{2} \frac{dE^2}{d\eta} \tag{37}$$

where

$$\omega_0{}^2 = \frac{n_0 e^2}{\varepsilon_0 m} \qquad \omega_p{}^2 = \frac{n_+ e^2}{\varepsilon_0 m} \tag{38}$$

Equation 37 integrates to

$$\left(\frac{E}{2\omega_0} \right)^2 = F = C_1 - C_2 \eta + \sum_1^\infty \frac{M_a \eta^{a+1}}{(a+1)!} \tag{39}$$

The constants C_1 and C_2 appear as constants of integration, and it is noted that the summation in Equation 39 is the same as that in the kinetic energy Equation 34. The quantity F must be positive. In order for $\eta = 0$ to fall within the range of the wave (see Figure 1), we choose $C_1 > 0$. All then depends on the sign of $dn_-/d\eta = n_0 F''$. For trapped electrons this is generally positive, because there are more energy levels above the bottom of the potential well, where η is largest, than elsewhere. For free electrons, it is always negative because the electrons go faster where the potential is positive. In the first case the curvature of $F(\eta)$ is positive and there is one or no root of the equation $F = 0$. In the second case, we can always choose C_1 and C_2 so that there are two roots, η_1 and η_2 (Figure 5). Roughly, C_1 determines the amplitude $\eta_2 - \eta_1$, and C_2 the position of $\eta = 0$ between η_1 and η_2. The field E is zero at the crest η_2 and trough η_1 of the wave, and no periodic wave is possible if there are less than two roots. Because $E = 0$ at both limits, there is no net charge between η_1 and η_2. Average space charge neutrality is automatically ensured.

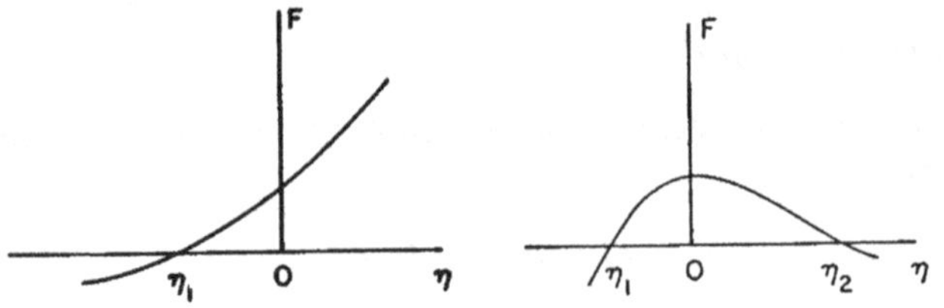

Figure 5. Determination of wave amplitude.

The electron flow Γ_0 fluctuates with n_- and its average value is given

$$\langle \Gamma_0 \rangle = (\langle n_- \rangle - n_0)u = (n_+ - n_0)u = C_2 n_0 u \tag{40}$$

Thus the constant C_2 is proportional to the dc electron flow. For a plasma in which there is no induced direct current other than the symmetrization current, Γ_s, it is zero, which means that the zero of potential falls at the inflection point of the potential profile (Figure 1).

Finally we must integrate Equation 39 for the phase

$$\omega_0 x = \int_{\eta_1}^{\eta \leq \eta_2} \frac{d\eta}{2\sqrt{F}} \tag{41}$$

This completes the exact solution for the wave form and gives the dispersion relation for the normal velocity distribution. We shall now carry out the integration for the limiting cases. However, the constants C_1 and C_2 will be replaced by two new constants A and b, algebraically related to C_1 and C_2 but more convenient in writing the solutions.

5. Limiting Solutions

(a) *Slow Waves.* The Boltzmann limit, Equation 30, is typical of trapped electrons: the curvature of F is such that there are no two solutions of $F = 0$ with $F > 0$ in between. Very slow electron waves do not exist, and we shall see that for small amplitudes the limit is $u^2 > 0.857$.

(b) *Cold Plasma.* In the cold plasma limit, the unscaled distribution is a delta function, and all electrons follow the same, open, trajectory. There are no trapped electrons. In this form the problem was solved by Akhiezer and Lyubarskii.[10] The scaled distribution remains finite, however, so that by use of Equations 31, 36, and 41 the following solution is obtained, in terms of a phase angle θ.

$$\omega_p x = bu(\theta + A \sin \theta)$$

$$\frac{\eta}{u^2} = b^2(1 + A \cos \theta)^2 - 1$$

$$E = 2\omega_p Abu \sin \theta \tag{42}$$

$$n_- = \frac{n_+}{1 + A \cos \theta}$$

$$n_+ = \frac{n_0}{b}$$

The first equation shows that the wave is periodic with scaled wavelength $2\pi bu/\omega_p$ and fundamental frequency

$$\omega = \frac{\omega_p}{b} \tag{43}$$

but there is a phase modulation $A \sin \theta$.

The crest and trough potentials are

$$\begin{aligned} \frac{\eta_2}{u^2} &= b^2(1 + A)^2 - 1 > 0 \\ \frac{\eta_1}{u^2} &= b^2(1 - A)^2 - 1 < 0 \end{aligned} \tag{44}$$

and the ascribed signs of η_1 and η_2 limit the range of b and hence the frequency range.

$$(1 - A)\omega_p < \omega < (1 + A)\omega_p \tag{45}$$

There is also the limit

$$A^2 < 1 \tag{46}$$

beyond which $dx/d\theta$ goes through zero. The condition $dx/d\theta = 0$ indicates overtaking, and at this point $n_- = \infty$. When $dx/d\theta$ changes sign, we find three values of η at the same x, which indicates that the wave has "broken" and the solution, Equation 42, is no longer valid. The maximum amplitude is $A = 1$. Two wave forms are shown in Figure 6. Equation 40 gives for the average current

$$\langle \Gamma_0 \rangle = (n_+ - n_0)u = (1 - b)n_+ u \tag{47}$$

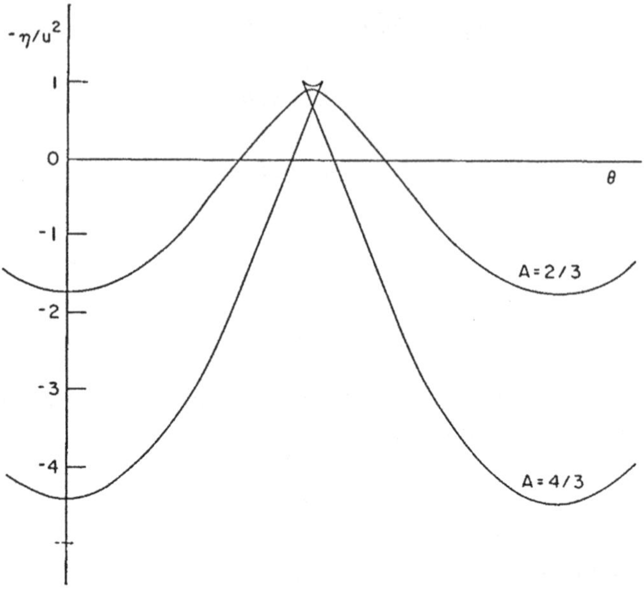

Figure 6. Cold plasma waves. The large wave has broken.

so that for no current $b = 1$ and the "dispersion" relation 43 reduces to the plasma frequency, as indeed we know it should.

(c) *Small Amplitudes.* If we keep the first three terms of the Series 26

$$n_- = n_0(1 + M_1\eta + \tfrac{1}{2}M_2\eta^2) \tag{48}$$

we are able to integrate Equation 41 in terms of elliptic functions

$$\eta = A(2sn^2\theta + b - 1) \tag{49}$$

where

$$\theta = \frac{K\omega x}{\pi u}$$

and

$$K = \frac{\pi}{2}\left[1 + \frac{m}{4} + \left(\frac{3m}{8}\right)^2 + \cdots\right]$$

is the complete elliptic integral of the first kind. The parameter m is usually written k^2 in terms of the modulus k of the elliptic function. Using Equation 41 and equating coefficients in Equation 39 then gives

$$\left(\frac{\pi u}{K}\frac{\omega_0}{\omega}\right)^2 C_2 = \frac{A[m(1+3b)(1-b)-4b]}{2}$$

$$\left(\frac{\pi u}{K}\frac{\omega_0}{\omega}\right)^2 M_1 = m(1-3b)-2 \tag{50}$$

$$\left(\frac{\pi u}{K}\frac{\omega_0}{\omega}\right)^2 M_2 = \frac{3m}{2A}$$

These yield

$$\frac{\omega^2}{\omega_0^{\,2}} = \frac{-M_1(\pi u/K)^2}{2+m(3b-1)}$$

$$A = \frac{-3M_1}{2M_2}\frac{m}{2+m(3b-1)} \tag{51}$$

and if there is no direct current, $C_2 = 0$, $\omega_0 = \omega_p$, and

$$m = \frac{4b}{(1-b)(1+3b)}$$

$$A = \frac{-M_1}{M_2}\frac{3b}{1+3b^2} \tag{52}$$

There is then only one independent variable for which it is convenient to use b. For periodicity, the parameter m must be less than one, so that

$$0 < b < \tfrac{1}{3} \tag{53}$$

The frequency depends on the amplitude both through the denominator $2 + m\,(3b - 1)$ and through the period $2K$ of the elliptic function. To first order, these two variations cancel, leaving

$$\frac{\omega^2}{\omega_p^{\,2}} = \frac{-2u^2M_1}{1+15m^2/16} \tag{54}$$

but as terms in m^2 are uncertain to the order of approximation in Equation 48, we can only say that no effect of the amplitude on the fundamental frequency has been proved. Of course the elliptic function in Equation 49 generates harmonics, and there is an effect caused by the current if C_2 is kept. The wave form for $m = 0.8$ is shown in Figure 1. No wave is possible unless M_1 is negative, or $u^2 > 0.857$.

(d) *Linear Limit.* Neglecting η^2 in Equation 48, we reach the linear limit whose solution is

$$\eta = A\sin\frac{\omega x}{u} - b \tag{55}$$

If we substitute this in Equation 41 and 36 we have

$$\left(\frac{\omega}{\omega_0}\right)^2 = -2u^2M_1$$

$$n_+ = n_0(1 - bM_1) \tag{56}$$

$$\left(\frac{\omega}{\omega_p}\right)^2 = \frac{-2u^2M_1}{1 - bM_1}$$

From Equation 40, we then find

$$\langle\Gamma_0\rangle = -bM_1n_0u \tag{57}$$

whence the dispersion relation

$$\left(\frac{\omega}{\omega_p}\right)^2 = -2u^2M_1\left(1 - \frac{\langle\Gamma_0\rangle}{n_+u}\right) \tag{58}$$

For $\langle\Gamma_0\rangle = 0$, this is precisely the Vlasov Formula 10 and yields the dispersion curve (Figure 7) given by him.[11]

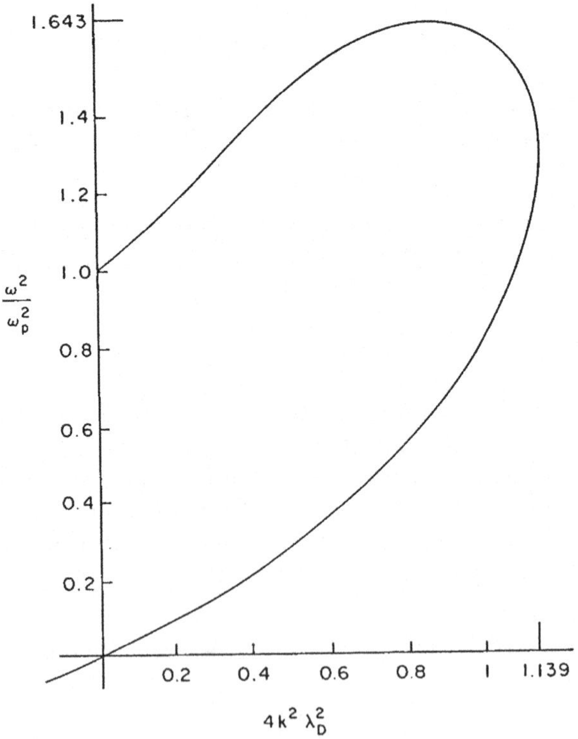

Figure 7. Dispersion curves for normally filled wells.

The dispersion formula 58 has exactly the form of Equation 13 given by Van Kampen, and we note that his arbitrary constant λ corresponds to the arbitrary current $\langle \Gamma_0 \rangle$ which appears in our theory.

Surprisingly, the current alters the frequency in the opposite direction here than in the cold plasma. (Equations 43 and 58).

6. *Trapped Electrons*

(a) *Antisymmetric Part.* Complex numbers are inappropriate in nonlinear theories. However, if the class of functions, 21, studied here is sufficiently broad to include linear waves it should be possible, in the linear limit, to introduce complex numbers and thus to show the relation between the real nonlinear theory and the complex Laplace transform theory. This is the intent of this section.

In computing the electron density (Equation 25), it was stated that the term in sinh $2uc$ was odd and therefore did not contribute to the integral. This is not quite true as can be seen by considering the complex plane of c, (Figure 8). For points at negative potential the branch points

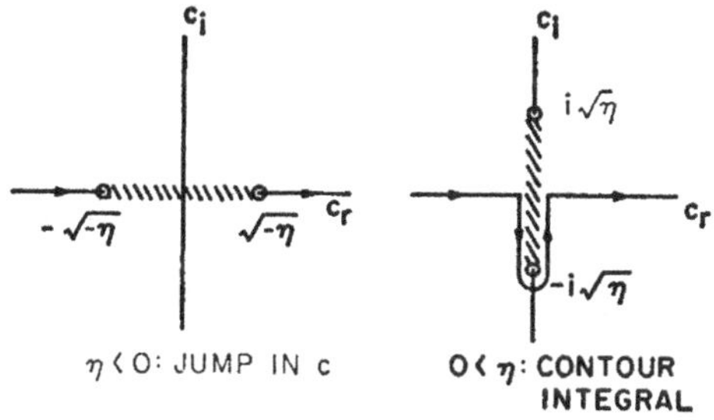

Figure 8. Complex c-plane.

at $c = \pm\sqrt{-\eta}$ are on the real axis, and the convention (Equation 20) for the sign of c introduces a discontinuity in c as w goes through zero. For points at positive potential the branch points are at $c = \pm i\sqrt{\eta}$, and for $|w| < \sqrt{\eta}$ it makes no sense to say that c has the sign of w. The path of integration must be around one of the branch points shown in Figure 8, and the result is

$$n_{t_0} = \frac{-n_0}{\sqrt{\pi}} e^{-u^2} \int_{-\sqrt{\eta}}^{+\eta} e^{-c^2} \sinh 2\,uc\,dw$$

$$= -i\sqrt{\pi}\,n_0\,ue^{-u^2}\eta\left(1 + \frac{3 - 2u^2}{4}\,n + \cdots\right) \tag{59}$$

and therefore in Equation 26 M_1 is to be replaced by

$$M_1 - i\sqrt{\pi}\, n_0\, u e^{-u^2} = -\tfrac{1}{2}Z'(u)_{\text{complex}} \tag{60}$$

which is the complete plasma dispersion function. The role of linearized theory is the remains of a branch cut between finite potentials, and damping comes entirely from the antisymmetric part of the trapped distribution. The Landau value of the damping coefficient comes from the sinh $2uc$ term in Equation 21 with the zero of potential taken at the top of the wells. This distribution is derived from an analytic continuation of the free distribution into the wells. However, analytic continuations are not entirely satisfactory, and it would be preferable to retain the Fokker–Planck terms in the Boltzmann equation so that the trapped distribution is physically determinate.[12]

(b) *Symmetric Part.* Returning to real variables, we obtained the number of trapped electrons by integrating Equation 25 between the limits $w = \pm\sqrt{\eta - \eta_1}$ and this yields

$$n_{t_r} = \frac{n_0}{\sqrt{\pi}} e^{-u^2} \int_{-\sqrt{\eta-\eta^2}}^{\sqrt{\eta-\eta_1}} e^{-c^2} \cosh 2uc\, dw$$

$$= 2n_0 \sqrt{\frac{\eta - \eta_1}{\pi}}\, e^{-u^2}$$

$$\times \left[1 + \frac{1 - 2u^2}{3} 2(\eta - \eta_1) + \frac{3 - 12u^2 + 4u^4}{3 \cdot 3 \cdot 5} 4(\eta - \eta_1)^2 + \cdots\right] \tag{61}$$

The appearance of $\sqrt{\eta - \eta_1}$ is as predicted by Bernstein, Greene, and Kruskal.[9] It leads to a function F of the form[13]

$$F = (M_1\eta_1 - C_2)(\eta - \eta_1) - C_3(\eta - \eta_1)^{3/2} + \tfrac{1}{2}M_1(\eta - \eta_1)^2 \cdots \tag{62}$$

where C_3 represents electrons subtracted from the normal distribution. As the number subtracted must be less than the total

$$C_3 \leqslant \frac{4}{3\sqrt{\pi}} e^{-u^2} \tag{63}$$

If we cut the series off at $(\eta - \eta_1)^2$ and substitute

$$y^2 = \eta - \eta_1 \tag{64}$$

the integral 41 can be evaluated and leads to

$$y = \sqrt{\eta - \eta_1} = A \sin \theta - B \qquad (65)$$
$$\theta = \frac{\omega x}{2u}$$

and if we compare coefficients with Equation 62, we have

$$\frac{\omega^2}{\omega_0{}^2} = -2u^2 M_1$$
$$C_3 = -BM_1 \qquad (66)$$
$$C_2 = \left(\eta_1 + \frac{A^2 - B^2}{2}\right) M_1 = \frac{\langle \Gamma_0 \rangle}{n_0 u}$$

Equation 40 remains valid so that for zero current $C_2 = 0$, $n_+ = n_0$, and Equation 66 gives the normal Vlasov dispersion. However we note the factor 2 in Equation 65. Unless $B = 0$, that is unless the potential wells have the "normal" number of trapped electrons, the subharmonic at $\omega/2$ will appear. The potential profile is shown in Figure 9.

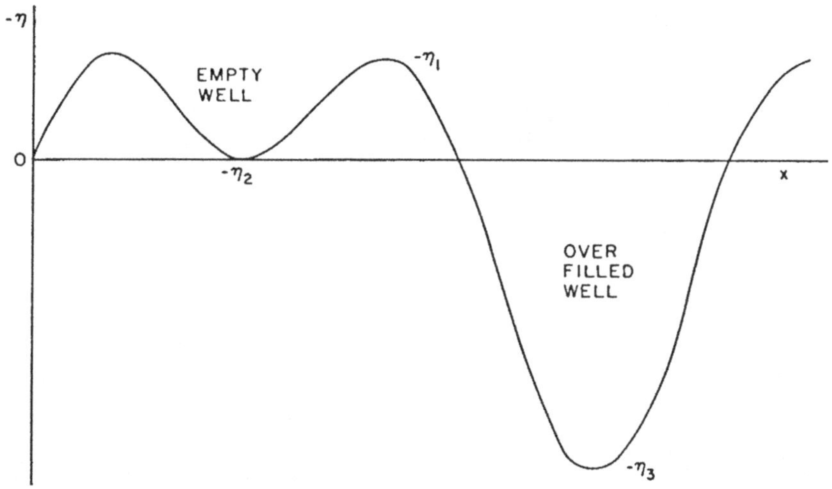

Figure 9. Wave with empty well for $A = 3B$.

As usual the position of potential zero is determined by the flow Γ_0. For $C_2 = 0$ we find the following three solutions to $F = 0$:

$$
\begin{aligned}
2\eta_1 &= (B + A)(B - A) < 0 \\
2\eta_2 &= (3B - A)(B - A) \\
2\eta_3 &= (3B + A)(B + A) > 0
\end{aligned}
\tag{67}
$$

As B is set by the number of electrons withdrawn (Equations 63 and 66), A is the amplitude factor, and the variation of η_1, η_2 and η_3 with A is shown in Figure 10. For $A < B$ the oscillation is between η_2 and η_3.

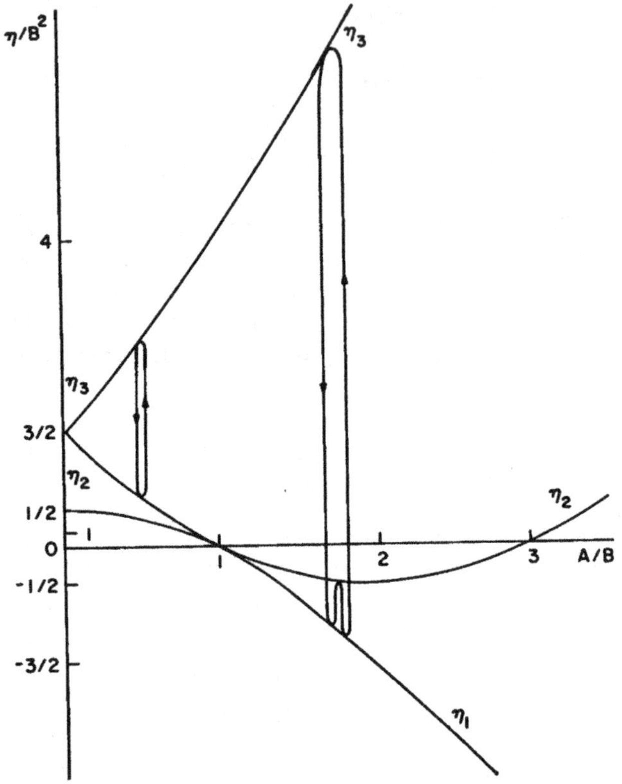

Figure 10. Oscillation amplitude variation for abnormally filled wells.

This is not a valid solution as we have, in Equation 61, subtracted electrons up to η_1. There is therefore a minimum amplitude to these oscillations.

For $B < A$ the variable y still varies between $y_3 = -A - B$ and $y_2 = A - B$ changing sign at $y_1 = 0$, thus producing the double well, shown in Figure 10, whose fundamental frequency is $\omega/2$. The deeper well corresponds to negative values of y and correspondingly negative values of $C_3(\eta - \eta_1)^{3/2}$. Thus subtracting electrons from one well results in additional electrons in the succeeding well.

The distribution 24 may be called "normal" because it gives the number of trapped electrons for which a small amplitude wave may propagate. Any deviation from this number will kill the wave if the amplitude is too small, or if the amplitude is adequate the deviation in one trough will automatically be compensated in the following one and result in the appearance of the subharmonic.

It is interesting to note that the trough to which electrons are added is the most positive. The effect of the free electrons, which go faster where the potential is positive, overcompensates for the negative charge which has been added.

It was expected that adding or subtracting trapped electrons would lead to results similar to those derived by Van Kampen from his assumption, Equation 12. This has not proved to be the case. Rather we find that the plasma rejects, or at least tries to compensate for, the delta function. Arbitrary dispersion relations such as Equation 13 are obtained from currents (Equation 58) not from trapped electrons. In any case the addition of trapped electrons leads to nonlinear waves and they cannot be simply added to form wave packets.

7. *Conclusions*

The usual perturbation expansion, Equation 5, of the distribution function diverges, and it is shown that this is equivalent to the series expansion Equation 23, of a square root in powers of η/w^2, which clearly diverges. This expansion is avoided by using a distribution function 21 which:

(1) Satisfies the appropriate boundary condition 22 in velocity space.
(2) Yields exactly the conventional solution 6 if linearized.
(3) Satisfies the time-independent, collisionless, Boltzmann Equation 4 in a frame moving with the velocity u.

The distribution is expanded in powers of the particle energy, Equation 19, and integration yields an absolutely convergent series, Equation 26, for the density in powers of η. Substitution in Poisson's equation then yields the wave form $\eta(x - ut)$ and dispersion relation $\omega(u)$ implicitly, Equation 41. Several limiting cases are studied in detail and it is shown that:

(a) There are no linear periodic waves below $u^2 = 0.857$ (Figure 4).

(b) The dispersion is of the undamped Vlasov type modified if a direct current exists, Equation 58.

(c) The dispersion is modified at most to terms in A^2 by large amplitudes, Equation 54.

(d) However, the wave may break at large amplitudes when overtaking occurs, Equation 46.

(e) A dispersion relation of the type of Equation 13 found by Van Kampen is shown to follow from the presence of an induced direct current.

Fortunately the method used here allows the effect of the distribution of trapped electrons to be studied. One finds that

(f) Landau damping results from continuing the antisymmetric part of the assumed distribution 21 into the potential wells.

(g) That the symmetric part of the trapped distribution Equation 24 is an equilibrium distribution both in the sense that it is the one left by phase mixing and because a reduction of this distribution in one well leads to a balancing augmentation in the following well, generating the subharmonic.

(h) Such a modified wave is nonlinear.

References

1. P. M. Morse, W. P. Allis, and E. S. Lamar, *Phys. Rev.*, **48**, p. 412 (1935).
2. A. A. Vlasov, *J. Exp. Theor. Phys. USSR*, **8**, p. 291 (1938).
3. B. D. Fried and S. D. Conte, *The Plasma Dispersion Function*, Academic Press Inc., New York 1961.
4. L. D. Landau, " On the Vibrations of the Electronic Plasma," *J. Phys. (USSR)* **10**, 25 (1946).
5. N. G. Van Kampen, *Physica*, **21**, p. 949 (1955); **23**, p. 641 (1957).
6. K. M. Case, *Ann. Phys.*, **7**, 352 (1959).
7. Ta-You Wu, *Kinetic Equations of Gases and Plasmas*, p. 143, Addison-Wesley Publishing Co. Inc., Reading, Mass. (1966).
8. T. O'Neil, " Collisonless Damping of Nonlinear Plasma Oscillations," *Phys. Fluids*, **8**, 2255 (1965).
9. I. B. Bernstein, J. M. Greene, and M. D. Kruskal, " Exact Nonlinear Plasma Oscillations,' *Phys. Rev.*, **108**, 546 (1957).
10. A. I. Akhiezer and G. Ya. Lyubarskii, *Dokl. Akad. Nauk. SSR*, **80**, p. 193 (1951).
11. A. A. Vlasov, *Many-Particle Theory and Its Application to Plasma*, Gordon and Breach, Science Publishers, Inc., N. Y. (1961).
12. W. P. Allis, *Proceedings of the Symposium on Electronic Waveguides*, p. 149, Polytechnic Institute of Brooklyn, N. Y. (1958).
13. Y. Ozawa, I. Kaji, and M. Kito, *Jour. Nucl. Energy C.*, **6**, p. 227 (1964). Study possible values of C_2 and C_3 for which this equation yields periodic solutions.

Leo L. Beranek
Bolt Beranek and Newman Inc.

4 Audience Absorption in Large Halls*

The ability to calculate reverberation times (R.T.) in large concert halls is important to the acoustical consultant because R.T. is one of the several parameters of a hall's musical quality which must be handled properly in design. The single most important absorption in determining R.T. in a concert hall is that provided by the audience and the seats. Many papers have appeared on the subject of audience absorption, since the beginnings of architectural acoustics, but inaccuracies in calculations have led to renewed interest in this subject especially in this decade.

Historically, the absorption introduced into a hall by an audience has been calculated on a per-person basis. Beranek postulated in 1960 that audience absorption is more accurately related to the floor area occupied by the audience, plus an edge correction to account for exposed sides and wave diffraction. Measurements in a reverberation chamber, on the other hand, suggest that the total sound absorption is neither a direct function of the number of people in an audience nor a direct function of the floor area which the audience occupies.[1] Other data indicate that the sound absorption of an audience is probably directly related to the total floor area it covers.[2] A further complicating factor is that the state of sound diffusion in a room varies according to room shape and, hence, the shape of an auditorium may affect the influence of the absorption of the audience area on the reverberation time.[3]

* Presented in part at the 71st Meeting of the Acoustical Society of America, Boston, Massachusetts, June 2, 1966, and published in the *Journal of the Acoustical Society of America*, January, 1969.

The purpose of this paper* is to study several alternate methods of accounting for audience absorption in the calculation of reverberation times in occupied concert halls, and to show which method yields the best results in practice. The halls selected for the study cover the range in the state of sound diffusion commonly in existence; they have been grouped for the purposes of analysis into several typical categories of shape and reverberation time, both of which have a significant effect on the state of sound diffusion in the room. Unusual designs of halls that might lead to decoupled spaces are avoided.[3]

1. Equivalent Audience Absorption Coefficient

Kosten[4] has introduced a simple method of calculating the reverberation time of an occupied concert hall using the Sabine reverberation equation,

$$T = \frac{0.049V}{S_T \alpha_{eq}} \tag{1}$$

where α_{eq} is called the *equivalent sound absorption coefficient* for the audience and musicians and S_T is the floor area occupied by them (see terminology below).

Kosten's α_{eq} includes: (a) the absorption of the audience; (b) the absorption of all other surfaces, chandeliers, statues, ventilating grilles and doors; (c) sound absorption in the air, which is significant at frequencies above 1500 Hz; and (d) large areas of sound absorbing materials, if any exist. All four items are handled in Equation 1 as though they were concentrated in the audience area S_T. This method of calculation of R.T. is convenient for early calculations and for rapid checks of measured data. The accuracy of prediction should be improved by allowing α_{eq} to include only items a through c and adding explicitly into the calculations the absorption of significant particular areas of added absorbing materials.

In this paper, new values for α_{eq} are determined which differ somewhat from Kosten's values. The data for determining α_{eq} in this paper are given in Tables 1 to 3. Three categories of shape of hall are presented—rectangular, "modern"† and wide-fan-plan. The symbols used in the tables are: T_{MID} is the average of the reverberation times at 500 and 1000 Hz; bass ratio $R = T_{LOW}/T_{MID}$, where T_{LOW} is the average of the

* The author is grateful to C. W. Kosten, H. Kuttruff, and T. J. Schultz for comments on earlier drafts of this paper.

† A "modern" hall generally is a moderate fan-plan hall with relatively a low and nonhorizontal ceiling. More generally, it is a hall that does not fall in the rectangular or wide-fan-plan categories.

reverberation times at 125 and 250 Hz; V in cubic feet is the volume of the hall; S_T in square feet is the floor area occupied by the audience, musicians, and chorus (if present when the reverberation time is to be known) and includes an edge effect at aisles equal in area to a strip one meter wide for an aisle bounded on both sides by audience and 0.5 meter wide for an aisle bounded on one side by audience, but not where the seating abuts walls or balcony fronts, nor around musicians; S_R = area in square feet of all surfaces of the hall, including under-balconies, not included in S_T ; N_A = number of people in the audience (all seats filled) ; and S_A = the audience area in square feet without orchestra, chorus, or edge corrections.

Kosten estimates that the accuracies of the components that go into determining the equivalent absorption coefficient are about $\pm 3\%$ for room volume; $\pm 5\%$ for reverberation time; $\pm 3\%$ for audience area (including edge corrections). Reverberation times are usually rounded off to tenths of a second, which adds another $\pm 3\%$. The rms error, therefore, is about $\pm 7\%$.

It must be emphasized, however, that not all the reverberation times given in Tables 1 to 3 are accurate to $\pm 3\%$. They were measured by various experimenters, using several different techniques, over a span of 30 years. (The sources of the data in these tables are given in the references.[2, 5, 6]) Thus we must deal with the medians of categories of halls in order to determine suitable values for α_{eq}. Whenever a value for bass ratio is not given in a table, it is because thin wooden walls or ceiling exist in the hall which produce significant low-frequency absorption, and this would prevent the tabulated R.T. from all being principally dependent on audience absorption. The wood should not significantly affect the sound absorption at mid-frequencies.

Inspection of the tables reveals that, in general, as T_{MID} increases, S_R/S_T increases and the bass ratio R decreases. Plots of the medians of these quantities *vs* the medians of T_{MID} are given in Figures 1 and 2. Architectural design considerations apparently have led to near-direct proportionality between S_R/S_T and T_{MID} for "modern" halls. For high values of T_{MID}, the values of S_R/S_T for the two shapes of room appear to converge. A simple calculation will show that a rectangular room tends to have a higher wall area for a given R.T. than a modern-shaped or fan-plan hall.

In using the data of this paper to calculate reverberation times, we expand Kosten's α_{eq} to the following form,

$$\text{Kosten's } \alpha_{eq} = \alpha_T + \frac{S_R \alpha_R}{S_T} + \frac{S_1 \alpha_1}{S_T} + \frac{S_2 \alpha_2}{S_T} + \cdots + \frac{4mV}{S_T} \qquad (2)$$

Table 1. Statistical Data for Nineteen Modern Shaped Halls

Name	T_{MID} (sec)	Bass Ratio R	S_T (ft²)	S_R (ft²)	S_R/S_T	V/S (ft)	V/S_T (ft)	N_A	S_A/N_A (ft²)	New α_{eq} MID	New α_{eq} LOW
		GROUP IA									
Berlin, Philharmonie	2.00	1.10	—	—	—	—	—	2350	—	—	—
Berlin. Freies	1.95	1.15	10,600	23,600	2.22	13.4	43.0	1120	7.7	1.08	0.94
Chicago, Arie	1.7	1.18	35,100	60,900	1.73	13.5	36.8	5081	6.5	1.06	0.90
Cleveland	1.7	1.18	15,000	45,000	3.00	9.3	36.9	1890	6.9	1.06	0.90
Montreal	1.7	1.22	25,800	70,700	2.74	9.7	36.2	2980	8.0	1.04	0.85
Gothenburg	1.7	—	10,350	24,050	2.32	12.2	40.6	1371	6.5	1.17	—
Lakeland, Florida	1.65	—	13,100	28,900	2.20	9.1	29.0	1814	6.8	0.86	—
Berlin MHS	1.65	—	9,850	22,570	2.30	10.5	34.6	1340	6.0	1.03	—
Stuttgart	1.62	—	16,500	38,500	2.34	10.3	34.2	2000	7.0	1.03	—
IA Median	1.7	**1.18**	14,000	33,700	2.31	10.4	36.5	1890	6.85	1.05	0.90
		GROUP IB									
Lafayette	**1.6**	**1.28**	**39,500**	**69,100**	**1.75**	**12.2**	**33.4**	**6107**	**6.1**	**1.02**	**0.80**
Turku	**1.6**	**1.14**	**9.750**	**26,250**	**2.70**	**9.5**	**34.9**	**1002**	**8.0**	**1.07**	**0.94**
Cambridge	**1.5**	—	**10,550**	**23,450**	**2.22**	**10.4**	**33.6**	**1238**	**7.5**	**1.10**	—
Salzburg	**1.5**	**1.08**	**16,900**	**31,600**	**1.87**	**11.3**	**32.4**	**2158**	**6.9**	**1.06**	**0.98**
Vancouver	**1.5**	**1.25**	**21,300**	**47,000**	**2.21**	**8.7**	**27.8**	**2800**	**6.9**	**0.91**	**0.73**
Liverpool	**1.5**	—	**15,300**	**32,500**	**2.12**	**10.0**	**31.3**	**1955**	**7.1**	**1.02**	—
London RFH	**1.5**	—	**23,100**	**53,900**	**2.35**	**10.1**	**33.6**	**3000**	**7.1**	**1.10**	—
IB Median	**1.5**	**1.20**	**16,900**	**32,500**	**2.21**	**10.1**	**33.4**	**2160**	**7.1**	**1.06**	**0.87**

			GROUP IC								
Chicago, O.H.	1.45	1.28	20,000	38,000	1.90	9.3	26.8	2582	7.0	0.91	0.71
Brussels	1.4	1.34	16,000	32,000	2.00	9.2	27.6	2150	6.5	0.97	0.72
Buffalo	1.3	1.27	23,200	42,000	1.81	9.9	27.8	2839	7.4	1.05	0.83
IC Median	1.4	1.28	20,000	38,000	1.90	9.3	27.6	2582	7.0	0.97	0.72
Group I Median	1.6	1.20	16,200	35,200	2.22	10.1	33.6	2150	7.0	1.04	0.85

Table 2. Statistical Data for Twelve Rectangular Halls

Name	T_{MID} (sec)	Bass Ratio R	S_T (ft²)	S_R (ft²)	S_R/S_T	V/S (ft)	V/S_T (ft)	N_A	S_A/N_A (ft²)	New α_{eq} MID	New α_{eq} LOW
		GROUP IIA									
N. Y. Philharmonic	**2.1**	**1.17**	**19,200**	**60,000**	**3.12**	**10.9**	**45.0**	**2700**	**6.5**	**1.05**	**0.90**
Indianapolis	**2.05**	**1.22**	**17,340**	**56,260**	**3.24**	**10.2**	**43.4**	**2100**	**7.4**	**1.04**	**0.85**
Vienna, MVS	**2.05**	**1.12**	**12,000**	**37,500**	**3.12**	**10.7**	**44.2**	**1680**	**6.3**	**1.06**	**0.95**
Amsterdam	**2.00**	**1.10**	**13,800**	**43,000**	**3.12**	**11.6**	**48.0**	**2206**	**5.5**	**1.18**	**1.07**
IIA Median	**2.05**	**1.15**	**15,500**	**50,000**	**3.12**	**10.8**	**44.6**	**2150**	**6.4**	**1.05**	**0.92**
		GROUP IIB									
Glasgow	**1.9**	—	**14,900**	**38,300**	**2.58**	**10.7**	**38.2**	**2133**	**6.3**	**0.98**	—
Boston	**1.8**	**1.17**	**16,600**	**45,600**	**2.74**	**10.6**	**39.9**	**2631**	**5.7**	**1.09**	**0.93**
Bristol	**1.7**	—	**13,460**	**36,540**	**2.74**	**9.5**	**35.3**	**2180**	**5.6**	**1.02**	—
La Chaux de Fonds	**1.7**	—	**8,350**	**24,050**	**2.88**	**8.6**	**33.3**	**1032**	**6.8**	**0.96**	—
IIB Median	1.75	—	14,100	37,200	2.74	10.0	36.7	2150	6.0	1.0	

		GROUP IIC									
Basel	1.7	1.24	9,650	28,850	2.99	9.6	38.3	1400	5.7	1.10	0.89
Zurich	1.6	1.22	11,000	29,300	2.66	10.0	36.6	1546	6.1	1.12	0.92
Leipzig	1.55	—	11,000	30,300	2.76	9.1	34.0	1560	6.3	1.07	—
Dartmouth	1.5	1.13	8,190	25,950	3.18	7.8	32.5	900	7.6	1.06	0.94
IIC Median	1.57	1.22	10,300	29,000	2.88	9.4	35.3	1470	6.2	1.08	0.92
Group II Median	1.75	1.17	12,700	37,000	2.94	10.1	38.2	1890	6.3	1.06	0.92

Table 3. Statistical Data for Seven Wide-Fan-Plan Halls

Name	T_{MID} (sec)	Bass Ratio R	S_T (ft²)	S_R (ft²)	S_R/S_T	V/S (ft)	V/S_T (ft)	N_A	S_A/N_A (ft²)	New α_{eq} MID	New α_{eq} LOW
Lenox	2.05	1.24	33,000	108,000	3.28	10.6	45.5	6000	5.1	1.09	0.88
Jerusalem	1.75	1.20	25,900	45,900	1.77	12.2	33.8	3142	7.3	0.95	0.79
Jacksonville	1.75	1.28	24,800	76,700	3.09	10.7	44.0	3200	7.0	1.23	0.96
San Diego	1.65	1.24	22,500	61,500	2.74	9.8	36.4	3000	6.8	1.07	0.86
Tel Aviv	1.55	—	20,800	51,700	2.48	10.3	36.0	2715	6.7	1.14	—
Detroit	1.55	1.27	22,100	44,100	2.00	10.1	30.6	2926	6.8	0.97	0.76
Edmonton	1.42		23,000	56,000	2.44	9.6	33.0	2731	7.7	1.14	—
Group III Median	1.65	1.24	23,000	56,000	2.48	10.3	36.0	3000	6.8	1.09	0.86

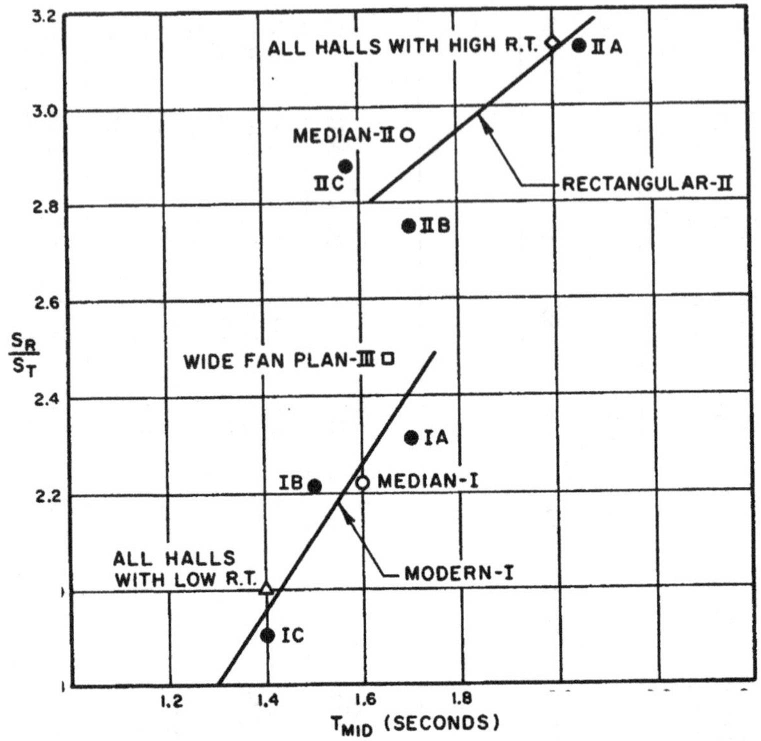

Figure 1. Variation of medians of S_R/S_T for "modern" and rectangular halls *vs* T_{MID}. Medians are also shown for groups of halls with high reverberation times, low reverberation times, and wide-fan plans.

The total area in the room equals

$$S = S_T + S_R + S_1 + S_2 + \dots \qquad (3)$$

Here, α_1 is the absorption coefficient for any substantial area S_1 of a particular sound absorbing material, α_2 and S_2 are the corresponding values for a second material, and so forth. The quantity m is the air absorption coefficient given in Table 4.[7] In the calculations of this paper, no hall is included for which there are significant areas of particular sound absorbing materials so the S_1 and S_2 terms of Equation 2 *can* be omitted. Thus, we have,

$$\text{New } \alpha_{\text{eq}} = \alpha_T + \frac{S_R \alpha_R}{S_T} + \frac{4mV}{S_T} \qquad (4)$$

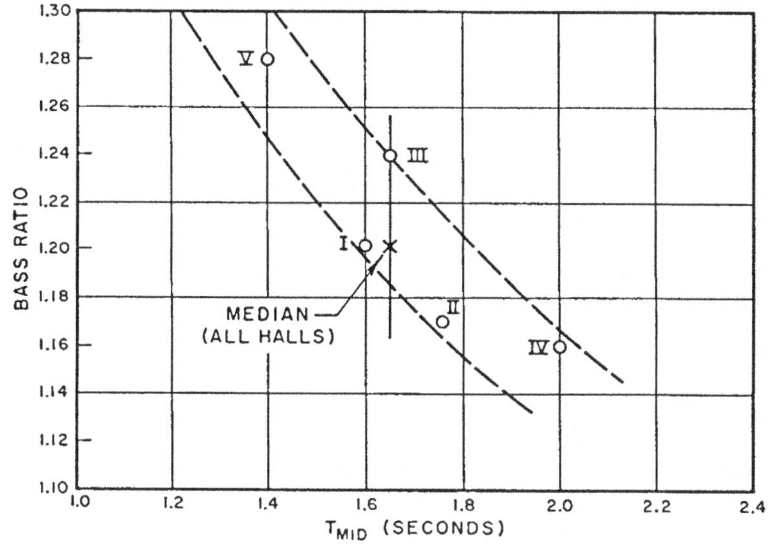

Figure 2. Variation of bass ratio with T_{MID} for several groupings of halls. The data are from Tables 1 to 3. Points I, II, and III are the medians of the three groups. Point IV is the median of halls with high reverberation times and Point V is the median of halls with low reverberation times. The dashed lines encompass the spread of the medians. The decrease in bass ratio with increase in T_{MID} is explainable from Equation 4 and the data shown in rows 1 and 3 of Table 6. For the seated area, the ratio of α_{MID} to $(S_R/S_T)\alpha_{LOW}$ is 0.72, for $S_R/S_T = 2.5$. For the unseated area, this ratio is 0.18. Thus, for halls with high values of S_R/S_T (high reverberation times) the bass ratio must decrease because the unseated area S_R becomes much more important relative to S_T in determining the reverberation time than is true for smaller values of S_R/S_T (low reverberation times).

Table 4. Value of the Air Absorption Coefficient m in Units of Inverse Feet. Normal Room Temperature is Assumed[7]

	Absorption Coefficient m (ft^{-1})			
Relative Humidity	2000 Hz	4000 Hz	6000 Hz	8000 Hz
30%	0.0010	0.0034	0.0057	0.0091
50%	0.0008	0.0023	0.0046	0.0067
70%	0.0006	0.0020	0.0039	0.0057

and the Sabine equation must include the missing terms when significant added absorbing areas are present as follows,

$$T = \frac{0.049V}{S_T\alpha_{eq} + S_1\alpha_1 + S_2\alpha_2 + \cdots} \tag{5}$$

In Table 5, values of New α_{eq} for occupied areas at individual frequencies and at MID and LOW frequencies are given, as calculated from the medians of data in Tables 1 to 3. These α_{eq} are lower, particularly at low frequencies, than those of Kosten. By taking account of the extra terms, $S_1\alpha_1$ etc. of Equation 2 and using the new values of α_{eq}, we will achieve greater accuracy of calculation for all halls than is obtained with Kosten's average α_{eq}. Of course, one could obtain a spectrum of α_{eq} from Kosten's curves by using the upper ends for halls with added absorbing materials and the lower ends for halls with plaster-on-concrete-block walls. But, because of the inaccuracies in measurement that underlie the data, this procedure would involve an element of judgement that couldn't be described adequately.

Table 5. Determination of New α_{eq} Using Equation (5) But Restricting the Halls to Those Without Appreciable Areas of Sound Absorption Other than the Audience, Musicians and Normal Ventilation Openings and Echo Control Materials, i.e., $S_1\alpha_1 + S_2\alpha_2 \cdots = 0$.

	New α_{eq} For Halls That Are:		
Frequency (Hz)	Occupied	Unoccupied (Occupied $-\Delta\alpha_{eq}$)	$\Delta\alpha_{eq}^*$
63	0.77	0.60	0.17
125	0.87	0.67	0.20
250	0.91	0.71	0.20
500	1.03	0.79	0.24
1000	1.09	0.82	0.27
2000	1.15	0.87	0.29
4000	1.29	1.01	0.28
6000	1.50	1.27	0.23
LOW: 125/250 av	0.89	0.69	0.20
MID: 500/1000 av	1.06	0.80	0.26

*$\Delta\alpha_{eq}$ is obtained from Figures 1 through 9 of Kosten[4] by entering an abscissa with the (occupied) New α_{eq} of column two above and determining a percentage value from the right-hand curve. The $\Delta\alpha_{eq}$ is equal to the separation of the two curves at that percentage value.

2. *Audience Absorption by Area*

Although it may be argued that New α_{eq} and knowledge of $(S_1\alpha_1 + S_2\alpha_2 \ldots)$, where necessary, are sufficient for the calculation of reverberation time within the accuracy needed for many design considerations, it is often desirable to separate the sound absorption into its components. Let us now proceed to do this.

First, we need to determine α_T, the absorption coefficient for areas occupied by audience, orchestra, and chorus, plus edge corrections, as defined earlier. It seems best to interpolate the values for α_T from the data of Kath and Kuhl,[8] Meyer *et al*,[3] and Beranek,[2] measured respectively in a large broadcasting studio, in a reverberation chamber and in halls, with people seated in well-upholstered seats. The interpolation is plotted as the upper curve of Figure 3.

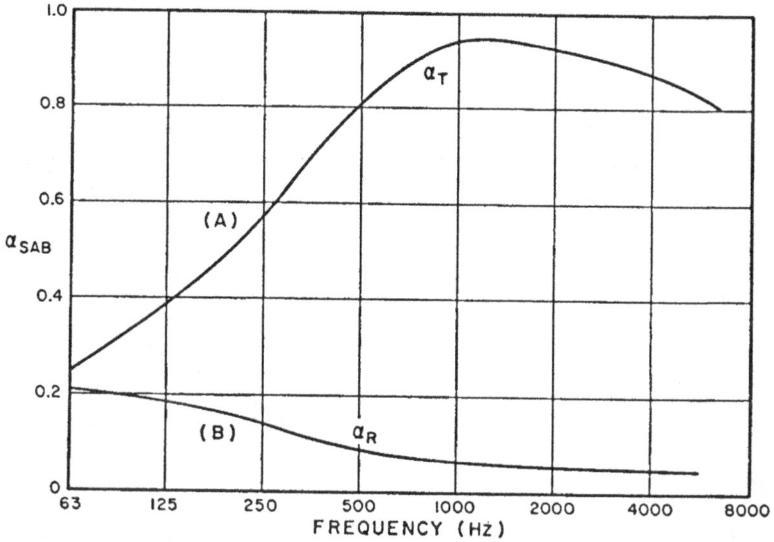

Figure 3. (A) Sabine absorption coefficients α_T for occupied audience, orchestra and chorus with areas S_T: This curve is interpolated from data of Kath and Kuhl[8], Meyer *et al.*[3], and Beranek[2]; (B) Sabine absorption coefficients α_R for unseated (remaining) areas S_R, in halls with approximately 1.0 inch plaster, or crack-free wood walls (more than 1.5 inch thick). The absorption coefficient α_R and the area S_R include all surfaces not included in S_T, including grilles, chandeliers, statues, doors, glass, and underbalcony soffits, except for areas of highly absorbing materials which must be added into the calculations according to Equation 5.

Now, we shall calculate α_R, the absorption coefficient for the unseated areas, including underbalconies. At frequencies above 1500 Hz, a relative

Table 6. Absorption coefficients of (1) occupied audience, orchestra, and chorus areas; (2) unoccupied "average," well-upholstered seating areas (seats with perforated bottoms and either wooden arms or wooden rear of seat back); (3) 1.0-in. thick plaster walls and ceiling or 1.5-in. thick (or more), crack-free, wooden walls; (4) plaster on concrete block; (5) thin (0.2 to 0.4-in.) wood-panelling areas with air-space behind. The numbers below are for use in the Sabine equation

$$T = \frac{0.049\,V}{(S_T\alpha_T + S_R\alpha_R + 4mV)}$$

Description (See Caption Above)	Absorption Coefficients, α_{Sab}, at Indicated Frequencies							
	63 Hz	125 Hz	250 Hz	500 Hz	1000 Hz	2000 Hz	4000 Hz	6000 Hz
(1) Audience, α_T†	0.25	0.39	0.57	0.80	0.94	0.92	0.87	0.80
(2) Upholstered Seats,† Unoccupied, α_S	0.08	0.19	0.37	0.56	0.67	0.61	0.59	0.57
(3) Plaster or Thick Wood Panelling, α_R*	0.21	0.19	0.14	0.09	0.06	0.06	0.05	0.04
(4) Plaster on Concrete Block, α_R*	0.16	0.12	0.09	0.07	0.05	0.05	0.04	0.04
(5) Thin Wood Panelling, α_P**	—	0.42	0.21	0.10	0.08	0.06	0.06	0.06

† The audience area must be calculated to include an edge effect at aisles equal in area to a strip one meter wide for an aisle bounded on both sides by audience and a strip 0.5 meter wide for an aisle bounded on only one side by audience. No edge effect is used where the seating abuts walls or balcony fronts. Orchestra areas include men, instruments, music racks and lighting. No edge effects are used around musicians. These coefficients are valid for areas per person (without aisles) of between 5 and 8 ft^2. The audience is assumed to be seated on upholstered chairs.

* These absorption coefficients have averaged in them the absorption of usual doors, ventilating grilles, statues, glass, chandeliers, and small areas of sound-absorbing material used for echo control. However, large areas of sound absorbing materials must be added, in addition, into the denominator of the Sabine Equation (see Equation 5).

** The absorption of thin wood panelling is added into the denominator of the Sabine Equation (see Equation 5) above by the term $S_P\alpha_P$.

humidity of 60% and a median value of $V/S_T = 35$ ft is used to determine $4mV/S_T$. The median value of S_R/S_T for all halls is 2.50. Given New α_{eq} from Table 5, α_T from Figure 3, $S_R/S_T = 2.5$, and the values of m from Table 4, α_R is calculated from Equation 4 and is plotted as the lower curve of Figure 3.

The values of α_T and α_R are also tabulated in Table 6. Absorption coefficients α_S for typical unoccupied, well-upholstered chairs are next calculated by subtracting $\Delta\alpha_{eq}$ of Kosten[4] from New α_{eq} for occupied halls to obtain New α_{eq} for unoccupied seating (see Table 5). Then using Equation 4, along with New α_{eq} from Table 5, $S_R/S_T = 2.5$, α_R from Table 6, $V/S_T = 35$ and m from Table 4, α_S for upholstered seats is calculated and tabulated in Table 6.

3. *Audience Absorption by Occupancy*

The Sabine equation for calculating reverberation time on a per-person or per-seat basis is,

$$T = \frac{0.049V}{N_A a + S_R \alpha_R + 4mV} \tag{6}$$

where a is the absorption per seat, occupied or unoccupied as desired, in sabins (ft^2).

To convert the absorption coefficients of rows 1 and 2 of Table 7 into absorption per person or per seat, a conversion factor of $(S_A/N_A \times 1.10)$ is used. The median value of S_A/N_A, the number of square feet per person in the audience or per seat (without edge corrections) is found from Tables 1 to 3 to be 6.8 ft^2. The 10% increase (1.10) accounts approximately for the contribution owing to the edge correction for seating area. Thus, the conversion factor is 7.5 ft^2. The absorption per seated person and per unoccupied seat in sabins (equal to 7.5 α_T and 7.5 α_S, respectively) are given in Table 7. These values are roughly comparable to those published by Furrer[9] and Lane and Botsford[10] and are lower at low frequencies than those published earlier by the author.[2,11]

4. *Analysis*

This paper combines the best currently known measured data to establish coefficients for determining the total sound absorption (in sabins) in concert halls, assuming large halls and an average state of sound diffusion. They supersede earlier coefficients of the author.[2,11]

Table 7. Absorption of Seated Persons and of Unoccupied Upholstered Chairs on a per Seat Basis. The numbers below are for use in the Sabine equation,

$$T = \frac{0.049\ V}{(N_A a + S_R \alpha_R + 4mV)}$$

Absorption *a* per person or per seat in sabins (ft^2)

Description	Frequency (Hz) 63	125	250	500	1000	2000	4000	6000
(1) Person in Upholstered Seat, a_T	1.9	2.9	4.3	6.0	7.0	6.9	6.5	6.0
(2) Unoccupied, "Average" Well-Upholstered Seat, a_S	0.6	1.4	2.8	4.2	5.0	4.6	4.4	4.3

Meyer *et al.*[3] have concluded that the absorption of an audience, when measured in a reverberation chamber, is neither directly proportional to the area which it occupies, nor to the number of persons present, but somewhere in-between. Their conclusion is not in agreement with the author's earlier postulate[11] and so needs further examination.

We shall now apply the per-person and the per-unit-area methods to particular large halls. Remember that the quantities in Table 6 were calculated for a median seating density of 6.8 ft^2 per person. We find, from Tables 1 to 3, five halls with high seating densities (5.1 to 5.7 ft^2 per person), and five halls with low seating densities (7.4 to 8.0 ft^2 per person). We may use these to compare reverberation times (calculated and measured) using α_T and a_T for the median group (6.8 ft^2 per person). The reverberation times, using Equations 5 and 6, Tables 6 and 7, and particular values of S_R, S_T and V/S_T from Tables 1 to 3 are calculated. The results are shown in the right-hand two columns of Table 8. The measured reverberation times are shown in the second column.

It is obvious that, for halls in which the seating density differs widely from the norm, calculation according to the audience absorption per-unit-area method is quite accurate while the per-person method yields reverberation time values that deviate considerably from those measured. Furthermore, the differences are in the expected direction: the per-person method yields reverberation times about 11% too low when S_A/N_A is about 12% below the norm of 6.8 ft^2 per person. The per-person method also yields reverberation times about 19% too high when S_A/N_A is about 13% larger than the norm. Similar results are obtained when calculating R.T. at low frequencies, as shown in Table 9.

5. *Conclusion*

The evidence presented in this paper further confirms the postulate of References 2 and 11 that in a large concert hall the absorbing power of a seated audience, chorus, and orchestra increases in proportion to the floor area they occupy, nearly independent of the number of seated persons in that area (uniform distribution of persons and well-upholstered seats assumed). This hypothesis is valid for seating densities in at least the range of 5 to 8 ft^2 of floor space per person and for concert halls with normally diffuse sound fields. No attempt has been made to extend the applicability of these data to small auditoriums or churches where the state of sound diffusion or the seating density may be substantially different.

Table 8. Comparison of Calculations of Reverberation Times at Mid-Frequencies by Use of Absorption Per Unit Area (Equations 4 and 5) and Absorption Per Person (Equation 6) with Measured R.T.

		Reverberation Time at Mid-Frequencies				
Hall	Area Per Person S_A/N_A (ft^2)	Measured (sec)	Data Taken By	Calculated, Per Unit Area Eqs. 4 and 5 (sec)	Calculated, Per Person Eq. 6 (sec)	Group
Lenox, Mass. Tanglewood	5.1	2.05	BBN	2.09	1.55	Low Area Per Person
Amsterdam	5.5	2.0	Geluk BBN	2.13	1.85	
Bristol	5.6	1.7	Parkin	1.63	1.38	
Boston	5.7	1.8	BBN	1.82	1.60	
Basel	5.7	1.7	Furrer	1.72	1.61	
Montreal	8.0	1.7	BBN	1.65	1.86	High Area Per Person
Turku	8.0	1.6	Arni	1.59	1.97	
Berlin, Freies	7.7	1.95	ITA	2.03	2.35	
Edmonton	7.7	1.42	Northwood	1.54	1.69	
Indianapolis	7.4	2.05	BBN	2.01	2.25	
Median Deviation of Upper Group				+0.02 sec	−0.20 sec	
Median Deviation of Lower Group				−0.01 sec	+0.27 sec	

Table 9. Comparison of Calculations of Reverberation Times at Low-Frequencies by Use of Absorption Per Unit Area (Equations 4 and 5) and Absorption Per Person (Equation 6) with Measured R.T.

		Reverberation Time at Low-Frequencies				
Hall	Area Per Person S_A/N_A (ft^2)	Measured (sec)	Data Taken By	Calculated, Per Unit Area Eqs. 4 and 5 (sec)	Calculated, Per Person Eq. 6 (sec)	Group
Lenox, Mass.						
Tanglewood	5.1	2.55	BBN	2.18	1.86	Low
Amsterdam	5.5	2.2	Geluk	2.36	2.17	Area
			BBN			Per
Boston	5.7	2.10	BBN	2.10	1.92	Person
Basel	5.7	2.1	Furrer	1.93	1.85	
Montreal	8.0	2.02	BBN	1.90	2.07	High
Turku	8.0	1.82	Arni	1.85	2.10	Area
Berlin, Freies	7.7	2.25	ITA	2.50	2.82	Per
Indianapolis	7.4	2.5	BBN	2.62	2.75	Person
Median Deviation of Upper Group				−0.08 sec	−0.22 sec	
Median Deviation of Lower Group				+0.07 sec	+0.26 sec	

References

1. E. Meyer, D. Kunstmann, and H. Kuttruff, " Measurements of Sound Absorption of Audience ", *Acustica*, **14**, 119–124 (1964).
2. L. L. Beranek, *Music, Acoustics, and Architecture*, John Wiley & Sons, Inc., New York (1962), pp. 541–554.
3. E. Meyer, H. Kuttruff, and N. Roy, " Room Acoustical Measurements in a Model of the 'Stadthall' in Goettingen," *Acustica*, **19**, 132–142 (1967/68).
4. C. W. Kosten, " New Method for the Calculation of the Reverberation Time of Halls for Public Assembly," *Acustica*, **16**, 325–330 (1965/66).
5. L. L. Beranek, and T. J. Schultz, " Design and Testing of Concert Halls with Suspended Panel Arrays," *Acustica*, **15**, 307–316 (1965).
6. L. Cremer, " The Acoustical Necessities in the New Berlin Philharmonie," *Deutsche Bauzeitung*, **10**, 3–7 (1965).
7. C. M. Harris, "Absorption of Sound in Air," *J. Acoust. Soc. Am.*, **35**, 11–17 (1963).
8. U. Kath and W. Kuhl, " Measurements of Sound Absorption of Upholstered Chairs, With and Without People." *Acustica*, **14**, 50–55 (1964).
9. W. Furrer, *Raum und Bauakustik*, Birkhäuser Verlag, Basel (1961) p. 91.
10. R. N. Lane and J. Botsford, " Total Sound Absorption for Upholstered Theatre Chairs," *J. Acoust. Soc. Am.*, **24**, 125–126 (1952).
11. L. L. Beranek, "Audience and Seat Absorption in Large Halls," *J. Acoust. Soc. Am.*, **32**, 661–670 (1960).

Cyril M. Harris
Columbia University

5 Measurements of the Absorption of Sound in Air

Introduction

The losses in acoustic energy that result when sound waves are transmitted through dry air may be predicted from classical theory. This *classical* absorption results primarily from viscosity, heat conduction and radiation. But the addition of water vapor or other impurities gives rise to additional losses, so called *anomalous* or *molecular* absorption which may be much larger in value than the classical absorption. In moist air, significant molecular absorption takes place as a result of the lag between internal temperatures of rotation and vibration and translational temperature.[1] This molecular absorption results in a peak in the curve of absorption *vs* humidity. The humidity at which such a peak occurs depends on frequency — in the audio frequency range the peak falls in the range from about 5 to 20% relative humidity, for normal atmospheric pressure.

In recent years more accurate air absorption data have been required, particularly in the low frequency range where experimental measurements had never been made, in order to predict the noise fields at some distance from jet and rocket engines. The required air attenuation data could not be calculated from theory because there is no theoretical model of the absorption process in moist air which is sufficiently close to the actual mechanism, although the general mechanism of the absorption process is well established. As a result, the experimental program described here was initiated — not only to provide the required engineering data for sound attenuation but also to provide basic information which may serve as a check on the validity of proposed theoretical models of the absorption process.

In approaching the measurement problem, it was only natural for the author, as a former student of Professor Morse's, to employ a technique which follows as a consequence of an assumption made in the now classical Chapter VIII of his book *Vibration and Sound*, i.e. that the air in a room be considered as an assemblage of resonators which result in "standing waves that can be set into vibration by a source and that will die out exponentially when the source is stopped."[2]

The rates of decay of the sound pressure of such standing waves in an enclosure can be used to determine values of sound absorption in air in the following way. The decay rates of the standing waves depend on two factors: (1) the boundary conditions, and (2) the absorptive characteristics of the air, i.e.

$$R_{\text{total}} = R_{\text{walls}} + R_{\text{air}} \qquad (1)$$

where R_{walls} is the rate of decay in dB/sec resulting from the absorption of acoustic energy at the boundaries, and R_{air} is the rate of decay in dB/sec caused by the absorption of sound in air. When decay rates are measured in an enclosure, it is the total rate of decay of sound pressure which is evaluated. In this study, a technique was employed in which the contribution of the wall losses is determined, thereby providing a means for evaluating the contribution caused by only air absorption.

1. Apparatus and Measurement Technique

A hollow sphere was selected as the enclosure whose acoustic properties were to be studied.[3] This shape was chosen and the walls were fabricated of thick steel in order to reduce to a minimum the losses of acoustic energy at the walls of the enclosure. In the low frequency range, the decay rate of the sound pressure of each radial mode of vibration was measured. (These modes have a particle velocity which is normal to the spherical surface.) In the high frequency range, the sphere was excited over a series of bands of frequencies and the decay rate of sound pressure was measured for each frequency band. This is the *statistical case* in which a large number of modes of vibration are excited.[4]

The control and measurement of humidity is one of the most important factors in determining the accuracy of air absorption data. In the measurement system described here, electric hygrometers were employed to advantage since they have relatively high accuracy at low values of humidity; greatest accuracy is required in this range in order to define peaks in the curves of absorption vs. frequency. In order to establish a state of equilibrium between the air in the sphere and the boundary surface which may either absorb or give off moisture, the system shown in Figure 1 was used to recirculate the air until such a state of equilibrium

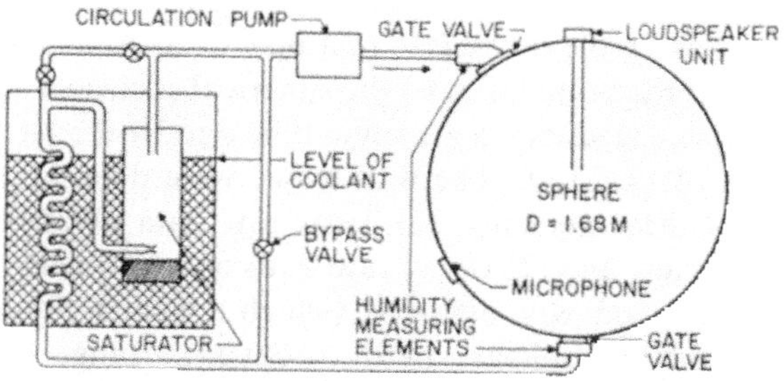

Figure 1. Simplified schematic diagram of the air-circulation system. Air is recirculated continuously through the spherical chamber. The saturator either takes away moisture from the air or adds moisture to it, depending on the relative temperatures of the spherical chamber and the saturator.

was achieved. In order to increase the humidity in the sphere, valves of the recirculation system were set so that the air passed through a *saturator chamber*, which contained a small amount of distilled water at an elevated temperature. In order to lower the humidity in the sphere, the temperature of the saturator was reduced to a value well below freezing, and the air from the sphere was recirculated through it, thereby removing moisture from the air and depositing it in the saturator. After changing the moisture content of the air, the saturator was bypassed and the air was again recirculated to establish a condition of moisture equilibrium within the sphere. At all times the temperature of the sphere during measurements was held to within 0.1°C since sound absorption in air is a function of temperature.

Because impurities in the air or contamination on the walls of the enclosure could affect the absorption of sound in air significantly, the entire system was evacuated for many days to allow contaminants to outgas. Then the sphere was filled with dry air having a carbon dioxide content of 300 parts per million.

Sound was introduced from a loudspeaker driving unit through a probe tube into the sphere. This point source was poorly matched to the spherical enclosure so that its contribution to the absorption of the chamber would be negligible. The length of the probe was selected so that the point source would be located near pressure maxima for the radial modes but of such a length as to discriminate against the excitation of nearby normal modes. After a steady-state sound level had been established by the sound source, the sound source was turned off. As indicated in Equation 1, the ensuing decay rate depends on the absorption of the bounding spherical surface as well as that caused by the dissipation of energy in the air.

The contribution to the decay rate that may be attributed to losses in acoustic energy at the boundaries of the spherical enclosure was evaluated by a technique suggested to me some time ago by the late Professor R. W. Leonard of U.C.L.A. The wall losses were determined directly from two sets of measurements: (1) decay rate data obtained with the sphere filled with air, and (2) decay rate data obtained with the sphere filled with pre-purified dry nitrogen (which exhibits no anomalous absorption caused by moisture in the frequency range of measurement).

Suppose the decay rate of sound in the sphere that one measures when it is filled with nitrogen is represented by

$$R_{\text{total}(N_2)} = R_{N_2} + R_{\text{wall}}\ dB/sec \tag{2}$$

where R_{wall} is the decay rate due to absorption at the boundaries of the sphere and where R_{N_2} is the decay rate owing only to the nitrogen gas (a value that can be calculated). Then the wall-loss term in Equations 1 and 2 is approximately the same because both gases are closely similar in molecular weight and characteristic acoustic impedance. By combining Equations 1 and 2, the rate of decay of sound in the spherical enclosure owing only to the absorption of sound in air is given by

$$R_{\text{air}} = R_{\text{total(air)}} - [R_{\text{total}(N_2)} - R_{N_2}]\ dB/sec \tag{3}$$

where the quantity within the brackets represents the contribution to the rate of decay of sound in a sphere that is caused by wall losses.

2. *Results*

Suppose the absorption of sound in air is expressed quantitatively in terms of the total attenuation coefficient m per meter, where m is the exponent in the equation for the diminution of intensity of progressive sound waves: $I = I_0 e^{-mx}$. The total attenuation coefficient $m = m_{\text{m}} + m_{\text{c}}$ (i.e., the sum of the attenuation coefficients due to molecular and classical attenuation) is related to the decay rate by the expression

$$m = R_{\text{air}}/(4.34c) \text{ meters}^{-1}$$

where c is the velocity of sound. Since the value of the classical absorption is known, the value of the molecular attenuation coefficient m_c may be determined.

Humidity at which Absorption is Maximum

At a given temperature, pressure, and frequency, the molecular absorption m_m varies with the moisture content of the air. The value of humidity at which the absorption is maximum, for a frequency f, is defined as h_{max}. This is illustrated in Figure 2 which shows data of

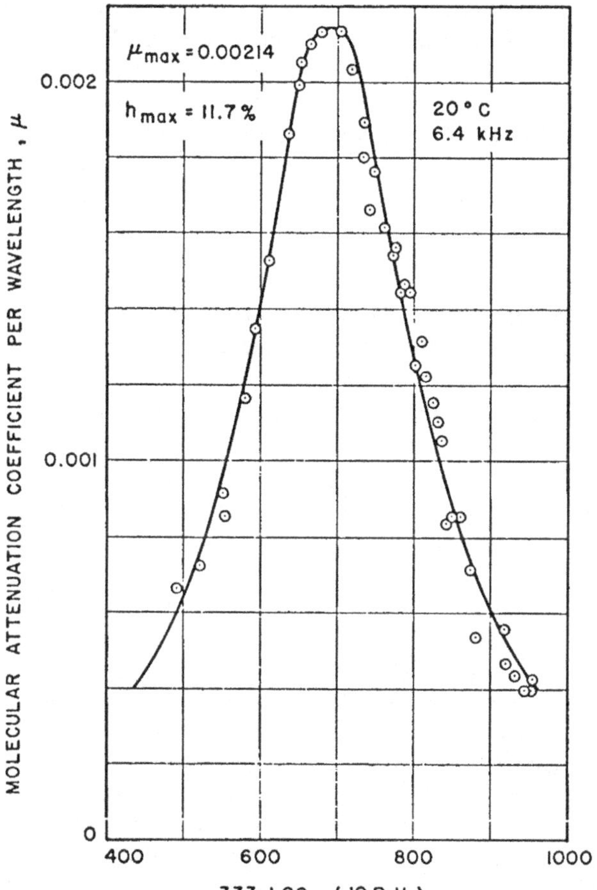

Figure 2. Example of a plot of experimental data of the molecular attenuation coefficient per wavelength in air as a function of humidity. These data are for a temperature of 20 °C and a frequency of 6400 Hz.

molecular absorption vs. humidity at a frequency of 6400 Hz (at 20 °C); in this example the absorption has a maximum value at a relative humidity of 11.7%. Experimental data obtained for other frequencies at this temperature, and normal atmospheric pressure, show the following relationship between f_{max} (the frequency in kilohertz) and the corresponding humidity at which the absorption peaks h (the moisture content of the air in grams per cubic centimeter):

$$f_{max} = 0.79h^2 + 1.47h - 0.15 \tag{4}$$

This is in contrast to the early theoretical prediction of Kneser[5] indicating that f_{max} varies as h^2, but it is in reasonably good agreement with the more recent theoretical work of Henderson and Herzfeld.[6]

At a given frequency, the effect of reducing the pressure of the air to a lower value is to shift the position of the absorption peak to a value which is lower than that given by Equation 4 at a fixed value of relative humidity. Above the peak (which is the usual range for most conditions out of doors) the effect of this shift is to lower the attenuation of sound in air at reduced pressures.[7]

Molecular Absorption versus Frequency. For a given temperature and frequency, the maximum value of molecular attenuation coefficient m_m as the humidity is varied is represented by m_{max}. According to theory, the maximum values of molecular absorption increase linearly with frequency.[1,5] Suppose we denote the attenuation coefficient per wavelength by μ. Since $m_{max} = \mu_{max}/\lambda$, we may obtain values of both μ_{max} and m_{max} from plots of μ versus humidity. A smooth curve drawn through the resulting values of μ_{max} yields values that are close to those predicted by theory.

For measurements at atmospheric pressure, the following values of μ_{max} were obtained: −10 °C, 0.00114; −5 °C, 0.00128; 0 °C, 0.00159; 10 °C, 0.00176; 15 °C, 0.00194; 20 °C, 0.00214; 25 °C, 0.00235; 30 °C, 0.00258. From these data the maximum value of the molecular absorption m_m for a frequency f was obtained by dividing the value of μ_{max} for the temperature by the corresponding wavelength. For example, for a temperature of 30 °C, the maximum absorption at 10,000 Hz is determined by dividing the value of μ_{max} for 30 °C (0.00258) by λ for this frequency and temperature, obtaining a value of m_{max} equal to 0.0740 m^{-1} is equal to 0.0074 m^{-1}.

In a recent study,[7] values of the molecular absorption maxima were determined at reduced pressures, for frequencies in the range from 2500

Hz to 8000 Hz, at 20 °C. Within experimental error, μ_{max} appeared to be independent of pressure in the range from 0.4 to 1.0 atm, the average value being 0.0022 for this temperature. This compares with the value for μ_{max} of 0.00214 obtained at atmospheric conditions and at 20 °C.

Oxygen Data as a Check on the Accuracy of the Measurement Technique. The accuracy of the method of measurement which has been employed here in evaluating the absorption of sound in air has been checked by the following means. The same measurement system was employed in another study in which the spherical enclosure was filled with oxygen instead of air. Then measurements of the decay rate of sound pressure were made, at various frequencies, as a function of water vapor content. From these data a relationship was obtained between the relaxation frequency for oxygen containing water vapor and the concentration of that water vapor. The oxygen data are of particular interest because absorption measurements have been made on this gas by a number of investigators using widely different measurement techniques which are subject to quite different types of measurement error. In a recent paper, Harlow and Nolan[8] compare their results for oxygen with the above data and with the data of Henderson,[9] and Knotzel.[10] There is excellent agreement between the results of these investigations.

3. Data Reduction and Preparation of Air Absorption Tables

In the above experiments, during a given run, the total absorption (classical plus molecular) was measured as a function of humidity — the frequency and temperature were held constant. By taking many data runs, families of such absorption *vs.* humidity curves were obtained at different frequencies and for different temperature conditions. It was possible to prepare tables of absorption of sound in air directly from such information. However, a more accurate data reduction procedure, described below, was used to generate the tabular values.

Consider a point on a given curve of total absorption vs. humidity. From its value of total absorption, the classical absorption was subtracted. Following a similar procedure for each point, a curve of molecular absorption vs. humidity was obtained. Then the following normalization procedure was employed. The value of molecular attenuation coefficient for each data point was divided by the maximum value of molecular attenuation coefficient (for any humidity, but for that frequency and temperature); the corresponding value of humidity for that data point was divided by the value of humidity at which maximum absorption occurs. Each data point, so normalized, for the different families of curves was replotted on a single graph, as shown in Figure 3.

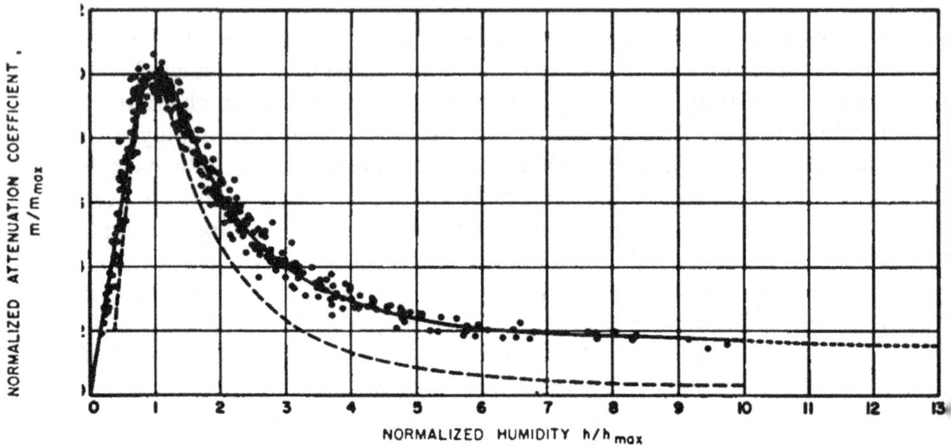

Figure 3. Plot of experimental data of the molecular attenuation coefficient in air versus humidity. These data are presented in the normalized form, m/m_{max} versus h/h_{max}. The theoretical relationship of Kneser is shown by the dashed line.

Note that this normalization procedure has the effect of collapsing a whole family of curves into a single curve of "normalized absorption" *vs.* "normalized humidity." These results are in contrast to the normalized relationship developed from theory by Kneser[5] which is indicated by the dashed curve. The discrepancy between theory and the experimentally derived relation, shown as the solid curve, results largely from the assumption in the theory that the angular relaxation frequency varies with h^2 whereas the data represented by Equation 4 show that the relationship between f_{max} and humidity is much more complex.

Making use of the solid curve of Figure 3, a computer program was written to generate values of the molecular absorption at any frequency, humidity, and temperature. The corresponding values of classical absorption also were computed, and added to the molecular absorption to yield the total absorption. This procedure has resulted in tables of sound attenuation in air for a wide range of frequencies, humidities, and temperatures.[11]

References

1. K. F. Herzfeld and T. A. Litovitz, *Absorption and Dispersion of Ultrasonic Waves*. Academic Press, Inc., New York (1959).
2. P. M. Morse, *Vibration and Sound*, Second Edition, McGraw-Hill Book Co., Inc., New York (1948).
3. C. M. Harris, *J. Acoust. Soc. Am.*, **40**, 148–159 (1966).
4. The statistical case of the decay of sound in the higher frequency range is described in Ref. 2, p. 382.
5. H. O. Kneser, *J. Acoust. Soc. Am.*, **5**, 122–126 (1933); also, *Akust.* Zh., **5**, 256–257 (1940); *Ergeb. Exakt. Naturw.*, **22**, 121–185 (1949).
6. M. C. Henderson and K. F. Herzfeld, *J. Acoust. Soc. Am.*, **37**, 986–988 (1965).
7. C. M. Harris, *J. Acoust. Soc. Am.*, **43**, 530–532 (1968).
8. R. G. Harlow and M. E. Nolan, *J. Acoust. Soc. Am.*, **43**, 1287–1290 (1968).
9. M. C. Henderson, A. V. Clark, and P. R. Lintz, *J. Acoust. Soc. Am.*, **37**, 457–463 (1965).
10. H. Knotzel and L. Knotzel, Ann. Physik., **2**, 393–403 (1948).
11. C. M. Harris, " Absorption of Sound in Air Versus Humidity and Temperature ", NASA Report CR-647, *Fed. Sci. and Tech. Inf.*, Springfield, Va (January 1967).

Robert H. Katyl and Uno Ingard
International Business Machines Corporation and
Massachusetts Institute of Technology

6 Light Scattering from Thermal Fluctuations of a Liquid Surface*

Introduction

Interest in Raman and Brillouin scattering has increased sharply in recent years as the advent of the laser has made light a versatile and powerful diagnostic tool in the study of properties of matter. New spectroscopic techniques have emerged which allow much higher resolution than before, and as a result many experimental studies which previously gave marginal results, or simply were out of range because of resolution problems, can now be handled successfully.

A typical example is the study of (thermal) phonons in liquids and solids by means of light scattering (Brillouin scattering[1]). The light scattered in a certain direction from the bulk of a liquid or solid is modulated by the fluctuations in density and contains the frequency ($\omega/2\pi$) of particular fluctuations or phonons that contribute to the scattering in the chosen direction. The scattered spectrum contains a (Brillouin) doublet, and the separation of the lines of the doublet is twice the frequency of the phonons involved. Furthermore, the spectral line broadening yields information about the lifetime of the phonons. From measurements of these quantities at different scattering angles, the dispersion relation for sound waves in the kilomegacycle region can be determined. Experiments of this type were attempted by Rao[2] as early as 1938. However, not only was the scattered light very weak,

* This work was supported principally by the U.S. Navy (ONR) under Contract No. N00014-67-A-0204-0019.

making the detection problem very difficult, but the resolution was not sufficient to enable measurement of the line broadening.

The problem of Brillouin scattering from the bulk of liquids and solids has recently been considered,[3, 4] using a laser source of light which made possible measurements not only of the separation of the Brillouin doublet but also the line broadening. Since then numerous studies have been carried out in liquids and solids yielding information about the *dynamic* compressibility and the dispersion of *hypersound*.[5]

At thermal equilibrium, all collective modes of motion of a liquid or solid are excited, and it occurred to us that it might be possible to observe Brillouin-type scattering also from thermally excited capillary waves or ripples on a liquid surface. According to classical hydrodynamics, the dispersion relation for ripples is determined solely by surface tension, shear viscosity, and the liquid density, and a determination of the dispersion relation by means of light scattering experiments would enable measurement of shear viscosity and surface tension under the conditions of the short wavelengths and high frequencies present in the thermal vibrations.

Actually, the possibility of finding diffuse light scattering from *thermal roughness* of a liquid surface without regard to the spectral composition of the light was pointed out by Smoluchowski[6] in 1908. Mandelshtam[7] considered the problem in more detail and obtained the wave kinematic relation for the scattering problem. Qualitative observations of diffuse scattering of sunlight focused on liquid surfaces were made by Raman and Ramdas.[8]

In the Brillouin scattering from phonons in the bulk of liquids or solids, the standard spectroscopic tool is a Fabry-Perot interferometer. Typically the *bandwidth* of such a spectrometer is of the order of 200 MHz for a measurable range of 10 GHz; such a resolution is usually quite adequate for studies of phonons in the bulk. The phonon frequency range involved in such measurements is usually centered about 10^3–10^4 MHz, and the bandwidth is of the order of 500 MHz. For ripples, on the other hand, the frequency is expected to be considerably lower, and it is not immediately clear that this spectroscopic technique can be successfully applied. To explore this question, it is necessary to study in some detail the dispersion relation for capillary waves.

Dispersion Relation

If we neglect the effect of viscous stresses, the general form of the dispersion relation for surface waves on a liquid can be obtained by the following simple considerations. Let us imagine a wavelike deformation of the liquid surface of wavelength $\lambda = 2\pi/k$. A deformation of the liquid surface is impeded by surface tension and by gravity. The restoring force per unit area resulting from surface tension σ is $\sigma\partial^2\zeta/\partial x^2 \simeq \sigma k^2\zeta$, that is, proportional to the curvature of the surface, ζ being the vertical surface displacement. The corresponding contribution to the restoring force from gravity is simply $\rho g\zeta$. In a wave motion of the surface it is well known that the motion penetrates into the bulk of the fluid to a depth of the order of a wavelength, and consequently the mass involved in the periodic motion of the surface per unit area is of the order of $\rho\lambda$. Thus, with a total restoring force $\sim(\sigma k^2 + \rho g)\zeta$, it follows then that the frequency of oscillation should be of the form

$$\omega \approx \left(\frac{\sigma k^2 + \rho g}{\rho\lambda}\right)^{1/2} \approx \left(\frac{\sigma}{\rho} k^3 + gk\right)^{1/2} \tag{1}$$

We note that the effect of gravity is negligible for sufficiently short wavelengths given by $k \gg \sqrt{\rho g/\sigma}$.[9] Actually, this is the range of interest here, and the effect of gravity can generally be ignored (with the possible exception of the neighborhood of the critical point where surface tension goes to zero). The dispersion relation then indicates that ω should be proportional to the $\frac{3}{2}$ power of the wave number. However, this relation cannot apply for arbitrarily large values of k, and to study this question, the role of shear viscosity must be explored. Using classical hydrodynamics for a viscous incompressible liquid, such a study yields a dispersion relation for capillary waves or ripples which can be written in the form[10]

$$(\gamma + 1)^2 + (gk + \omega_0^2)\tau^2 - \sqrt{1 + 2\gamma} = 0 \tag{2}$$

where $\gamma = -i\omega\tau$, $\tau = \rho/2\mu k^2$, $\omega_0^2 = \sigma k^3/\rho$, σ = surface tension, ρ = density, and $k = 2\pi/\lambda$. If one wishes to introduce a correction for compressibility, the third term in the equation should be multiplied by $(1 + \kappa 4\gamma^2\mu^2k^2/\rho)^{1/2}$, where κ is the compressibility.

In thermal fluctuations we are interested in the temporal behavior of ripples of a given wavelength. This means that in Equation 2 the quantity k is regarded as real. Thus, solving for $\omega = \omega(k)$, the frequency of oscillation is given by $\mathrm{Re}(\omega)$ and the damping rate by $\mathrm{Im}(\omega)$. In the long wavelength limit $k \to 0$, we see that the dispersion relation reduces to Equation 1, and the damping rate is given by $\mathrm{Im}(\omega) = 1/\tau$. However, as the wavelength decreases, the decay becomes nonoscillatory for λ smaller than a certain critical value given by

$$\lambda_c = 8\pi y_c \frac{\mu^2}{\sigma\rho} \simeq 3.65 \frac{\mu^2}{\sigma\rho} \tag{3}$$

where the factor $y_c \simeq 0.145$ is obtained from a numerical study of Equation 2. The values of λ_c for some common liquids are given in Table 1.

Table 1. Critical Wavelengths and Maximum Frequencies of Surface Waves.

Liquid		λ_c, cm	k_c, cm^{-1}	f_m, Hz
Methanol	(20 °C)	$6.98 \cdot 10^{-5}$	$9.00 \cdot 10^{5}$	$1.49 \cdot 10^{8}$
Ethanol	(20 °C)	$2.86 \cdot 10^{-5}$	$2.19 \cdot 10^{5}$	$1.75 \cdot 10^{7}$
Isopropanol	(20 °C)	$1.19 \cdot 10^{-4}$	$5.30 \cdot 10^{4}$	$2.09 \cdot 10^{6}$
Benzene	(20 °C)	$6.22 \cdot 10^{-6}$	$1.01 \cdot 10^{6}$	$1.87 \cdot 10^{8}$
Water	(20 °C)	$5.02 \cdot 10^{-6}$	$1.25 \cdot 10^{6}$	$3.89 \cdot 10^{8}$
Carbon dioxide	(20 °C)	$1.46 \cdot 10^{-6}$	$4.30 \cdot 10^{6}$	$2.62 \cdot 10^{8}$
Mercury	(20 °C)	$1.31 \cdot 10^{-7}$	$4.80 \cdot 10^{7}$	$6.59 \cdot 10^{10}$
Argon	(−180 °C)	$1.22 \cdot 10^{-6}$	$5.14 \cdot 10^{6}$	$1.24 \cdot 10^{9}$
Glycol	(20 °C)	$2.73 \cdot 10^{-3}$	$2.29 \cdot 10^{3}$	$2.35 \cdot 10^{4}$

At a value $k = 0.744k_c$, where $k_c = 2\pi/\lambda_c$, the frequency of oscillation reaches a maximum value f_m which, according to Equation 2, is found to be

$$f_m \simeq 0.0736 \frac{\sigma^2\rho}{\mu^3} \tag{4}$$

This maximum frequency $f_m = \omega_m/2\pi$ is strongly dependent on the value of the shear viscosity μ of the liquid; values of f_m for some common liquids are given in Table 1.

Introducing the critical wave number $k_c = 2\pi/\lambda_c$ and the maximum frequency $\omega_m = 2\pi f_m$, we can present Equation 2 in normalized form, ω/ω_m vs k/k_c, as shown in Figure 1. In the overdamped region $k/k_c > 1$, we have two decay rates $\mathrm{Im}(\omega)$ consistent with the behavior of any overdamped harmonic oscillator, and in the long wavelength limit $k/k_c \ll 1$

$$\mathrm{Re}\left(\frac{\omega}{\omega_m}\right) \simeq 4.88\left(\frac{k}{k_c}\right)^{3/2}$$

$$\mathrm{Im}\left(\frac{\omega}{\omega_m}\right) \simeq 12.8\left(\frac{k}{k_c}\right)^{2}$$

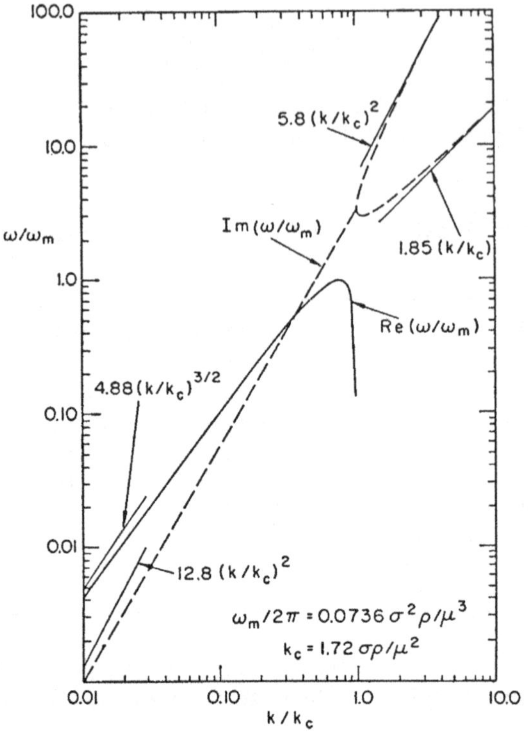

Figure 1. Dispersion relation for surface waves on a liquid $\omega = \omega(k)$. Here $\omega_m = 0.462\ \sigma^2\rho/\mu^3$ and $k_c = 1.72\ \sigma\rho/\mu^2$.

It follows from Table 1 that the line separation $2f$ in a Brillouin doublet in the scattered light from the liquid surface is expected to be less than about 100 MHz, and in order to resolve this doublet, the effective bandwidth of the spectrometer should be less than this value. Consequently, the Fabry-Perot interferometer, the standard tool for Brillouin scattering from the bulk, appears to be only marginal for surface wave scattering. Nevertheless, it was decided to attempt an experiment using a Fabry-Perot interferometer.

The Scattered Light

As a result of the thermal fluctuations of a liquid surface, there will be partial diffuse scattering in addition to the regular specular reflection. The scattered light observed in a certain direction is contributed by capillary waves of a particular wavelength consistent with the wave kinematical conditions of constructive interference. This condition obviously comes out of a classical analysis of diffraction but can most conveniently be obtained if we regard the scattering as an interaction between a photon and *ripplon*. The ripplon is then specified as a quasi-particle with momentum $\hbar k$ along the liquid surface and an energy $\hbar\omega$. The conservation of momentum along the surface in the collision between an incident photon and a ripplon then leads to the following relation, which is illustrated schematically in Figure 2,

$$(k/K)^2 = \sin^2 \theta + \sin^2 \theta_0 - 2 \sin \theta \sin \theta_0 \cos \varphi \qquad (5)$$

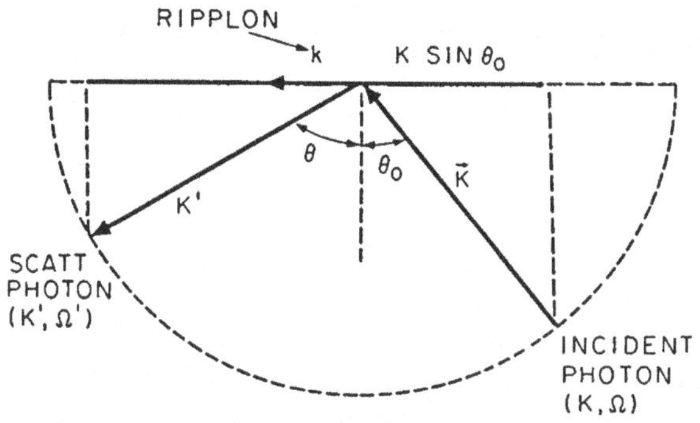

Figure 2. The relation between the scattering angles (θ, φ) and the ripple wavelength ($2\pi/k$), responsible for scattering in this direction, is obtained conveniently from momentum conservation in photon-ripplon collision. $K = 2\pi/\Lambda$ refers to the photon and $k = 2\pi/\lambda$ to the ripplon. Here $|K'| \simeq K$ and θ_0 = angle of incidence.

where the polar and azimuthal angles (θ, φ) define the scattering direction and θ_0 the incident direction as shown. Energy conservation requires that the frequencies of the scattered light are

$$\Omega' = \Omega \pm \omega$$

corresponding to absorption or creation of a ripplon.

As we have seen, the decay of the ripples can be either oscillatory or nonoscillatory. In the former case, the scattered light spectrum will consist of a Brillouin doublet. It is interesting to note that in this case there should be no central or unshifted spectral component corresponding to the Rayleigh line in the scattering from the bulk. Only if the decay is nonoscillatory will there be an unshifted component, but then the doublet is absent.

The intensity of the scattered light can be obtained conveniently from a classical analysis of wave scattering from a rough surface. Actually, an extensive analysis of wave scattering from a stationary rough surface was done in 1924 by Gans,[11] and his results readily can be extended to account for the motion and the corresponding frequency shift of the scattered light. The intensity ordinarily is quite weak and difficult to analyze unless the incident and scattered beam directions are chosen properly. The intensity is greatest at scattering angles in the neighborhood of the specular direction. This is related in part to the fact that the mean-square surface displacement is proportional to λ^2, and as we approach the specular direction, longer and longer wavelengths are involved in the scattering (the specularly reflected light can be regarded as a contribution from $k = 0$). Also, we find that in transparent liquids the scattering is enhanced if the light beam is totally reflected from the surface.

Preliminary Experiments with Fabry-Perot Interferometer

Guided by such observations, we performed preliminary experiments with two liquids, methanol and isopropanol, choosing our scattering angles and corresponding ripple wavelengths such that the oscillations would be underdamped for methanol and overdamped for isopropanol.[12] The incident light beam was obtained from a stabilized single-mode helium-neon laser at a wavelength of 6328 Å, and the angle of incidence chosen so that the total internal reflection was obtained as already illustrated schematically in Figure 2. The scattered light in a certain direction (θ, φ) was analyzed with a 10-cm flat-mirror Fabry-Perot interferometer, having a half intensity width of the order of 27 MHz.[13] The interferometer was pressure-scanned either by a slow-leak system or a mechanically driven piston. The detector, a photomultiplier tube with a pulse-discrimination system, permitted observation of light intensities as low as 1 photoelectron/sec with a room-temperature dark counting rate of approximately 2 photoelectrons/sec (Figure 3).

As indicated earlier, the preliminary measurements involved scattering from methanol and isopropanol. If the incident angle is chosen to be $\theta_0 = 80°$ and the scattering angle is given by $\theta = 70°$ and $\varphi = 50°$, the wavelength λ of the ripples is very nearly equal to the wavelength Λ of light, $\lambda \simeq 1.25\ \Lambda/n$, where n is the index of refraction. Using static values of the shear viscosity for the two liquids, we find from Equation 2 that at this wavelength methanol is underdamped, whereas isopropanol is overdamped. The expected separation of the Brillouin lines for methanol is found to be 42 MHz and the line width 23 MHz. For isopropanol, a doublet is not expected. Instead the line should consist of

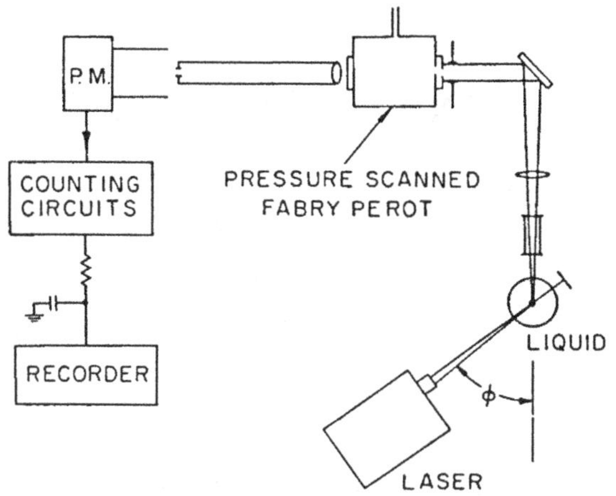

Figure 3. Experimental arrangement in spectral analysis of the scattered light from thermal fluctuations of a liquid surface, using Fabry-Perot interferometer.

two superimposed unshifted lines of widths 18 MHz and 81 MHz; if in thermal equilibrium these decaying modes are excited with the same average energy, their peak intensities should be inversely proportional to their widths. Thus, because of the occurrence of the doublet in one liquid and not in the other, the total spectrum width in the two cases should be considerably larger for methanol than for isopropanol.

The calculated ideal scattered spectra for the two liquids are shown in Figure 4. The solid lines represent the ideal spectra, assuming negligible instrumental bandwidth, whereas the dashed line is the result obtained when we account for the finite bandwidth of the Fabry-Perot interferometer. We note that the doublet for methanol should be just resolvable. However, some stray light from the Rayleigh scattering in the bulk of the liquid just below the surface can conceivably also contribute to the observed spectrum, and to get an idea of its magnitude, this spectrum was recorded from a position in the bulk of the liquid just below the surface. This unshifted spectral line has considerably smaller intensity than the surface scattered light, but its presence will somewhat modify the total observed spectrum. The modification of the spectral line shape is illustrated in Figure 5, where we have shown separately the Rayleigh scattering, the total measured spectrum, and also the expected line shape accounting for both the instrumental bandwidth and the contribution from the Rayleigh scattering.

As can be seen, this is in good agreement with the experimental data for both methanol and isopropanol, and the former spectrum is considerably

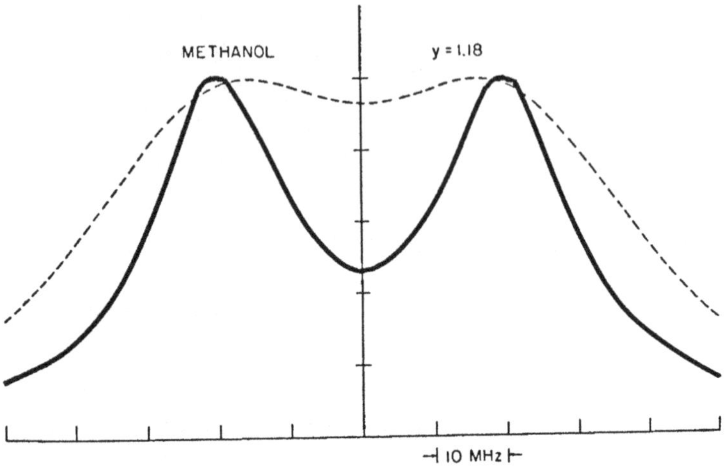

(a)

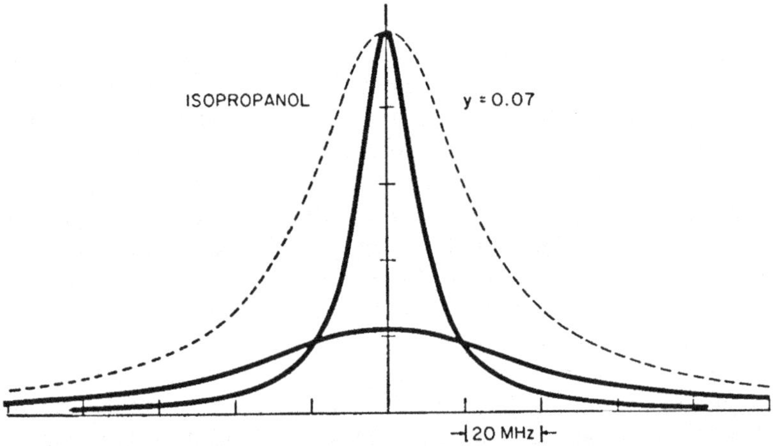

(b)

Figure 4. (a) and (b). Calculated scattering spectra for methanol and isopropanol for $\theta_0 = 80°$, $\theta = 70°$, $\varphi = 50°$ (see Figure 2). For methanol, being underdamped, the spectrum consists of a Brillouin doublet, and for isopropanol, being overdamped, it consists of two superimposed unshifted lines of different bandwidth. The dashed curves are the expected output spectrum when the finite bandwidth (23 MHz) of the Fabry-Perot interferometer is accounted for.

broader than the latter. Although this difference in width only gives an indirect proof of the presence and absence of a doublet in methanol and isopropanol, respectively, the results were sufficiently promising to

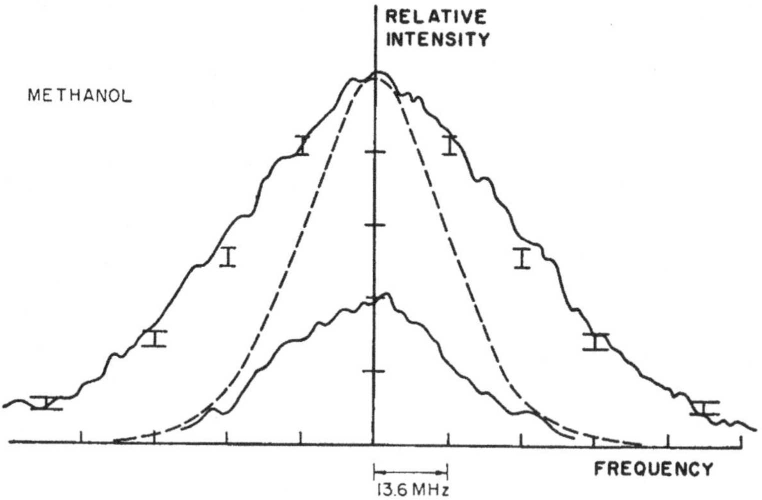

(a)

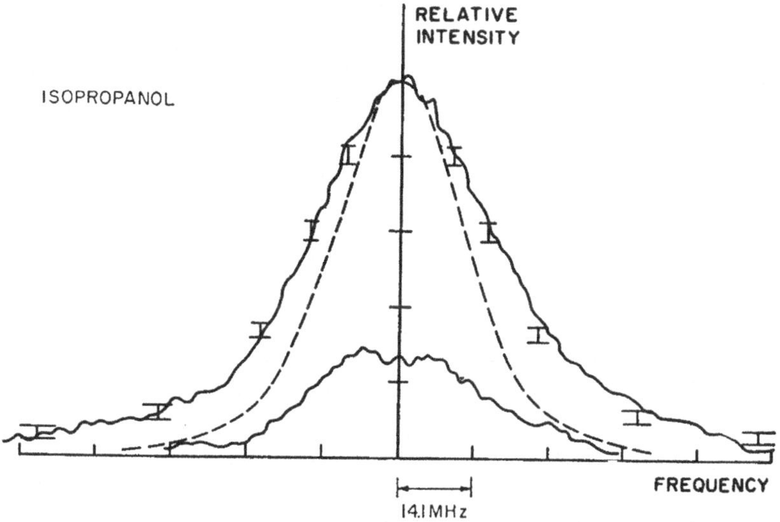

(b)

Figure 5. (a) Results for methanol. (b) Results for isopropanol. Upper solid curve is surface-scattered light; lower solid curve is Rayleigh scattering in bulk of fluid near surface; points (I) indicate calculated value of surface plus bulk scattering convolved with instrument response; dashed curve is instrument response. The measured peak intensity for the surface scattering in both liquids is approximately 30 photoelectrons/sec, using a 0.1 mW laser. Data were taken at approximately 25 °C. (From *Phys. Rev. Letters*, **19**, 64, July 10, 1967.)

justify further studies of the problem. Actually, subsequent measurements on water (which is considerably more difficult to work with both with respect to surface contamination and signal-to-noise ratio) gave a direct although marginal indication of the presence of a doublet.

Heterodyne Spectroscopy

In heterodyne spectroscopy the frequency shift in scattered light is mixed with a *local oscillator* beam at a photomultiplier. Because of the nonlinear response of the photomultiplier to the total incident electric field, the output current of the photomultiplier contains both the difference and the sum frequencies of the scattered and the local oscillator light. Considering the difference frequency, we note that this frequency is identical with the frequency of the capillary waves, and the corresponding spectrum representing the Brillouin doublet is thus downshifted to a value of zero for the center frequency. Consequently, this portion of the spectrum then can be analyzed by a conventional ultrasonic spectrum analyzer.

In our first heterodyne experiment[14] the incident light was obtained from one end of an argon-ion laser (operating in the principal transverse mode at 4880 Å at a power of ~5 mW) and the local oscillator beam from the other end, as indicated schematically in Figure 6. The surface

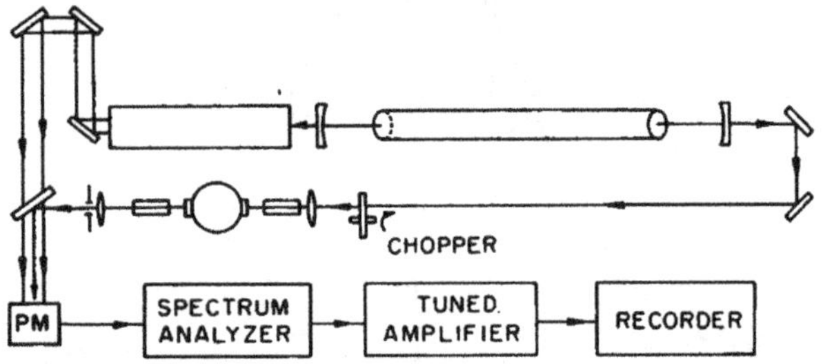

Figure 6. Experimental arrangement in heterodyne spectroscopy.

scattered light is collimated and combined with the local oscillator beam at a beam splitter and directed onto a photomultiplier. In a subsequent version of the experimental setup, we used the specularly reflected light as the oscillator beam, which proved advantageous in attempting to reduce the influence of background vibrations.

The photocurrent, which should contain the downshifted scattered light spectrum, is analyzed by means of an ultrasonic spectrum analyzer.

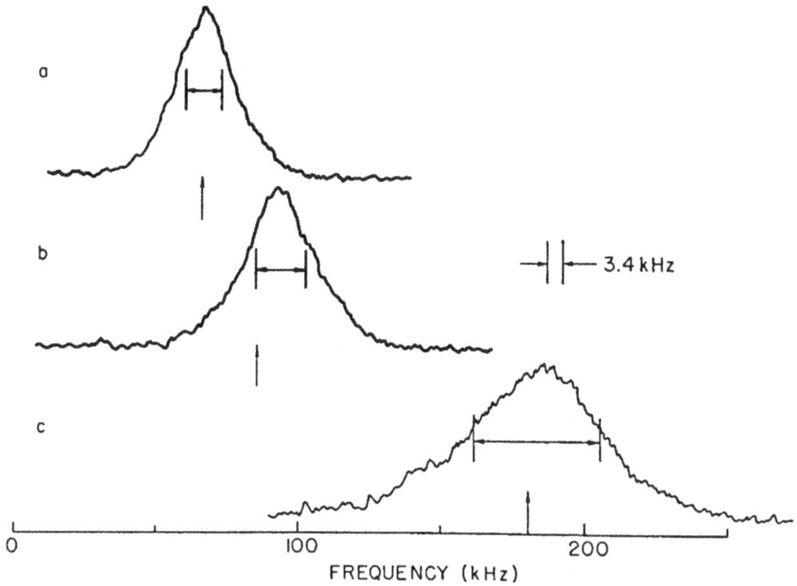

Figure 7. Experimental spectra of light scattered from thermal surface vibrations on a liquid (methanol at 25 °C) at three different surface wavelengths: (a) $k = 1860$ cm^{-1}, (b) $k = 2200$ cm^{-1}, and (c) $k = 3600$ cm^{-1}. The arrows and brackets indicate the frequencies and line widths calculated from classical theory of surface vibrations, with static values of surface tension and viscosity used. The instrumental bandwidth, including aperture broadening, is estimated to be 3.4 kHz, which is also indicated. (From *Phys. Rev. Letters*, **20**, 248, February 5, 1968.)

Typical records of spectra thus obtained are shown in Figure 7. The three spectral lines shown here refer to three different scattering angles and corresponding ripple wavelengths (see Equation 4). The center frequency of each line equals the ripple frequency of oscillation, and the width of each line is inversely proportional to the damping rate. The results clearly indicate that the frequency of oscillation, as well as the damping rate, increases with the wave number k.

From an analysis of data of this kind, obtained at different scattering angles, the dispersion relation for the thermally excited capillary waves has been determined for several liquids. Results obtained for some different liquids are shown in Figure 8, together with the theoretical curves corresponding to Equation 2, using static values for surface tension and viscosity. The results for these liquids indicate that, on the whole, the dispersion of capillary waves, even at these very short wavelengths considered in this experiment, can be satisfactorily described by classical hydrodynamics. However, it would not be surprising if for

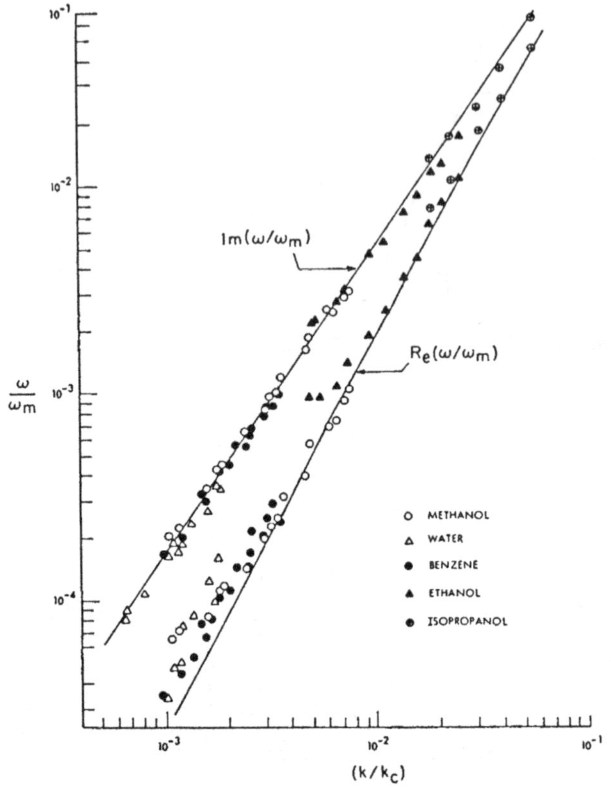

Figure 8. Experimental values of (normalized) frequency of oscillation $\mathrm{Re}(\omega/\omega_m)$ and decay rate $\mathrm{Im}(\omega/\omega_m)$ for thermally excited surface waves in some liquid surfaces as a function of (normalized) wave number k/k_c ($k = 2\pi/\lambda$). Here $k_c = 1.72\ \sigma\rho/\mu^2$, and $\omega_m = 0.462\ \sigma^2\rho/\mu^3$.

some liquid such a description were inadequate. This would be the case, for example, if there were a frequency relaxation in viscosity or surface tension, or if there were a difference between the value of the viscosity in the surface layer and the standard value obtained from measurements in the bulk. However, considerably more work must be done before any definite conclusions along these lines can be drawn.

Liquid-Liquid Interface

The light scattering technique lends itself equally well to the study of fluctuations of a free liquid surface as a liquid-liquid interface and liquid films. We shall describe here an experiment designed to study the temperature dependence of the interfacial surface tension between methanol and hexane in the neighborhood of the critical mixing temperature. The experimental arrangement is much the same as in the study of the free liquid surface; the incident beam is reflected from the interface, strikes the interface from below, and the scattered light in a certain direction is mixed with the specularly reflected light beam at the photomultiplier. The liquid samples were filtered into a double flask assembly connected by an annealed copper pinch-off seal. After evacuation, the solution was repeatedly distilled into the scattering flask side of the double flask to remove all dust. Then the scattering flask was pinched off.

Examples of recorded beat spectra (actually the square root of the beat power spectra) obtained at three different scattering angles are shown in Figure 9. It is interesting to note that as the scattering angle (and the wave number) increases, the scattered spectrum changes markedly in character from a Brillouin doublet to a single line. At low

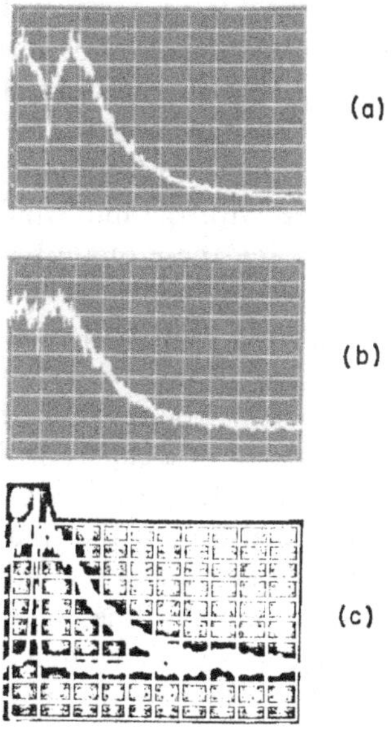

Figure 9. Examples of recorded spectra (the output is proportional to the square root of the power spectrum) of light scattered from thermally excited surface waves in a methanol-hexane interface at different scattering angles and corresponding wave numbers. (a) $k = 750\ \text{cm}^{-1}$; (b) $k = 1750\ \text{cm}^{-1}$; (c) $k = 2330\ \text{cm}^{-1}$.

scattering angles corresponding to long wavelengths, there is a distinct Brillouin doublet, but as the scattering angle increases (wavelength decreases), the doublet gradually turns into a single line. The same thing occurs at a given wavelength when the temperature is increased. Measurements of this sort were carried out at different temperatures ranging from 25 °C to 35 °C, and at each temperature several scattering angles corresponding to different ripple wavelengths were considered.

To interpret these experimental results, we again used classical hydrodynamic theory of surface waves adapted to the present problem, assuming no slippage between the two liquids at the interface. The resulting dispersion relation is found to be of the form

$$(\gamma + 1)^2 + (\omega_1 \tau)^2 - a\sqrt{1 + \frac{2a}{b}\gamma} - (1 - a)\sqrt{1 + \frac{2(1 - b)}{(1 - a)}\gamma} = 0 \quad (6)$$

where $\tau = (\rho_1 + \rho_2)/2(\mu_1 + \mu_2)k^2, \omega_1{}^2 = \sigma k^3/(\rho_1 + \rho_2), a = \mu_1/(\mu_1 + \mu_2)$, $b = \rho_1/(\rho_1 + \rho_2)$, and $\mu_{1,2}$ and $\rho_{1,2}$ are the viscosities and densities of each phase. With this binary system (n-hexane-methanol) we have a, $b \simeq 0.45$ to 0.50, and the normalized form of the dispersion relation then does not greatly differ from that for free surface waves in Figure 1.

It should be pointed out that in the analysis of the experimental spectra the frequency shift and the line width of each Brillouin line cannot be read directly from the recorded spectra unless the separation of the lines is large compared with the line width. Since in our experiment this is not always the case, these quantities can be obtained only by comparison of the scattered spectra with the spectral lines computed for different values of frequency shift and line width, or in the overdamped case for different values of the two line-widths Δ_1 and Δ_2. Such computed spectra are shown in Figures 10a and 10b. In the underdamped case, a spectrum consists of a superposition of two Lorentzians of equal width Δ, separated by a frequency 2ω, but two maxima in the resulting spectrum are obtained only if the frequency shift ω is larger than $\Delta/\sqrt{3}$. In the overdamped case, the scattered spectrum is a superposition of two unshifted Lorentzian lines of different line widths Δ_1 and Δ_2 with peak values inversely proportional to the line widths (since in thermal equilibrium each mode is excited with equal total energy).

In our experiment, we encountered scattered spectra of both the underdamped and overdamped case, corresponding to Figures 10a and 10b, respectively. For example, at a temperature of 25 °C and a ripple wavenumber $2\pi/\lambda = k = 750\ \text{cm}^{-1}$, the scattered spectrum closely matched the shape of that corresponding to $(\omega/\Delta) = 1.0$ in Figure 10a, whereas

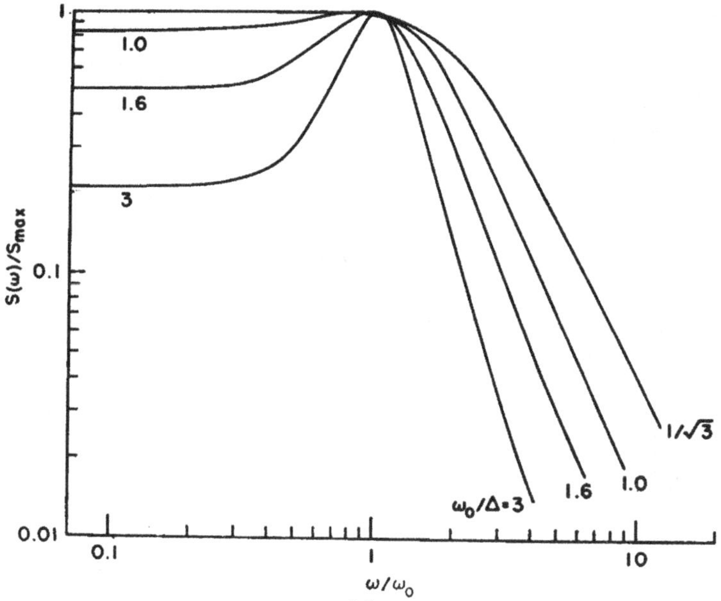

(a)

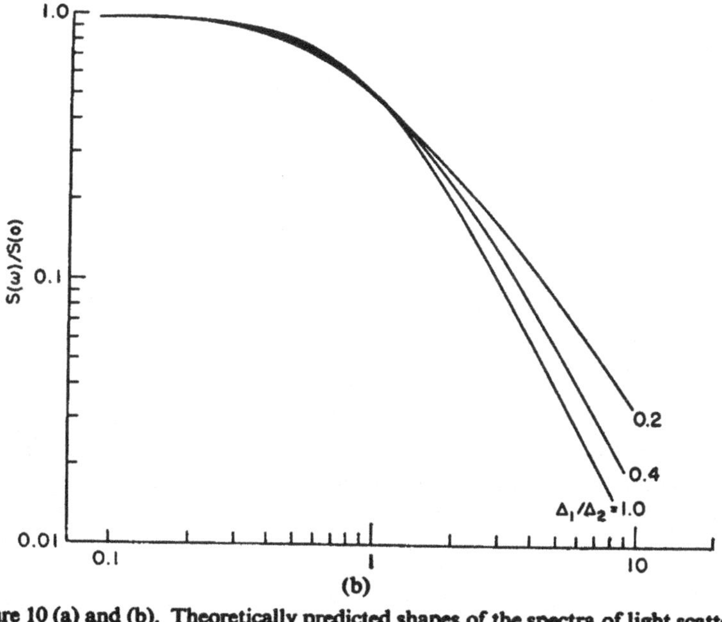

(b)

Figure 10 (a) and (b). Theoretically predicted shapes of the spectra of light scattered from underdamped and overdamped surface waves on liquid. Here $\omega_0/2\pi =$ frequency at the maximum of the spectrum from underdamped waves, and $\Delta =$ line width (Δ_1 and Δ_2 are the line widths of the two components in the spectrum from overdamped waves).

at $k = 2330\ \text{cm}^{-1}$ the scattered spectrum was in closest agreement with that of an unshifted line composed of two Lorentzians with $\Delta_1 = 0.5\,\Delta_2$ as in Figure 10b.

Having obtained the real and imaginary parts of $\gamma = -i\omega\tau$ in this manner, the corresponding value of the interfacial surface tension σ can be obtained from Equation 6. In fact, for sufficiently large values of the ripple wavelength, the surface tension is given simply by $(\rho_1 + \rho_2) \cdot(\omega^2/k^3)$, where ω is the frequency shift.

The values of surface tension obtained from such measurements are shown as a function of temperature in Figure 11. The critical mixing

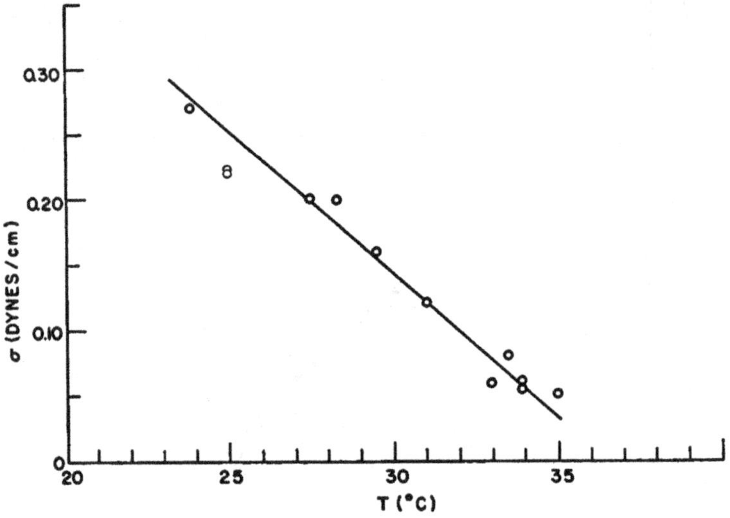

Figure 11. Experimental values of surface tension *vs* temperature for n-hexane-methanol interface.

temperature of the liquids is known to be 36.5 °C, and it is interesting to note that our experiments indicate linear decrease of surface tension with temperature as the critical temperature is approached. It is important to point out in this context that our present method of measurement allows measurement of surface tensions as low as 0.01 dyn/cm.

References

1. L. Brillouin, *Ann. Phys. (Paris)*, **17**, 88 (1922).
2. R. Rao, *Proc. Ind. Acad. Sci.*, **7**, 163 (1938).
3. E. Garmire, and C. H. Townes, *Appl. Phys. Letters*, **5**, 84 (1964).
4. G. B. Benedek *et al.*, *J. Opt. Soc. Am.*, **54**, 1284 (1964).

5. N. R. Goldblatt, and T. A. Litovitz, *J. Acoust. Soc. Am.*, **41**, 1301 (1967).
6. M. V. Smoluchowski, *Ann. Physik*, **25**, 225 (1908).
7. L. I. Mandelshtam, *Ann. Physik*, **41**, 609 (1913).
8. C. V. Raman, and L. A. Ramdas, *Proc. Roy. Soc.* (*London*), **A108**, 561 (1924); **A109**, 150, 272 (1925).
9. For water $\sqrt{\rho g/\sigma} = 3.67\ \text{cm}^{-1}$, corresponding to a wavelength of 1.72 cm.
10. V. G. Levich, *Physicochemical Hydrodynamics*, translated by Scripta Technica, Inc., Prentice-Hall, Inc., Englewood Cliffs, N. Y., 1962, p. 603.
11. R. Gans, *Ann. Physik*, **74**, 231 (1924); **79**, 204 (1926).
12. R. H. Katyl, and U. Ingard, *Phys. Rev. Letters*, **19**, 64 (1967).
13. R. H. Katyl, *Proc. IEEE*, **56**, 1109 (1968).
14. R. H. Katyl, and U. Ingard, *Phys. Rev. Letters*, **20**, 248 (1968).

George L. Lamb, Jr.
United Aircraft Research Laboratories

7 Propagation of Ultrashort Optical Pulses*

1. Introduction

Recent advances in laser technology have led to the production of coherent optical pulses having durations from the nanosecond regime down to 10^{-12} seconds.[1] Such time intervals are comparable to or shorter than the relaxation times associated with the energy levels of many atomic systems. The interaction of radiation and matter on such short time scales gives rise to new phenomena which can not be described by the rate equation analysis used previously[2] for much longer pulses.

Whenever it becomes necessary to extend the range of a theory to encompass new phenomena, it is useful to seek limiting cases of the new formalism[3,4,5] which will admit of exact solution. The purpose of the present paper is to exhibit a model which is exactly soluble in regard to certain aspects of ultrashort pulse propagation. The model describes a number of effects which have been observed both experimentally[4,6] and as output from machine computations based upon more complete theoretical descriptions.[4,5] Hence it is felt that the model preserves many essential features of the processes which take place.

Although the coupling between radiation and matter is too strong to be treated by perturbation theory, a fairly extensive analysis of the model is still possible since it expresses this interaction in terms of a

* This research is part of project DEFENDER under the joint sponsorship of the Advanced Research Projects Agency, the Office of Naval Research, and the Department of Defense.

single nonlinear partial differential equation which arose long ago in differential geometry. Therefore, techniques developed about the turn of the century for the solution of such equations may be employed here to great advantage.

2. Basic Equations

We begin by summarizing the semiclassical description of the interaction of radiation with an assembly of two-level systems. The electric field $E(\mathbf{r}, t)$ of a plane polarized optical pulse satisfies the usual wave equation

$$\nabla^2 E - \frac{1}{c^2}\frac{\partial^2 E}{\partial t^2} = \frac{4\pi n_0}{c^2}\frac{\partial^2 p}{\partial t^2} \tag{1}$$

where c is the velocity of light in the medium, p is the polarization of an individual two-level system and n_0 is the number density of such systems.

The response of a two-level system to the optical field is described by equations for the density matrix of a two-level system which are similar to the Bloch equations of nuclear magnetic resonance,[7,8] *i.e.*,

$$i\hbar\left(\frac{\partial \rho_{aa}}{\partial t} + \gamma_a \rho_{aa}\right) = V(\rho_{ba} - \rho_{ab}) \tag{2a}$$

$$i\hbar\left(\frac{\partial \rho_{ab}}{\partial t} + \gamma_{ab} \rho_{ab}\right) - \hbar\omega_{ab}\rho_{ab} = V(\rho_{bb} - \rho_{aa}) \tag{2b}$$

plus another pair of equations that are obtained from Equations 2 by interchanging a and b. In Equations 2, γ_a is the decay rate or longitudinal relaxation time of the upper laser level, γ_{ab} is the transverse relaxation time of dephasing time for levels a and b, ω_{ab} is the transition frequency and ρ_{aa}, ρ_{ab}, etc., are the elements of a density matrix for a single two-level system. It is assumed that the optical field is coupled to the two-level systems through the usual dipole approximation so that V, the matrix element of the interaction potential between states a and b, is of the form

$$V = -\wp E \tag{3}$$

where $\wp$ is the dipole matrix element connecting these two levels.

The polarization of a two-level system is expressed in terms of the off-diagonal elements of the density matrix by[8]

$$p = -\wp(\rho_{ab} + \rho_{ba}) \tag{4}$$

while the population difference is $n_0 \rho$ where ρ is related to the diagonal elements through

$$\rho = \rho_{aa} - \rho_{bb} \tag{5}$$

From Equations 2 and the definitions given in Equations 4 and 5, one finds that p and ρ are governed by the equations[9]

$$\frac{\partial \rho}{\partial t} = -\frac{2V}{\hbar \omega_{ab} \wp} \frac{\partial p}{\partial t} \tag{6}$$

$$\frac{\partial^2 p}{\partial t^2} + \omega_{ab}^2 p = \frac{2\omega_{ab} \wp}{\hbar} V \rho \tag{7}$$

The relaxation rates γ_a, γ_b, γ_{ab} have been set equal to zero in these last two equations since we shall henceforth specialize the analysis to the ultrashort pulse limit in which pulse durations are short compared to all relaxation times.

However, since even the shortest pulses produced to date contain many optical cycles, it is appropriate to write the electric field in terms of a carrier wave, an envelope function $\mathscr{E}(\mathbf{r}, t)$ which varies slowly on the length and time scales of the carrier wave and a characteristic amplitude E_0. Hence, we write

$$E(\mathbf{r}, t) = E_0 \mathscr{E}(\mathbf{r}, t) \cos(\mathbf{k}_0 \cdot \mathbf{r} - \omega_0 t) \tag{8}$$

An additional unknown phase term could be included in the carrier wave. However, it is found[10] to be unnecessary for many aspects of the problem and will be neglected here. Because of the assumption that $\mathscr{E}(\mathbf{r}, t)$ varies slowly compared to the carrier wave, only first derivatives of $\mathscr{E}(\mathbf{r}, t)$ need be retained on the left-hand side of Equation 1 when it is expressed in terms of the envelope function. On the right-hand side of Equation 1, one may replace $\partial^2 p/\partial t^2$ by $-\omega_0{}^2 p$. In general, it is appropriate to consider not a single transition frequency ω_{ab} but a spread in transition frequencies about ω_{ab}. It is also convenient to assume that this distribution is symmetric about ω_{ab} and that the carrier frequency of the optical pulse is at this frequency, *i.e.*, $\omega_0 = \omega_{ab}$.

Equation 1, specialized to a plane wave travelling in a positive x direction, now becomes

$$\left(\frac{\partial \mathscr{E}}{\partial t} + c \frac{\partial \mathscr{E}}{\partial x}\right) \sin(k_0 x - \omega_0 t) = \frac{2\pi \omega_0 n_0}{E_0} \int_{-\infty}^{\infty} d\Delta\omega\, g(\Delta\omega) p(\Delta\omega, x, t) \tag{9}$$

where $\Delta\omega = \omega - \omega_0$ and

$$\int_{-\infty}^{\infty} d\,\Delta\omega g(\Delta\omega) = 1 \tag{10}$$

The formal solution of Equation 7, with ω_{ab} now replaced by ω, may be written

$$p(\Delta\omega, x, t) = -2\wp\Omega \int_{-\infty}^{t} dt' \sin\omega(t - t') \times \rho(\Delta\omega, x, t')\mathscr{E}(x, t')\cos(k_0 x - \omega_0 t') \tag{11}$$

where $\Omega = E_0\,\wp/\hbar$. This expression for the polarization may be readily decomposed into parts which are in phase and $\pi/2$ out of phase with the electric field. One finds

$$p = \wp[\mathscr{P}(\Delta\omega, x, t)\sin(k_0 x - \omega_0 t) + \mathscr{Q}(\Delta\omega, x, t)\cos(k_0 x - \omega_0 t)] \tag{12}$$

in which

$$\mathscr{P} = \Omega \int_{-\infty}^{t} dt'\mathscr{E}(x, t')\rho(\Delta\omega, x, t')\cos\Delta\omega(t - t') \tag{13}$$

$$\mathscr{Q} = -\Omega \int_{-\infty}^{t} dt'\mathscr{E}(x, t')\rho(\Delta\omega, x, t')\sin\Delta\omega(t - t') \tag{14}$$

In obtaining this result, terms near the second harmonic of ω_0 have been discarded. The function $\mathscr{P}$ and $\mathscr{Q}$ are readily shown to satisfy the differential equations

$$\frac{\partial\mathscr{P}}{\partial t} = \Omega\mathscr{E}\rho + \Delta\omega\mathscr{Q} \tag{15}$$

$$\frac{\partial\mathscr{Q}}{\partial t} = -\Delta\omega\mathscr{P} \tag{16}$$

When time dependence near the second harmonic of ω_0 is neglected in Equation 6, it takes the form

$$\frac{\partial\rho}{\partial t} = -\Omega\mathscr{E}\mathscr{P} \tag{17}$$

With the transformation

$$\tau = \Omega\left(\frac{t - x}{c}\right) \qquad \xi = \left(\frac{\delta\Omega}{c}\right)x \tag{18}$$

where

$$\delta \equiv \frac{n_0\,\hbar\omega_0}{(E_0{}^2/2\pi)} \tag{19}$$

Equations 9, 15, 16, and 17 become

$$\frac{\partial \mathscr{E}}{\partial \xi} = \int_{-\infty}^{\infty} d\Delta\omega g(\Delta\omega)\mathscr{P}(\Delta\omega, \xi, \tau) \tag{20}$$

$$\frac{\partial \rho}{\partial \tau} = -\mathscr{E}\mathscr{P} \tag{21a}$$

$$\frac{\partial \mathscr{P}}{\partial \tau} = \mathscr{E}\rho + \frac{\Delta\omega}{\Omega}\mathscr{Q} \tag{21b}$$

$$\frac{\partial \mathscr{Q}}{\partial \tau} = -\frac{\Delta\omega}{\Omega}\mathscr{P} \tag{21c}$$

Equations 21 describe how the electric field $\mathscr{E}$ determines $\mathscr{P}$, $\mathscr{Q}$ and ρ for a two-level system that is off resonance by an amount $\Delta\omega$. Equation 20 shows how the polarization caused by a distribution of such systems reacts back on the electric field.

3. *Reduction of the Basic Equations*

It is interesting to note that Equations 21 are a set of scalar equations that have exactly the same structure as the Frenet-Serret equations of differential geometry.[11] It is known that the solution to such a set of equations can be made to rest upon the solution of a Riccati equation.[12] To show this, one first observes that an integral of Equation 21 is readily obtained in the form

$$\rho^2 + \mathscr{P}^2 + \mathscr{Q}^2 = 1 \tag{22}$$

in which an arbitrary function of ξ has been set equal to unity since before the arrival of the pulse $\mathscr{P} = \mathscr{Q} = 0$ and $\rho^2 = 1$.

Two new functions are now introduced by writing

$$\frac{\rho + i\mathscr{P}}{1 - \mathscr{Q}} = \frac{1 + \mathscr{Q}}{\rho - i\mathscr{P}} = \varphi \tag{23}$$

$$\frac{\rho - i\mathscr{P}}{1 - \mathscr{Q}} = \frac{1 + \mathscr{Q}}{\rho + i\mathscr{P}} = -\frac{1}{\psi} = \varphi^* \tag{24}$$

One then finds

$$\rho = \frac{1 - \varphi\psi}{\varphi - \psi} \tag{25a}$$

$$\mathscr{P} = i\frac{1 + \varphi\psi}{\varphi - \psi} \tag{25b}$$

$$\mathcal{Q} = \frac{\varphi + \psi}{\varphi - \psi} \tag{25c}$$

Equations governing the time dependence of φ and ψ are readily deduced by inserting Equations 25 into Equations 21. It is found that φ satisfies the Riccati equation

$$\frac{\partial\varphi}{\partial\tau} = i\mathscr{E}\varphi + \frac{i}{2}\frac{\Delta\omega}{\Omega}(\varphi^2 - 1) \tag{26}$$

and that ψ also satisfies this equation.

It was shown by McCall and Hahn[4] that one may obtain simple analytical expressions for the response of a two-level system to a field profile having the shape of a hyperbolic secant. This may be seen within the present formalism since the equation

$$\frac{\partial\varphi}{\partial\tau} = i(2\,\mathrm{sech}\,\tau)\varphi + \frac{i}{2}\frac{\Delta\omega}{\Omega}(\varphi^2 - 1) \tag{27}$$

has the solution

$$\varphi = \frac{f^2 + e^{i\sigma}}{2f\sin\frac{\sigma}{2} - (1 + f^2)} \tag{28}$$

where $f = \Delta\omega/\Omega$ and

$$\sigma = 2\int_{-\infty}^{\tau} d\tau'\,\mathrm{sech}\,\tau = 4\tan^{-1} e^{\tau} \tag{29}$$

According to Equations 25, the response of the system is

$$\mathscr{P} = -\frac{\sin\sigma}{1 + f^2} \tag{30a}$$

$$\mathcal{Q} = \frac{2f\sin(\sigma/2)}{1 + f^2} \tag{30b}$$

$$\rho = \frac{2\sin^2(\sigma/2)}{1 + f^2} - 1 \tag{30c}$$

In general, one may carry out the usual transformation of the Riccati equation to a second-order linear equation. The problem may then be reduced to solving the equation

$$\ddot{w} + \tfrac{1}{4}(f^2 + \mathscr{E}^2 + 2i\dot{\mathscr{E}})w = 0 \tag{31}$$

where w is related to φ through the transformations

$$w = u\exp\left(-i/2\int_{-\infty}^{\tau} d\tau'\,\mathscr{E}\right),\quad \varphi = (2i/f)d(\ln u)/d\tau.$$

Equation 31 is particularly instructive since it puts power broadening in evidence.

However, from the well-known properties of these equations, it follows that in general it is impossible to write φ or ψ explicitly in terms of quadratures of $\mathscr{E}$.

4. *Specialization to a Soluble Model*

If we consider the case of vanishing bandwidth, *i.e.*,

$$g(\Delta\omega) = \delta(\Delta\omega) \tag{32}$$

then only the case $\Delta\omega = 0$ need be considered in Equation 26. Pulse propagation in a medium having a vanishing bandwidth is still possible because of the presence of power broadening.[13]

The equation for φ is now linear and the solution is

$$\varphi = \pm e^{i\sigma} \tag{33}$$

where σ is given by

$$\mathscr{E} = \frac{\partial \sigma}{\partial \tau} \tag{34}$$

The ambiguity of sign in Equation 33 is related to the two possible initial conditions $\rho(\xi, -\infty) = \pm 1$. The lower sign is associated with a medium that is initially in the lower level and is the one to be treated subsequently. This minus sign will be suppressed by incorporating it into the constant defined in Equation 19.

From Equation 25b, there follows

$$\mathscr{P} = \sin \sigma \tag{35}$$

and Equation 20 takes the form

$$\frac{\partial^2 \sigma}{\partial \xi \partial \tau} = \sin \sigma \tag{36}$$

This nonlinear partial differential equation is the fundamental equation of the model.[14] Fortunately, it has already been studied extensively since it arose long ago in the theory of surfaces of constant negative curvature.[15] The general solution of the equation is not known. However, particular solutions may be obtained since the equation is of a type for which a Baecklund transformation[16] may be devised. This transformation theory enables one to relate pairs of solutions of certain partial differential equations. As is shown in the Appendix, the transformation equations for Equation 36 are

$$\frac{1}{2}\frac{\partial}{\partial\tau}(\sigma_1 - \sigma_0) = a_1 \sin\left(\frac{\sigma_1 + \sigma_0}{2}\right) \tag{37a}$$

$$\frac{1}{2}\frac{\partial}{\partial\xi}(\sigma_1 + \sigma_0) = \frac{1}{a_1} \sin\left(\frac{\sigma_1 - \sigma_0}{2}\right) \tag{37b}$$

where a_1 is an arbitrary constant. According to Equations 36, if σ_0 is a known solution of Equation 36, another solution σ_1 may be obtained by solving this pair of first-order equations.

5. *Particular Solutions*

By inspection one sees that a solution of Equation 37 is

$$\sigma_0 = 0 \tag{38}$$

It then follows from Equations 37 that

$$\sigma_1 = 4 \tan^{-1} e^{a_1\tau + \xi/a_1} \tag{39}$$

is another particular solution. By Equation 34, the pulse envelope has the form

$$\mathscr{E}_1 = \operatorname{sech} \frac{1}{2}\Omega\left(t - \frac{x}{v}\right) \tag{40}$$

where

$$v = \frac{c}{1 - 4\delta} \tag{41}$$

The constant a_1 has been chosen so as to normalize $\mathscr{E}$.

It has been shown[4] that an extremely useful function in the study of ultrashort pulses is the area function $\theta(x)$ which is defined as

$$\theta(x) \equiv \Omega \int_{-\infty}^{\infty} dt\, \mathscr{E}_1(x, t) \tag{42}$$

For the envelope obtained above, one finds $\theta = 2\pi$. It has become customary to borrow from the terminology of nuclear magnetic resonance and refer to $\mathscr{E}_1$ as a 2π pulse. It is now also known[4] that the area function for a 2π pulse is not a stable quantity unless the two-level systems are initially in the ground state. In this case the pulse shape given in Equation 40 corresponds to the recently discovered self-induced transparency effect.[4]

The determination of σ_1 from σ_0 required the integration of two first-order equations. It is known[17] from the theory of surfaces of constant negative curvature that there exists a relationship among four solutions of Equation 36 which does not involve quadratures. Although this relationship is usually derived by appeal to the geometric significance of the analysis, for present purposes it may be more conveniently obtained by introducing the following symbolic representation of the transformation Equations 37.

A transformation from a solution σ_i to a solution σ_j with a constant a_k will be represented as shown in Figure 1. Now the above-mentioned

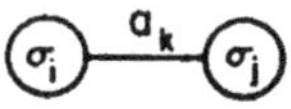

Figure 1. Schematic representation for the Baecklund transformation given by Equations 37.

relationship involving four solutions is

$$\tan\left(\frac{\sigma_3 - \sigma_0}{4}\right) = \frac{a_1 + a_2}{a_1 - a_2} \tan\left(\frac{\sigma_1 - \sigma_2}{4}\right) \tag{43}$$

It may be shown quite easily that the equations corresponding to the sequence of transformations depicted in Figure 2 can be combined to yield this purely trigonometric relation.

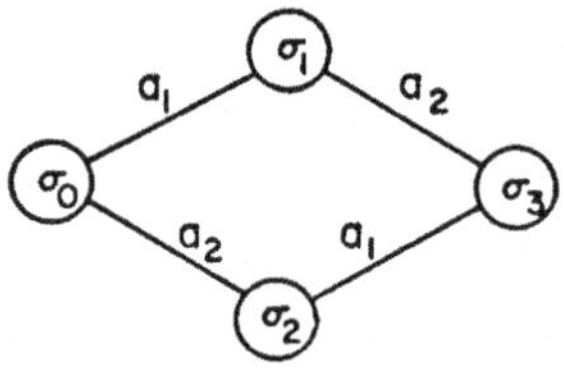

Figure 2. Diagram for sequence of transformations giving 4π pulse.

If one again assumes $\sigma_0 = 0$, the final solution σ_3 in Equation 43 may be made to vary from -2π to 2π as τ varies from $-\infty$ to ∞. One need only require $a_2 < 0$. Such a solution will represent a 4π pulse. In order to obtain a pulse shape which contains only one peak as a function of time at $x = 0$, one may impose the additional requirement $\partial^2\mathscr{E}/\partial\tau^2 < 0$ at $\xi = \tau = 0$. This leads to the inequality

$$(a_1 + a_2)(a_1{}^2 + a_2{}^2 + 3a_1a_2) > 0 \tag{44}$$

If one also requires $\mathscr{E}(0, 0) > 0$ and arbitrarily sets $a_2 = -1$, then the inequality is equivalent to $\frac{1}{2}(3+\sqrt{5}) \leq a_1 < \infty$. In terms of the angle defined by $a_1 = (1+\sin\theta)/\cos\theta$, this condition takes the form $\cos^{-1}(\frac{2}{3}) \leq \theta < \pi/2$.

The pulse shape is given by[18]

$$\mathscr{E}(x,t) = \frac{2\sin\theta[\operatorname{sech}\Omega(t-x/v) + a\operatorname{sech}a\Omega(t-x/v')]}{1-\cos\theta[\tanh\Omega(t-x/v)\tanh a\Omega(t-x/v') - \operatorname{sech}\Omega(t-x/v)\operatorname{sech}a\Omega(t-x/v')]} \tag{45}$$

where $v' = c/(1-\delta/a^2)$, $v = c/(1-\delta)$. An example for $\theta = \cos^{-1}(\frac{2}{3})$, $\delta = -\frac{1}{2}$ is shown in Figure 3. It is of interest to note that when the two

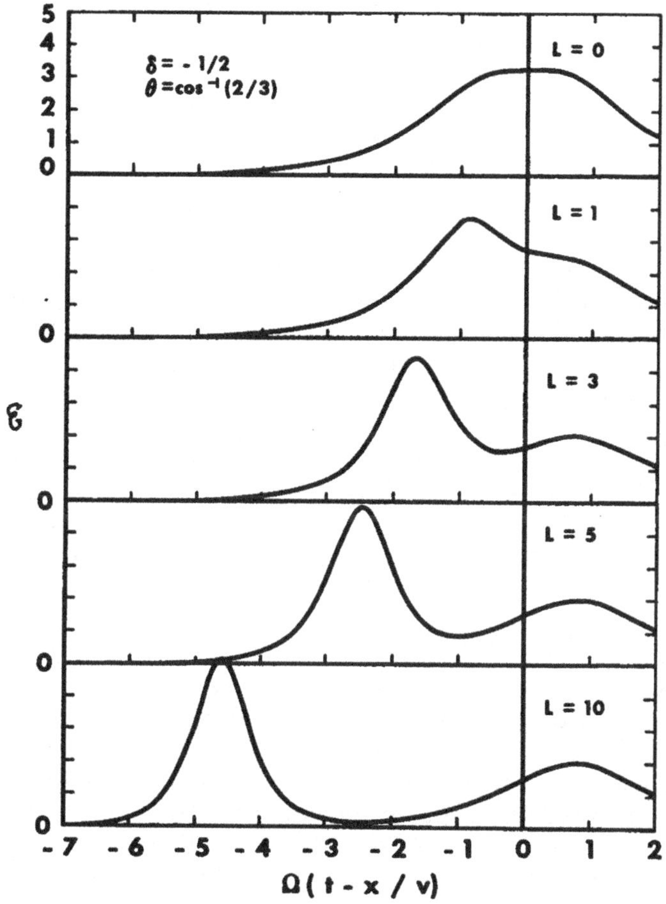

Figure 3. Breakup of 4π pulse predicted by Equation 45 as a function of the dimensionless distance $L = \Omega x/c$. Direction of propagation is to the left in the figure.

pulses are completely separated, only the leading one retains any information about the initial pulse shape.

The procedure outlined above may be generalized to yield expressions for any $2n\pi$ pulse where n is a positive integer. As an example, the 6π pulse is obtained from the sequence of transformations depicted in Figure 4. Certain constraints may be immediately imposed upon the

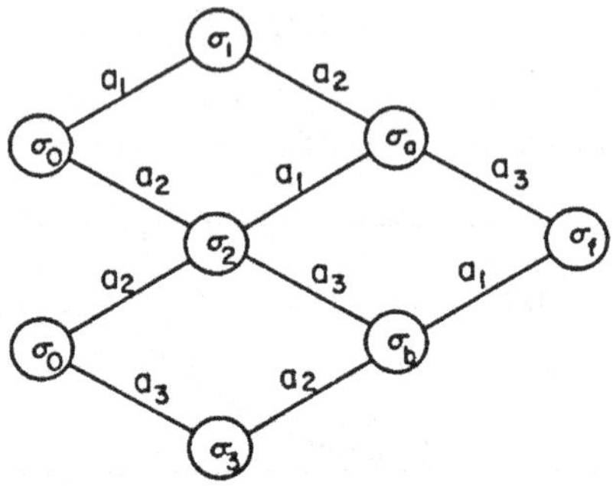

Figure 4. Diagram for sequence of transformations giving 6π pulse.

triad of constants a_1, a_2, a_3. Firstly, for the envelope function corresponding to σ_2 to be positive, one requires $a_2 > 0$. One then proceeds to make σ_a a 4π pulse which in addition requires $a_1 < 0$ and σ_b is made a zero π pulse which is obtained by requiring $0 < a_3 < a_2$. Additional restrictions are required in order to assure a pulse shape which consists of a single peak at $x = 0$. The inequality is much more complicated in this case than in Equation 44 and has not been analyzed in detail. By a simple trial-and-error method the case shown in Figure 5 has been obtained. Again it is found that the last pulse is insensitive to variations in the initial pulse shape. This effect has been observed experimentally.[19]

6. Stability of Pulse Shapes

The pulse shapes calculated above will not bear any relation to those realizable in nature unless they are stable with respect to perturbations about these shapes. In the present section a specific example will be used to show that an infinitesimal amount of phenomenological damping is necessary in order for the pulse shape to be stable.

To exhibit the instability of the undamped solutions, we first consider a perturbation to the solution given in Equation 39 and write

$$\sigma(u, v) = 4 \tan^{-1} e^u + \sigma^{(1)}(u, v) \tag{46}$$

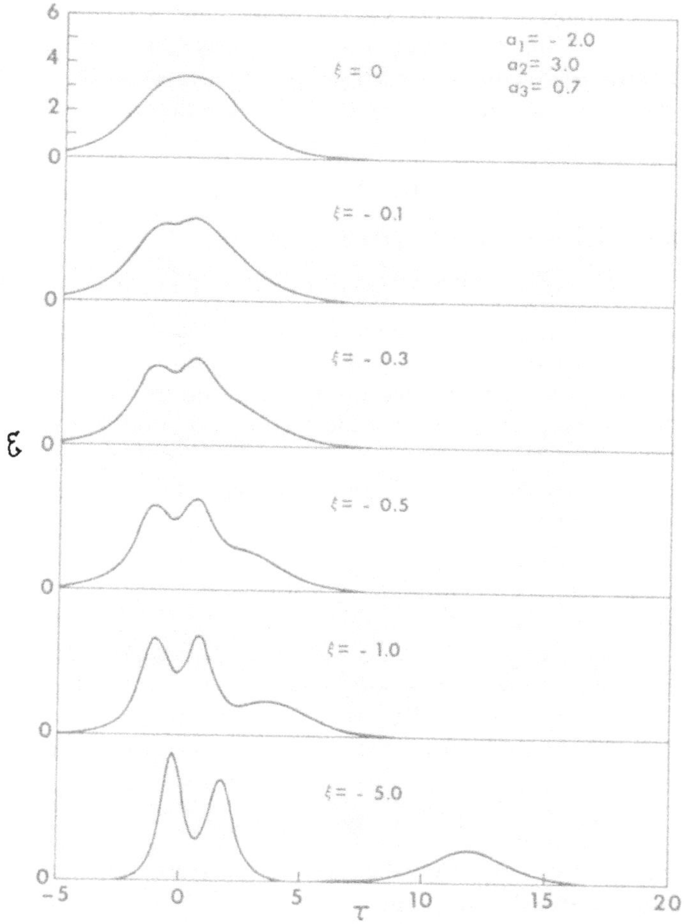

Figure 5. Breakup of 6π pulse as derived from sequence of transformations shown in Figure 4. Direction of propagation is to the left in the figure.

where new variables

$$u = a\tau + \xi/a$$
$$v = a\tau - \xi/a \tag{47}$$

have been introduced. We shall perform only a linear stability analysis which leads to the following equation for the perturbation $\sigma^{(1)}$

$$\frac{\partial^2 \sigma^{(1)}}{\partial u^2} - \frac{\partial^2 \sigma^{(1)}}{\partial v^2} - (1 - 2\,\text{sech}^2\, u)\sigma^{(1)} = -2a\,\text{sech}\, u \tag{48}$$

Particular solutions of this equation are readily obtained in the product form $\sigma^{(1)} = U(u)V(v)$. The solution of the equation for V is immediate, while the form of the equation for U is that of a Schroedinger equation for a sech^2 potential. Such an equation may be analyzed completely.[20] The simplest eigenfunction is

$$U(u) = \text{sech}\, u \tag{49}$$

One then finds that a possible perturbation is

$$\sigma^{(1)} = (A + Bv)\text{sech}\, u = [A + B(u - 2x)]\ \text{sech}\, u \tag{50}$$

This perturbation may be interpreted as a disturbance which propagates at the velocity of the zero-order pulse but contains a contribution that grows linearly with distance. Hence the undamped solution is seen to be unstable although the growth of the perturbation in this case is only linear.

The stability of solutions in the presence of damping may be indicated by considering the equation

$$\frac{\partial^2 \sigma}{\partial \xi \partial \tau} + \kappa \frac{\partial \sigma}{\partial \tau} = \sin \sigma \tag{51}$$

which is the counterpart of Equation 36 when damping is included in Equation 1. While the steady-state solution will now differ slightly from the solution given in Equation 39, the perturbation may be analyzed as before. It is now found that the term $V(v)$ satisfies the equation

$$(V'' - \kappa v')/v = \text{const} \tag{52}$$

which gives rise to an exponential decay of the perturbation as it propagates in the positive dx direction.

Appendix: Derivation of the Baecklund Transformation

The usual[21] derivation of the Baecklund transformation given in Equations 37 relies to a great extent upon the geometric significance of Equation 36. Any attempt to derive these equations *ab initio* by an application of the transformation theory as it is usually developed for the purpose of solving differential equations is frustrated by the fact that these treatments are nearly always restricted to the case in which the transformed dependent variable (σ_1 in Equations 37) does not appear explicitly in the transformation equations. The only presentation of the theory which the author has found that is sufficiently general to permit the derivation of Equations 37 is contained in a relatively inaccessible paper by M. J. Clairin.[16] Since a technique for deriving such transformation equations (when they exist!) is essential in certain attempts to extend the theory of the present paper, or to apply the method to other problems, it was felt that a derivation of the transformation equations based upon Clairin's method would be in order.

Rather than present the theory in unnecessary generality, attention will be confined to transformations of the form

$$p = f(p_1, z, z_1) \tag{A1}$$

$$q = \varphi(q_1, z, z_1) \tag{A2}$$

where $p = \partial z/\partial x$, $q = \partial z/\partial y$, $p_1 = \partial z_1/\partial x$, $q_1 = \partial z_1/\partial y$ and z_1 is a solution of the equation $\partial^2 z_1/\partial x\, \partial y = \sin z_1$. It will turn out that z satisfies the same equation as z_1 but this is not true in all applications of the theory. Then

$$\frac{dp}{dy} = \frac{\partial f}{\partial p_1} s_1 + \frac{\partial f}{\partial z} q + \frac{\partial f}{\partial z_1} q_1 \tag{A3}$$

$$\frac{dq}{dx} = \frac{\partial \varphi}{\partial q_1} s_1 + \frac{\partial \varphi}{\partial z} p + \frac{\partial \varphi}{\partial z_1} p_1 \tag{A4}$$

where $s_1 = \partial^2 z_1/\partial x\, \partial y$. The requirement that

$$\frac{dp}{dy} = \frac{dq}{dx} \tag{A5}$$

yields

$$s_1\left(\frac{\partial f}{\partial p_1} - \frac{\partial \varphi}{\partial q_1}\right) - p_1 \frac{\partial \varphi}{\partial z_1} + q_1 \frac{\partial f}{\partial z_1} + \varphi \frac{\partial f}{\partial z_1} - f \frac{\partial \varphi}{\partial z} = 0 \tag{A6}$$

A function $\Omega(p_1, q_1, z, z_1)$ is now introduced by the definition

$$\Omega(p_1, q_1, z, z_1) = \sin z_1\left(\frac{\partial f}{\partial p_1} - \frac{\partial \varphi}{\partial q_1}\right) - p_1 \frac{\partial \varphi}{\partial z_1} + q_1 \frac{\partial f}{\partial z_1} + \varphi \frac{\partial f}{\partial z} - f \frac{\partial \varphi}{\partial z} = 0 \tag{A7}$$

and successive derivatives of Ω are calculated as shown below.

$$\frac{\partial \Omega}{\partial p_1} = \sin z_1 \frac{\partial^2 f}{\partial p_1{}^2} - \frac{\partial \varphi}{\partial z_1} + q_1 \frac{\partial^2 f}{\partial p_1\, \partial z_1} + \varphi \frac{\partial^2 f}{\partial p_1\, \partial z} - \frac{\partial f}{\partial p_1} \frac{\partial \varphi}{\partial z} = 0 \tag{A8}$$

$$\frac{\partial\Omega}{\partial q_1} = -\sin z_1 \frac{\partial^2\varphi}{\partial q_1{}^2} - p_1 \frac{\partial^2\varphi}{\partial q_1\,\partial z_1} + \frac{\partial f}{\partial z_1} + \frac{\partial\varphi}{\partial q_1}\frac{\partial f}{\partial z} - f\frac{\partial^2\varphi}{\partial q_1\,\partial z} = 0 \tag{A9}$$

$$\frac{\partial^2\Omega}{\partial p_1\,\partial q_1} = -\frac{\partial^2\varphi}{\partial q_1\,\partial z_1} + \frac{\partial^2 f}{\partial p_1\,\partial z_1} + \frac{\partial\varphi}{\partial q_1}\frac{\partial^2 f}{\partial p_1\,\partial z} - \frac{\partial f}{\partial p_1}\frac{\partial^2\varphi}{\partial q_1\,\partial z} = 0 \tag{A10}$$

$$\frac{\partial^3\Omega}{\partial p_1{}^2\,\partial q_1} = \frac{\partial^3 f}{\partial p_1{}^2\,\partial z_1} + \frac{\partial\varphi}{\partial q_1}\frac{\partial^3 f}{\partial p_1{}^2\,\partial z} - \frac{\partial^2 f}{\partial p_1{}^2}\frac{\partial^2\varphi}{\partial q_1\,\partial z} = 0 \tag{A11}$$

$$\frac{\partial^3\Omega}{\partial p_1\,\partial q_1{}^2} = -\frac{\partial^3\varphi}{\partial q_1{}^2\,\partial z_1} + \frac{\partial^2\varphi}{\partial q_1{}^2}\frac{\partial^2 f}{\partial p_1\,\partial z} - \frac{\partial f}{\partial p_1}\frac{\partial^3\varphi}{\partial q_1{}^2\,\partial z} = 0 \tag{A12}$$

$$\frac{\partial^4\Omega}{\partial p_1{}^2\,\partial q_1{}^2} = \frac{\partial^2\varphi}{\partial q_1{}^2}\frac{\partial^3 f}{\partial p_1{}^2\,\partial z} - \frac{\partial^2 f}{\partial p_1{}^2}\frac{\partial^3\varphi}{\partial q_1{}^2\,\partial z} = 0 \tag{A13}$$

One now proceeds to obtain a solution of Equation A13. The arbitrary functions involved are then determined so that Equations A8 to A12 are also satisfied. The resulting solution will constitute the desired Baecklund transformation.

In Equation A13 it is possible to separate the p_1 and q_1 dependence and write

$$\frac{\partial^3\varphi/\partial q_1{}^2\,\partial z}{\partial^2\varphi/\partial q_1{}^2} = \frac{\partial^3 f/\partial p_1{}^2\,\partial z}{\partial^2 f/\partial p_1{}^2} = \frac{\partial \mathrm{H}(z, z_1)/\partial z}{\mathrm{H}(z, z_1)} \tag{A14}$$

These equations may be integrated to give

$$f(p_1, z, z_1) = g(p_1, z_1)\mathrm{H}(z, z_1) + \ell(z, z_1)p_1 + m(z, z_1) \tag{A15}$$

$$\varphi(q_1, z, z_1) = \theta(q_1, z_1)\mathrm{H}(z, z_1) + \lambda(z, z_1)q_1 + \mu(z, z_1) \tag{A16}$$

It is evident that $\mathrm{H}(z, z_1)$ may be set equal to unity without changing the structure of these equations. Substitution of Equations A15 and A16 into Equation A12 then yields

$$\frac{\partial^3\theta}{\partial z_1\,\partial q_1{}^2} - \frac{\partial^2\theta}{\partial q_1{}^2}\frac{\partial\ell}{\partial z} = 0 \tag{A17}$$

which shows[22] that $\partial\ell/\partial z$ is independent of z. One may then integrate Equation A17 to see that $\theta(q_1, z_1)$ must be of the form

$$\theta(q_1, z_1) = e^{u(z_1)}v(q_1) + r(z_1)q_1 + w(z_1) \tag{A18}$$

where $u(z_1)$ is related to $\ell(z, z_1)$ by

$$\ell(z, z_1) = z \frac{\partial u}{\partial z_1} + \text{const} \tag{A19}$$

and $v(q_1)$, $r(z_1)$ and $w(z_1)$ are arbitrary functions of their arguments. When θ and ℓ are substituted into Equation A16, one finds that there is no loss of generality in setting $w(z_1)$ and the constant in Equation A19 equal to zero. We shall further restrict the calculation by setting $u = 0$, and obtain

$$\varphi(q_1, z, z_1) = v(q_1) + r(z_1)q_1 + \mu(z, z_1) \tag{A20}$$

A similar analysis of Equation A11 leads to

$$f(p_1, z, z_1) = \bar{v}(p_1) + \bar{r}(z_1)p_1 + m(z, z_1) \tag{A21}$$

Substitution into Equation A10 yields

$$\frac{\partial}{\partial z_1}(r - \bar{r}) = 0 \tag{A22}$$

Turning now to Equation A9, one finds that it imposes the restriction

$$-\frac{\partial^2 v}{\partial {q_1}^2} \sin z_1 + \frac{\partial m}{\partial z}\frac{\partial v}{\partial q_1} = -\frac{\partial m}{\partial z_1} - r\frac{\partial m}{\partial z} \tag{A23}$$

This equation may be solved to find the general form of $v(q_1)$. One finds that unless $\partial m/\partial z = \sin z_1$, one must have $v(q_1) = \text{const}$ and $v(q_1)$ may then be assimilated into $\mu(z, z_1)$. We shall follow up only this case.

Equation A23 now simplifies to

$$\frac{\partial m}{\partial z_1} + r(z_1)\frac{\partial m}{\partial z} = 0 \tag{A24}$$

and φ is of the form

$$\varphi(q_1, z, z_1) = r(z_1)q_1 + \mu(z, z_1) \tag{A25}$$

Application of a similar procedure to Equation A8 leads to

$$\frac{\partial \mu}{\partial z_1} + \bar{r}(z_1)\frac{\partial \mu}{\partial z} = 0 \tag{A26}$$

and

$$f(p_1, z, z_1) = \bar{r}(z_1)p_1 + m(z, z_1) \tag{A27}$$

Finally, Equation A7 yields

$$(\bar{r} - r)\sin z_1 + \mu\frac{\partial m}{\partial z} - m\frac{\partial \mu}{\partial z} = 0 \tag{A28}$$

If one sets $\bar{r} = 1$, $r = -1$, Equations A24 and A26 imply

$$m = m(z + z_1) \qquad \mu = \mu(z - z_1) \tag{A.29}$$

and Equation A28 is now

$$2 \sin z_1 + \frac{\partial}{\partial z_1}[m(z + z_1)\mu(z - z_1)] = 0 \tag{A30}$$

This functional differential equation is easily solved. Differentiation with respect to z yields

$$\frac{m''(z + z_1)}{m(z + z_1)} = \frac{\mu''(z - z_1)}{\mu(z - z_1)} = -K^2 \tag{A31}$$

where the primes indicate differentiation with respect to argument. From Equation A31,

$$\begin{aligned} m(t) &= a \cos Kt + b \sin Kt \\ \mu(t) &= \alpha \cos Kt + \beta \sin Kt \end{aligned} \tag{A32}$$

When these solutions are substituted into Equation A30, evaluated at $z = 0$ and odd and even functions of z_1 are separately equated to zero, one finds

$$\begin{aligned} \alpha b - \beta a &= 0 \\ \alpha a + \beta b &= 4 \\ K &= \tfrac{1}{2} \end{aligned} \tag{A33}$$

which may be satisfied by setting $a = \alpha = 0$, $b\beta = 4$ or $b/2 = 2/\beta = A$ where A is a constant. Equations A25 and A27 now yield

$$\begin{aligned} \tfrac{1}{2}(p - p_1) &= A \sin\left(\frac{z + z_1}{2}\right) \\ \tfrac{1}{2}(q + q_1) &= \frac{1}{A} \sin\left(\frac{z - z_1}{2}\right) \end{aligned} \tag{A34}$$

These are the desired transformation equations.

References

1. A. J. DeMaria, D. A. Stetser, and W. H. Glenn, Jr., *Science*, **156**, 1557 (1967); G. Magyar, *Nature*, **218**, 16 (1968). Both of these survey articles contain extensive reference to the original research publications.
2. R. Bellman, G. Birnbaum, and W. G. Wagner, *J. Appl. Phys.*, **34**, 750 (1963); L. M. Frantz, and J. S. Nodvik, *J. Appl. Phys.*, **34**, 2346 (1963); J. P. Wittke, and P. J. Warter, *J. Appl. Phys.*, **35**, 1668 (1964).

3. C. L. Tang, and B. D. Silverman, in *Physics of Quantum Electronics*, edited by P. L. Kelley, B. Lax, and P. E. Tannenwald, McGraw-Hill Book Co., Inc., New York (1966) p. 280. Although this paper is not addressed to short-pulse propagation, the formalism is adequate for such considerations.
4. S. L. McCall, and E. L. Hahn, *Phys. Rev. Letters*, **18**, 908 (1967).
5. F. Hopf, and M. O. Scully, to be published in *Phys. Rev.*
6. C. K. N. Patel, and R. E. Slusher, *Phys. Rev. Letters*, **19**, 1019 (1967).
7. F. Bloch, *Phys. Rev.*, **70**, 460 (1946).
8. W. E. Lamb, Jr., in *Proceedings of the International School of Physics "Enrico Fermi," Course XXXI*, edited by P. A. Miles, Academic Press, Inc., New York (1964); *Phys. Rev.*, **134**, A1429 (1964).
9. N. G. Basov, *et al.*, *Soviet Physics — JETP*, **23**, 16 (1966).
10. Private communication from M. O. Scully, Massachusetts Institute of Technology, Cambridge, Mass.
11. L. P. Eisenhart, *A Treatise on the Differential Geometry of Curves and Surfaces*, Dover Publications, Inc., New York (1960) p. 23.
12. Reference 11, p. 25.
13. The author is indebted to E. B. Treacy for this observation.
14. The relevance of this equation to optical pulse propagation was apparently first discovered by C. Cercignagni and A. Ghirardi and is referred to by F. T. Arecchi and R. Bonifacio, *IEEE J. Quant. Electr.*, **1**, 169 (1965) where a summary of their calculation is presented. Discussion of this equation in relation to other physical problems of contemporary interest may be found in A. Seeger, *et al.*, *Z. Physik*, **134**, 173 (1953); P. Lebwohl, and M. J. Stephen, *Phys. Rev.*, **163**, 376 (1967).
15. Reference 11, p. 280 *et seq.*
16. A. R. Forsyth, *Theory of Differential Equations*, Vol. *VI*, Chap. 21, Dover Publications, Inc., New York (1959). Additional references are cited herein. M. J. Clairin, *Annales de Toulouse*, 2ᵉ Sér. **5**, 437 (1903).
17. Reference 11, p. 286.
18. G. L. Lamb, Jr., *Phys. Lett.*, **25A**, 181 (1967). In this work the analysis was improperly applied to the amplifier case $\delta > 0$.
19. C. K. N. Patel, and R. E. Slusher, to be published.
20. P. M. Morse, and H. Feshbach, *Methods of Theoretical Physics*, Vol. *I*, pp. 734 and 768 and Vol. *II*, p. 1651. McGraw-Hill Book Co., Inc., New York (1953).
21. Reference 11, p. 284.
22. An alternate approach is to set $\theta = 0$.

David Mintzer and W. James Hadden, Jr.
Northwestern University

8 Two-Point Correlations for Random-Medium Acoustic Propagation

Introduction

The propagation of a series of sound pulses through a medium whose refractive index varies randomly as a function of space and time has been the subject of extensive theoretical research[1,2] and experiments in the ocean[3] and the laboratory[4,5]. At the present time, it can be said that the case of propagation from a source to a single receiver is well understood, with the coefficient of variation (standard deviation of a series of sound pulses, expressed as a fraction of the mean) of the received pulses being related to the correlation function of the refractive index variations; the laboratory experiments,[4] especially, have shown the close agreement between the theoretical predictions and the measurements over wide ranges of distance from source to receiver, and of frequency.

There has been, on the other hand, only a small amount of research on two-point correlations of the acoustic field, and how they are related to the temperature (or refractive index) microstructure. Calculations have been done for the cases of a plane wave propagated to two receivers oriented so that the line between them is parallel to or perpendicular to the wave front[6]; and for a spherical source with two arbitrarily placed receivers, but in the ray-theory approximation.[7] It may be expected, however, that the correlation of the signals received at two different points in the medium will give additional information concerning the microstructure which may well be difficult to obtain otherwise.

General Analysis

We proceed along the lines set forth in Reference 1(I); the single-scattering approximation for the pressure received at r_1 due to a source at the origin is

$$p(\mathbf{r}_1) = \frac{e^{ik_0 r_1}}{r_1} + \frac{k_0{}^2 \alpha}{2\pi} \int_V n(\mathbf{r}') \frac{e^{ik_0|\mathbf{r}_1 - \mathbf{r}'|}}{|\mathbf{r}_1 - \mathbf{r}'|} \frac{e^{ik_0 r'}}{r'} \, dV' \tag{1}$$

where $k_0 = 2\pi f/c_0$, with c_0 being the mean sound speed. The refractive index is taken as

$$\mu(\mathbf{r}) = 1 + \alpha n(\mathbf{r}) \qquad \alpha \ll 1 \tag{2}$$

where $n(\mathbf{r})$ is the refractive index variation, random in space and time, whose rms value has been normalized to unity; and α is the rms value of the refractive index variations. Since the time scale of the variations of the refractive index is much greater than the pulse length, that variable is ignored in Equation 1; it is the variations between pulses which cause the pulse-to-pulse amplitude variation. The volume integrated over is that portion of the medium which gives rise to the scattered waves arriving at the receiver during the pulse reception.

We change the integration variables $\mathbf{r}' = (x', y', z')$ to prolate spheroidal coordinates (ξ', η', φ'):

$$x' = \left(\frac{r_1}{2}\right)\xi'\eta' \qquad y' = \left(\frac{r_1}{2}\right)[(\xi'^2 - 1)(1 - \eta'^2)]^{1/2} \cos\varphi$$

$$z' = \left(\frac{r_1}{2}\right)[(\xi'^2 - 1)(1 - \eta'^2)]^{1/2} \sin\varphi'$$

$$\xi' = \left(\frac{1}{r_1}\right)[r' + |\mathbf{r}_1 - \mathbf{r}'|] \qquad \eta' = \left(\frac{1}{r_1}\right)[r' - |\mathbf{r}_1 - \mathbf{r}'|] \tag{3}$$

$$dV' = \left(\frac{r_1{}^3}{8}\right)(\xi'^2 - \eta'^2)\, d\xi'\, d\eta'\, d\varphi'$$

$$1 \le \xi' \le \infty \qquad -1 \le \eta' \le 1 \qquad 0 \le \varphi' \le 2\pi$$

The scattered waves which reach the observation point at time t have come from volume elements satisfying the inequality

$$c_0(t - T) \le [r' + |\mathbf{r}_1 - \mathbf{r}'|] \le c_0 t,$$

where T is the pulse length; the spheroidal coordinates allow these inequalities to be used in the limits of integration of Equation 1. We will be interested in the correlation of the pressure amplitudes[8], each averaged over the pulse length, received at the points $\mathbf{r}_1$ and $\mathbf{r}_2$:

$$\mathscr{P} = \frac{\langle [\overline{|p(\mathbf{r}_1)|} - \langle \overline{|p(\mathbf{r}_1)|} \rangle][\overline{|p(\mathbf{r}_2)|} - \langle \overline{|p(\mathbf{r}_2)|} \rangle] \rangle}{\langle [\overline{|p(\mathbf{r}_1)|} - \langle \overline{|p(\mathbf{r}_1)|} \rangle]^2 \rangle} \tag{4}$$

$$\equiv \frac{I(\mathbf{r}_1, \mathbf{r}_2)}{I(\mathbf{r}_1, \mathbf{r}_1)} \tag{4'}$$

The ensemble average, $\langle\rangle$,i nvolves averaging over many different distributions of refractive index variations.[1] If we use Equations 3 in Equation 1, expand to first order in the small parameter α (second, and higher order, terms in α drop out when we keep only second order terms in the numerator and denominator of Equation 4), and average over the pulse length, we obtain

$$I(\mathbf{r}_1, \mathbf{r}_2) = \langle \mathrm{Re}(e^{-ik_0 r_1} \bar{P}_1) \mathrm{Re}(e^{-ik_0 r_2} \bar{P}_2) \rangle \tag{5}$$

where

$$\bar{P}_1 = \frac{\alpha k_0^{\,2} r_1}{4\pi} \int_0^{2\pi} d\varphi' \int_{-1}^{1} d\eta' \int_1^{1 + c_0 T/r_1} \times d\xi' \left(1 + \frac{r_1}{c_0 T} - \frac{r_1}{c_0 T} \xi'\right) e^{ik_0 r_1 \xi'} n(\xi', \eta', \varphi') \tag{5'}$$

In Equation 5′ we have taken the axis of the prolate spheroidal coordinate system along the line $\mathbf{r}_1$, that is, along the line from source to receiver. Thus,

$$\begin{aligned} I(\mathbf{r}_1, \mathbf{r}_2) = \frac{\alpha^2 k_0^{\,4} r_1 r_2}{32\pi^2} & \int d\xi'() \int d\xi''() N(|\mathbf{r}' - \mathbf{r}''|) \\ & \cdot \mathrm{Re}\{\exp[ik_0 r_1(\xi' - 1) - ik_0 r_2(\xi'' - 1)] \\ & + \exp[ik_0 r_1(\xi' - 1) + ik_0 r_2(\xi'' - 1)]\} \end{aligned} \tag{6}$$

where $\int d\xi'()$ is symbolic for the appropriate expressions of Equation 5′; similarly for $\int d\xi''()$; and the autocorrelation functions of the refractive index variations is

$$N(x' - x'', y' - y'', z' - z'') = \langle n(x', y', z') n(x'', y'', z'') \rangle \tag{7}$$

We introduce the three-dimensional Fourier transform of the autocorrelation:

$$\mathscr{N}(k_1, k_2, k_3) = \frac{1}{(2\pi)^{3/2}} \iiint_{-\infty}^{\infty} N(x_1, x_2, x_3) e^{-i(k_1 x_1 + k_2 x_2 + k_3 x_3)} \, dx_1 \, dx_2 \, dx_3 \tag{8}$$

We let

$$I(\mathbf{r}_1, \mathbf{r}_2) = \frac{\alpha^2 k_0^4 r_1 r_2}{32\pi^2} \operatorname{Re}[e^{-ik_0(r_1 - r_2)} T_1 + e^{-ik_0(r_1 + r_2)} T_2] \tag{9}$$

so that

$$T_1 = \frac{1}{(2\pi)^{3/2}} \int dk_x\, dk_y\, dk_z \int d\xi'(\,) \int d\xi''(\,) e^{ik_0(r_1\xi' - r_2\xi'')} e^{i\mathbf{k}\cdot(\mathbf{r}' - \mathbf{r}'')} \cdot \mathcal{N}(k_x, k_y, k_z) \tag{10}$$

The geometrical situation is as shown in Figure 1, where we take an

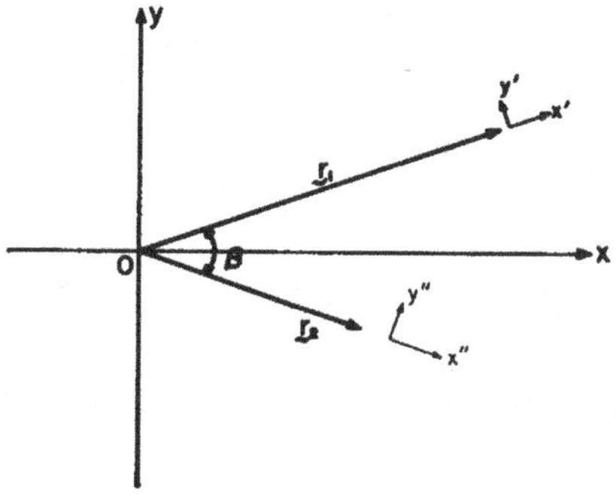

Figure 1. Geometry of source and receivers; source at origin, receivers at $\mathbf{r}_1$ and $\mathbf{r}_2$. Axes for coordinate systems used in integrations are also shown.

x-axis bisecting the angle β between the vectors $\mathbf{r}_1$ and $\mathbf{r}_2$ from the source to the receivers; a y-axis perpendicular to it in the plane of the vectors; and an appropriate z-axis. It is to be noted that, in Equation 10, the $\xi' = 0$ and $\xi'' = 0$ lines are along $\mathbf{r}_1$ and $\mathbf{r}_2$, respectively. The last exponent of Equation 10 is to be put in terms of the spheroidal coordinates, therefore, whose major axes are at angles β with each other. It is simplest to do the ξ'- and ξ''-integrations by decomposing the vector $\mathbf{k}$ into the primed and double-primed coordinate systems:

$$\begin{aligned} \mathbf{k} \cdot \mathbf{r}' &= k_x' x' + k_y' y' + k_z' z' \\ &= \left(\frac{r_1}{2}\right)[k_x' \xi' \eta' + k_y' (\xi'^2 - 1)^{1/2}\, 1 - \eta'^2)^{1/2} \cos\varphi' \\ &\quad + k_z'(\xi'^2 - 1)^{1/2}(1 - \eta'^2)^{1/2} \sin\varphi'] \end{aligned} \tag{11}$$

where

$$k_x' = k_x \cos\left(\frac{\beta}{2}\right) + k_y \sin\left(\frac{\beta}{2}\right) \qquad k_y' = -k_x \sin\left(\frac{\beta}{2}\right) + k_y \cos\left(\frac{\beta}{2}\right)$$

$$k_z' = k_z \tag{11'}$$

with similar expressions in the double-primed system (with $-\beta$ replacing β). Finally, as in Reference 1(I), we express the $\mathbf{k}'$ and $\mathbf{k}''$ in spherical coordinates (k, θ', ψ') and (k, θ'', ψ''), using the x'- and x''-axes as the polar axes:

$$\mathbf{k}\cdot(\mathbf{r}'-\mathbf{r}'') = \tfrac{1}{2}kr_1[\xi'\eta'\cos\theta' + (\xi'^2-1)^{1/2}(1-\eta'^2)^{1/2}\cdot\sin\theta'\cos(\varphi'-\psi')] - \tfrac{1}{2}kr_2[\text{primes}\rightarrow\text{double-primes}] \tag{12}$$

For simplicity of notation, we use $\mathbf{k} = (k_x, k_y, k_z)$, in the transform, in terms of the spherical coordinates (k, θ, ψ) based on the x-axis as polar axis. Thus,

$$T_1 = (2\pi)^{-3/2}\int d\mathbf{k}\mathcal{N}(k,\theta,\psi) \left\{\int d\xi'(\,)\cdot\exp(ik_0r_1\xi')\exp\left(\frac{1}{2}ikr_1\xi'\eta'\cos\theta'\right)\cdot\exp\left[\frac{1}{2}ikr_1(\xi'^2-1)^{1/2}(1-\eta'^2)^{1/2}\sin\theta'\cos(\varphi'-\psi')\right]\right\}\cdot\{r_1\rightarrow r_2;\ \text{prime}\rightarrow\text{double-prime};\ \text{complex conjugate}\} \tag{13}$$

The ξ' and ξ'' integrations can be evaluated as in Reference 1(I), Equations 15 to 18, for $k_0r_1 \gg 1$, $k_0r_2 \gg 1$:

$$T_1 = \frac{(2\pi)^{1/2}}{k_0^2r_1r_2}\exp[ik_0(r_1-r_2)]\int d\mathbf{k}\mathcal{N}(k,\theta,\psi)\int_0^\pi \sin\lambda'\,d\lambda'\int_0^\pi \sin\lambda''\,d\lambda'' \cdot\exp\left[-i\left(\frac{k^2}{8k_0}\right)(r_1-r_2)\right]\cdot\exp\left[\frac{1}{2}ik(r_1\cos\theta'\cos\lambda'-r_2\cos\theta''\cos\lambda'')\right]\cdot\exp\left[i\left(\frac{k^2}{8k_0}\right)(r_1\cos^2\lambda'-r_2\cos^2\lambda'')\right] \tag{14}$$

We return to rectangular coordinates in $\mathbf{k}$, using Equations 11′, and define

$$\tau' = \left(\frac{r_1}{a}\right)\cos\lambda' \qquad \tau'' = \left(\frac{r_2}{a}\right)\cos\lambda''$$

$$\frac{1}{2}ka = \lambda \qquad D_i = \frac{r_i}{k_0a^2} \qquad \Delta_i = \frac{1}{k_0r_i} \tag{15}$$

where a is a correlation length for the refractive index variations (which need not be the same in all three directions); the D_i are the Fresnel zone parameters see Reference 1(II), and Reference 7); and the $\Delta_i \ll 1$. We finally obtain

$$T_1 = \frac{8(2\pi)^{1/2}}{k_0{}^2 a r_1{}^2 r_2{}^2} \exp[ik_0(r_1 - r_2)] \int d\lambda \mathcal{N}(\lambda_x, \lambda_y, \lambda_z)$$

$$\cdot \int_{-r_1/a}^{r_1/a} d\tau' \int_{-r_2/a}^{r_2/a} d\tau''$$

$$\cdot \exp\left[-\frac{1}{2} i\lambda^2(D_1 - D_2)\right] \exp\left[i\lambda_x(\tau' - \tau'')\cos\left(\frac{\beta}{2}\right)\right.$$

$$\left. + i\lambda_y(\tau' + \tau'')\sin\left(\frac{\beta}{2}\right)\right] \exp\left[\frac{1}{2} i\lambda^2(\Delta_1\tau'^2 - \Delta_2\lambda''^2)\right] \qquad (16)$$

In a similar manner, we find

$$T_2 = -\frac{8(2\pi)^{1/2}}{k_0{}^2 a r_1{}^2 r_2{}^2} \exp[ik_0(r_1 + r_2)] \int d\lambda \mathcal{N}(\lambda_x, \lambda_y, \lambda_z)$$

$$\int_{-r_1/a}^{r_1/a} d\tau' \int_{-r_2/a}^{r_2/a} d\tau''$$

$$\cdot \exp\left[-\frac{1}{2} i\lambda^2(D_1 + D_2)\right]$$

$$\cdot \exp\left[i\lambda_x(\tau' - \tau'')\cos\left(\frac{\beta}{2}\right)\right.$$

$$\left. + i\lambda_y(\tau' + \tau'')\sin\left(\frac{\beta}{2}\right)\right]$$

$$\cdot \exp\left[\frac{1}{2} i\lambda^2(\Delta_1\tau'^2 + \Delta_2\tau''^2)\right] \qquad (16')$$

Using Equations 16 and 16′ in Equation 9, we have

$$I(\mathbf{r}_1, \mathbf{r}_2) = (2\pi)^{-3/2}\left(\frac{\alpha^2 k_0{}^2}{a r_1 r_2}\right)\mathrm{Re}\left(\int d\lambda \mathcal{N}(\lambda) \int_{-r_1/a}^{r_1/a} d\tau' \int_{-r_2/a}^{r_2/a} d\tau''\right.$$

$$\cdot \exp\left[i\lambda_x(\tau' - \tau'')\cos\left(\frac{\beta}{2}\right) + i\lambda_y(\tau' + \tau'')\sin\left(\frac{\beta}{2}\right)\right]$$

$$\cdot \left\{\exp\left[\frac{1}{2} i\lambda^2(\Delta_1\tau'^2 - \Delta_2\tau''^2 + D_2 - D_1\right]\right. \qquad (17)$$

$$\left.\left. - \exp\left[\frac{1}{2} i\lambda^2(\Delta_1\tau'^2 + \Delta_2\tau''^2 - D_1 - D_2)\right]\right\}\right)$$

A Particular Case

Rather than evaluate Equation 17 in general, which leads to rather unwieldy results, we particularize to the case of $r_1 = r_2 \equiv r$ (equidistant receivers), and a Gaussian correlation function

$$N(r) = e^{-r^2/a^2} \tag{18}$$

We let

$$\rho = \tau' - \tau'' \qquad \sigma = \tau' + \tau'' \tag{19}$$

so that Equation 17 becomes

$$I(r, \beta) = \left(\frac{\alpha^2 {k_0}^2 a^2}{4\pi^{3/2} r^2}\right) \mathrm{Re} \iiint_{-\infty}^{\infty} d\lambda_x \, d\lambda_y \, d\lambda_z \int_0^{2r/a} d\rho \int_0^{2r/a-\rho} d\sigma$$
$$\cdot e^{-\lambda^2} e^{i[\lambda_x \rho \cos(\beta/2) + \lambda_y \sigma \sin(\beta/2)]}$$
$$\cdot \left\{ \exp\left[\frac{1}{2} i\Delta\rho\sigma\lambda^2\right] - \exp\left[\frac{1}{4} i\lambda^2(\Delta\sigma^2 + \Delta\rho^2 - 4D)\right]\right\} \tag{20}$$

where $\Delta = 1/k_0 r$ and $D = r/k_0 a^2$. We first integrate over the λ-variables:

$$I(r, \beta) = \left(\frac{\alpha^2 {k_0}^2 a^2}{4r^2}\right) \mathrm{Re} \int_0^{2r/a} d\rho \int_0^{2r/a-\rho} d\sigma \left\{\left[1 - \frac{1}{2} i\,\Delta\rho\sigma\right]^{-3/2}\right.$$
$$\cdot \exp\left[\frac{-(\rho^2 \cos^2 \beta/2 + \sigma^2 \sin^2 \beta/2)}{4\left(1 - \frac{1}{2} i\,\Delta\rho\sigma\right)}\right]$$
$$- \left[1 - \frac{1}{4} i(\Delta\sigma^2 + \Delta\rho^2 - 4D)\right]^{-3/2}$$
$$\left.\cdot \exp\left[\frac{-(\rho^2 \cos^2 \beta/2 + \sigma^2 \sin^2 \beta/2)}{4\left(1 - \frac{1}{4} i(\Delta\sigma^2 + \Delta\rho^2 - 4D)\right)}\right]\right\} \tag{21}$$

For the case of $r \sin(\beta/2) \ll a$, the maximum value of σ is of the order of r/a, while the maximum value of ρ which contributes to the integrals is approximately unity; thus $\max(\Delta\rho\sigma) \simeq 1/k_0 a \ll 1$. We may therefore neglect the $\Delta\rho\sigma$-terms in the first term of the integrand. In the second term, the $\Delta\rho^2$-terms may be neglected, but the $\Delta\sigma^2$-term is of the order of D. If $D \ll 1$, the $(\Delta\sigma^2 - 4D)$-terms may be neglected (or expanded to obtain correction terms); if $D \gg 1$, then, as will be seen, the second term is small compared with the first one; we shall therefore neglect the $(\Delta\sigma^2 - 4D)$-terms. Thus, the case of $\beta \simeq 0$ (and similarly for $\beta \simeq \pi$) and $D \simeq 1$ will not be covered; for $\beta = 0$, the result for arbitrary D is found to be[9] $I(r, 0) = \alpha^2 k_0{}^2 (a/r)[1 - (1/\pi D)\tan^{-1}(\pi D)]$.

We therefore approximate Equation 21 for $r/a \gg 1$, by

$$I(r, \beta) \simeq (\alpha^2 k_0{}^2 a^2/4r^2)\mathrm{Re}\int_0^{2r/a} d\rho \int_0^{2r/a} \cdot d\sigma\left\{\exp\left[-\frac{1}{4}(\rho^2\cos^2\beta/2 + \sigma^2\sin^2\beta/2)\right] - (1 + iD)^{-3/2}\exp\left[\frac{-\frac{1}{4}(\rho^2\cos^2\beta/2 + \sigma^2\sin^2\beta/2)}{1 + iD}\right]\right\} \tag{22}$$

This may be evaluated to give, for $k_0 r \gg 1$, $k_0 a \gg 1$, and *both* $\beta \simeq 0, \pi$ and $D \simeq 1$ simultaneously not allowed:

$$I(r, \beta) \simeq \frac{\pi\alpha^2 k_0{}^2 a^2}{2r^2 \sin\beta}\mathrm{Re}\{\mathrm{erf}[r\cos(\beta/2)/a]\mathrm{erf}[r\sin(\beta/2)/a] - (1 + iD)^{-1/2}\mathrm{erf}[r\cos(\beta/2)/a(1 + iD)^{1/2}] \cdot \mathrm{erf}[r\sin(\beta/2)/a(1 + iD)^{1/2}]\} \tag{23}$$

where $D = r/k_0 a^2$, and the error function is $\mathrm{erf}(x) = (2/\sqrt{\pi})\int_0^x e^{-t^2}\,dt$.

It is interesting to note the limiting cases. For $D \gg 1$, the second term in the brackets (which came from the second term in the integral in Equation 21) goes to zero; for the case treated by Mintzer (I),[1] $r/a \gg 1$, and $\beta = 0$, we obtain

$$I(r, 0) \simeq \frac{\sqrt{\pi}\alpha^2 k_0{}^2 a}{2r} \qquad D \gg 1 \tag{24}$$

Since the coefficient of variation is given by $V^2 = r^2 I(r, 0)$, we see that this agrees with Mintzer's result, as modified by Potter and Murphy.[8] Moreover, for the case of $D \ll 1$, $r/a \gg 1$, and $\beta = 0$ we obtain

$$I(r, 0) \simeq \frac{(15/16)\sqrt{\pi}\alpha^2 r}{a^3} \qquad D \ll 1 \tag{25}$$

This agrees with the results of Bergmann[7] and of Mintzer[1] (II) (although the correct numerical factor should be 8/15).

Experiments have been performed by Sederowitz and Favret[5] with 500 kHz sound pulses, and $L \equiv r\cos(\beta/2)$ maintained at 60″; they have plotted the covariance $\mathscr{P}$, see Equation 4, as a function of β. We have, assuming small D,

$$\mathscr{P}(r, \beta) = \frac{\sqrt{\pi}a\,\text{erf}[L/a]\text{erf}[L\tan(\beta/2)/a]}{2L\sin(\beta/a)\text{erf}[L/a\cos(\beta/2)]} \tag{26}$$

We may estimate the value of a, which they do not give, by noting that for 50° all of the error functions are unity for any reasonable size of a (several inches). Using their experimental values of $P(60'', 50°) = 0.366$, we obtain a equal to 10.2″; we may then plot $\mathscr{P}$ for any value of β. Their values are shown as points in Figure 2 (they note that the value at $\beta = 12°$

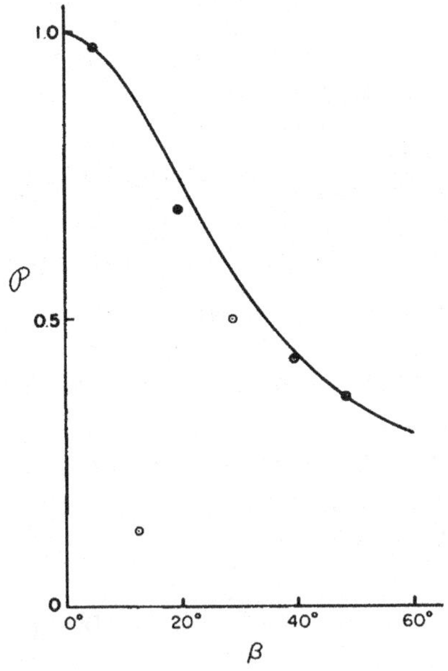

Figure 2. Covariance, $\mathscr{P}$, of the acoustic pulses received at points equidistant from source; as a function of angular separation β. Points are data of Sederowitz and Favret (Ref. 5); line is plot of Equation 26 fitted to the data at $\beta = 50°$.

is probably incorrect); Equation 26 is given by the line. The agreement seems very good. The value of a determined here seems quite large; however, it seems probable that there is a large-scale temperature microstructure which is made dominant over the small-scale structure by the process of averaging the data over a long time. It is interesting to note that, to within the approximations for which Equation 23 is valid, the covariance is independent of frequency. This point, among others, should be very interesting to experimentally substantiate.

References

1. D. Mintzer, *J. Acoust. Soc. Am.*, **25**, 922 (1953); **25**, 1107 (1953); **26**, 186 (1954); these papers will be referred to in the text as I, II, III, respectively.
2. L. A. Chernov, *Wave Propagation in a Random Medium*, McGraw-Hill Book Company, Inc., New York (1960); has extensive references to the Soviet literature.
3. M. J. Sheehy, *J. Acoust. Soc. Am.*, **22**, 24 (1950); F. H. Sagar, *J. Acoust. Soc. Am.*, **27**, 1092 (1955); D. C. Whitmarsh, E. J. Skudrzyk, and R. J. Urick, *J. Acoust. Soc. Am.*, **29**, 1124 (1957); G. R. Garrison, S. R. Murphy, and D. L. Potter, *J. Acoust. Soc. Am.*, **29**, 186 (1959); F. H. Sagar, *J. Acoust. Soc. Am.*, **32**, 112 (1960).
4. R. G. Stone and D. Mintzer, *J. Acoust. Soc. Am.*, **34**, 647 (1962); E. O. LaCasce, Jr., R. G. Stone, and D. Mintzer, *J. Appl. Phys.*, **33**, 2710 (1962); R. G. Stone, and D. Mintzer, *J. Acoust. Soc. Am.*, **38**, 843 (1965).
5. W. J. Sederowitz and A. G. Favret, *J. Acoust. Soc. Am.* (to be published). We wish to thank Dr. Favret for sending us a preprint of this paper, and allowing us to use this data in our Figure 2.
6. Reference 2, pp. 89–110.
7. P. G. Bergmann, *Phys. Rev.*, **70**, 486 (1946).
8. We use here the averages of the pressure amplitude rather than the amplitude of the averages, as suggested in D. S. Potter and S. R. Murphy, *J. Acoust. Soc. Am.*, **29**, 197 (1957).
9. *C.f.*, V. A. Krasil'nikov and A. M. Obukhov, *Sov. Phys. — Acoust.*, **2**, 103 (1956).

M. R. Schroeder
Bell Telephone Laboratories

9 Acoustics Research at Bell Telephone Laboratories: The Impact of Digital Computers

Preface

Acoustics, speech, and hearing have been active fields of study in Bell Laboratories for many years. Harvey Fletcher, in the preface to his classic book, *Speech and Hearing* (First Edition 1929) wrote:

> Some fifteen years ago [i.e., about 1914] the Research Laboratories of the Bell Telephone System undertook a comprehensive survey of speech and hearing to obtain the fundamental facts on which to base the design of apparatus and systems for telephone use. The art of analyzing electrical currents into their component frequencies and accurately measuring them, and the correlated art of describing and measuring the characteristics of mechanically vibrating systems, had reached an advanced state of development. It was apparent that great advantages would come from similarly analyzing speech and hearing, for if we could accurately describe every part of the system from the voice through the telephone instruments to and including the ear, we could engineer the parts at our disposal with greater intelligence.
>
> A greater plan for a prolonged laboratory investigation was evolved and has been in operation since that time. The attack was first launched most vigorously on the constitution of speech in an effort to establish a reasonable description of average speech, and to find to what extent small imperfections and variations in speech affected intelligibility. It is obvious that this work could not go far without entailing a study of the organs of speech and the organs and mechanisms of hearing, and its scope came to be extended as well to some of the abnormalities in these faculties.

As the work progressed it became apparent that better and more precise instruments must be developed than were available and a considerable part of the effort has been devoted to the matter of securing devices which would convert sound waves into electrical form and reconvert them again to sound with the least possible distortion. Out of this have come unexpected rewards to the telephone and phonograph arts, for as these devices were perfetced they found very immediate application to the great advantage of those industries.

Out of this work and what followed in the early 1930's came a succession of important scientific and technological advances which brought a better understanding of the mechanisms of speech and hearing and vastly improved instrumentation for their measurement. Contributions, some of which grew into new industries, flowed as by-products of Bell Laboratories research and development: high fidelity recording and reproducing systems, sound motion pictures, radio broadcasting equipment, the artificial larynx, simple reliable instruments for measuring hearing acuity, hearing aids.

Many scientists and engineers in the Engineering Department of the Western Electric Company and, later, Bell Telephone Laboratories pioneered in these advances: Irving B. Crandall (analysis of speech sounds by oscillographic techniques; analysis of energy distribution in speech; *Theory of Vibrating Systems and Sound*); J. P. Maxfield (acoustic detection of aircraft; research on loud speakers, electrical sound recording and reproduction); R. L. Wegel (mathematical analysis of the dynamics of the ear and vocal cords, studies of masking effects of pure tones); E. C. Wente (condenser microphone, ribbon light valve for recording sound on film, stereophonic transmission and reproduction; anechoic rooms; high quality magnetic tape recording, high quality loud speakers); L. J. Sivian (techniques for determination of speech power, measurement of peak amplitudes, frequency of occurrence and distribution in the spectrum; sound propagation; reference systems); N. R. French (articulation tests, relation of subjective tests to characteristics of circuits, speech, hearing and noise); John C. Steinberg (articulation testing methods, effects of distortion on speech recognition, studies of quality of music and speech, audiometry, underwater sound, speech privacy systems, visible speech); Robert R. Riesz (studies of differential sensitivity of the ear for pure tones; the artificial larynx; sound spectrograph; the cardio-spectrogram); W. A. Munson (basic studies in subjective acoustics: articulation testing, loudness measurement and reference levels, masking effects of tones and noise).

This list of early Bell System contributors to the science of acoustics is not complete; perhaps at best, it is representative of the scope of the work then being carried on and the depth of its penetration. The Director of Research at Bell Laboratories until 1933, H. D. Arnold, was a guiding spirit in the conception and initiation of acoustic research as described by Fletcher in his Preface. Arnold clearly recognized the need for precise measurement, a basic transmission reference system with components as nearly perfect as possible, and an excursion into psychological and physiological aspects of speech and hearing. He wrote in the Introduction to Fletcher's classic:

> The work is not complete — indeed some parts of it are hardly more than started; yet its results have been so great, both for the original purpose which was planned and for the many issues which have since arisen, that it presents a unique exemplification of the worth of systematic and sustained research.

And, indeed, "systematic and sustained research" is a good description of the activity which has followed, currently under the leadership of Manfred Schroeder. His paper, which follows, deals importantly with three of the fields to which Professor Morse has contributed importantly: acoustics, mathematical analysis, and computer sciences.

J. B. Fisk
Bell Telephone Laboratories

Introduction

In 1936, in the Preface to the first edition of *Vibration and Sound*,* Professor Morse noted that the vacuum tube and other applications of electronics, developed mainly during the preceding decade, had provided "immensely powerful tools for the measurement, recording, and reproduction of sound; tools which have revolutionized acoustic technique."

The dozen or so years immediately preceding this *Festschrift* have likewise seen significant advances in the frontiers of acoustics. The powerful tools that Morse mentioned in 1936 have matured, and through the additional developments of solid-state and quantum electronics, one can now cope with problems of measurement, signal processing, and acoustic wave generation and detection that but recently seemed forbidden. For example, the physical acoustics literature of the last decade has recorded such topics as the achievement of coherent elastic wave generation at frequencies of 10^9 to 10^{11} Hz, the use of Brillouin scattering to study the propagation characteristics of thermally

* McGraw-Hill Book Co., Inc., New York (1936).

excited waves at still higher frequencies, studies of heat pulse propagation and the observation of second sound, investigations of acoustic interactions with lattice defects, charge carriers, electron spins and nuclear spins, and the significant improvement of measurement techniques to obtain accurate data on the stress dependence of wave velocities in solids — information needed in evaluating lattice anharmonicity.

While there are a variety of relatively new tools and techniques that have made an impact on practically all branches of acoustics, I wish to discuss here just one, the high-speed digital computer, and its impact on research in acoustics at Bell Laboratories.

The superior power of digital computers for analysis, modeling and the control of experiments has stimulated much new research, including many topics on which Professor Morse has made pioneering contributions.[1,2,3,4] Because of the wide impact of digital computers, Professor Morse's abiding interest in the application of physical principles to acoustics has come to full fruition. The speed and accuracy of today's computers has made possible the direct application of the fundamental laws of physics to a wide variety of practical problems. Thus, the geometrical shapes of resonant cavities, such as the human vocal tract, have been calculated from their resonance frequencies using Ehrenfest's adiabatic invariance principle. Computers have even enabled us to determine the approximate shapes of unknown and inaccessible objects from their diffraction patterns by inverting the diffraction process, as it were, using well-known wave diffraction principles.

In architectural acoustics and underwater sound, where approximations based on *geometrical acoustics* have played such a useful role over the years, the precision afforded by digital computers in calculating ray paths has given new insights into such important phenomena as reverberation in concert halls and sound ray propagation in the ocean. Severe shortcomings in the classical reverberation formulas of Sabine, Eyring, and others have recently been exposed in this manner.

In psychoacoustics and hearing, computer modeling of the mechanical and neural activities of the ear has led to a deeper understanding of human perception and its relation to the physical properties of acoustic signals. In speech research, new concepts of the human speech process are put to the test by digital computers whose programs incorporate the postulated human behavior. Current research activity promises computers capable of understanding spoken instructions, recognizing individual voices, and answering questions in a human-sounding tongue.

Modern digital computers allow the realization of signal processing schemes of great complexity and sophistication — from the coding of speech signals to the evaluation of the acoustic properties of concert halls. Because of the versatility of modern computers, powerful new methods for analyzing and displaying physical and subjective data have become feasible. The application of these methods is having a profound influence on our understanding of human behavior and the correlation between perception and underlying physical and physiological structures.

Speech Processing of Digital Simulation

One of the earliest applications of computers in acoustics occurred in speech processing. Vocoders[5] and other speech processing devices had become so complex that it often took months or even years to implement a new device. When the implementation was finally completed and the result unsatisfactory (as was frequently the case), the question arose "What does the result mean? Is the original idea bad? Or are the often numerous approximations in its realization responsible?" Many new and ingenious ideas could not be put to the test because instrumentation would have been too time consuming.

All this was changed by the emergence of large digital computers and the introduction of *digital simulation.*[6] The device to be simulated, if it was not already in digital form, was first represented by a sampled-data equivalent amenable to digital simulation. The bridge between the real world of analog signals and the computer realm of numbers was established by analog-to-digital and digital-to-analog converters which translate analog sample values into binary representations and *vice versa.*[7] Early programs had to be written in machine language or available assemblers and compilers, but a special block-diagram compiler ("BLØDI"[8]) soon facilitated the task of programming digital simulations of problems representable by sampled-data block diagrams.[9,10] Engineers could learn the use of the BLØDI-compiler within a few hours. Considerable advances have since been made in speech processing for a variety of practical applications.[11] Spectrum channel vocoders and formant vocoders, both pitch and voice excited, were substantially improved and new ideas for syllabic-rate changers[12,13] and the restoration of *helium speech* of deep-sea divers[14] were successfully tested. New principles for speech processing, based on analytic signal representation[15] and linear predictability[16] of speech signals, were formulated and demonstrated by means of digital simulation.

A new and even more powerful version of the block-diagram compiler (BLØDI-B) allowed still more complex simulations.[17]

The impressive applications of digital simulation in speech processing, aided by block-diagram compilers, soon aroused interest in allied fields, particularly in speech analysis.

Speech Analysis

Digital simulation allowed the realization of a long-standing desire of speech researchers: pitch-synchronous (period-by-period) spectrum analysis of speech.[18] This led to improved methods for inverse filtering to recover the excitation wave form[19] and to track formant frequencies.[20] Notable gains were made in the extraction of formant frequencies and in the measurement of fundamental frequency by *cepstrum*,[21] period histogram,[22] and other methods.[23] Perceptually significant aperiodicities were found in voiced speech segments by high-resolution spectrograms obtained on the computer. The concept of *spectrum liftering* facilitated spectrum envelope measurement[24] and the Fast-Fourier Transform (FFT)[25,26] allowed spectral analysis of high resolution in near real time. Special plotting programs[27] made possible the rendition of half-tone pictures on microfilm plotters, thereby allowing sound spectrograms to be both efficiently computed and plotted by digital machines. New methods for 3-dimensional displays[28] of speech spectra with continuously adjustable resolutions in time and frequency may lead to refined insights into the structure of speech signals and their spectra.

Interesting results were also obtained by computer analysis of the sounds of musical instruments such as the trumpet and the violin.[29]

Although notable successes in automatic word recognition have been achieved with analog devices based on mechanical and optical principles, future progress will strongly depend upon computers.

Automatic Speech and Speaker Recognition

Various methods for automatic word recognition have been proposed and tested by digital simulation.[30] Automatic voice classification and speaker identification, of great practical potential in future banking operations and limited-access information retrieval, is being explored using spectral data of isolated words[31] and fundamental-frequency contours of connected utterances.[23]

In all these studies, the reported *correct* scores are high, but the inventories are still limited to relatively few words and speakers. However, even with such limitations, useful applications can be foreseen in such diverse activities as postal sorting, spacecraft maneuvering and surgery.

New methods of analyzing speech spectra combined with deeper insights into the relations between vocal tract shape and acoustic parameters led to articulatory analysis based on acoustic measurements.

Articulatory Analysis Based on Acoustic Measurements

The formant frequencies of a wide variety of vocal-tract shapes were computed by iterative solution of Webster's horn equation and analyzed in terms of articulatory parameters. The *inverse* problem of determining (spatially *band limited*) vocal tract shapes from formant frequencies[32] and lip-impedance functions[33] was also solved. The crucial uniqueness problem ("Is there more than one (band limited) shape that could have given rise to the measured formants or impedance singularities?"), for which no mathematical proof is available, was answered empirically by a near exhaustive, and happily unsuccessful, computer search for multiple solutions.

Speech Synthesis by Rule

While these advances in speech analysis were still in progress, speech synthesis by rule was already attempted by the more ambitious speech researchers. Speech synthesis by rule (the *talking computer*) has been one of the practical goals of speech research for some time. Once it had been shown that natural-sounding speech can indeed by synthesized by electrical networks or simulations of such networks, the way to ultimate success seemed open. But it became soon apparent that electromechanical analogies and analog-digital equivalences were only one of many prerequisites for successful automatic speech synthesis. An early speech synthesizer[34] included simple rule for formant transitions and synthesized speech with simulated analogs of the vocal tract. Intelligibility was perfect (if one knew beforehand what was said) and signal quality was often amusing (especially in the *singing* mode).

Of great importance in the history of speech research was the emergence of a new laboratory tool: the small *dedicated* computer for on-line experimentation.[35] The small computer (which has since grown considerably but is still dedicated) allows the manipulation of synthesis parameters and linguistic rules and an immediate subjective evaluation of their auditory effects. The experienced experimenter, using typewriter and graphic input devices to modify the program according to his auditory evaluation, and the computer, with speech synthesizer and oscilloscopes attached, form a tightly coupled loop. This most advanced (and expensive) version of trial-and-error methods is uniquely fitted to surmount many of the remaining obstacles on the way to better synthetic speech.

A particularly promising approach is speech synthesis by on-line computer based on an *articulatory* model[36] of the speech process. The computer displays (simplified) outlines of the tongue and lips as well as the corresponding area function and formant frequencies. Articulatory *target* shapes and rules for transition are entered into the program by the experimenter. The almost immediately available acoustic output signal allows rapid subjective evaluation and continual refinement of the rules and the model itself.

One of the advantages of a small owned (or rented) computer which has been paid for (or is being paid for at a fixed rate) is the possibility of processing large volumes of data, a frequent requirement in speech research, at low cost. The saving can be as much as a factor of 10 compared to large central computers whose extensive arsenal of hardware and software is often not needed but still has to be paid for. Thus, once *free* access to a small machine is obtained, many problems can be run more economically on an existing on-line facility. A realistic model of the human vocal cords[37] and their interaction with the vocal tract was thus successfully tested and refined on a small Honeywell DDP-516 computer using a powerful General Electric GE-645 only for compiling and debugging the (Fortran IV) program. The same basic program was used to imitate the sound of a Flügelhorn in which the vibrating lips of the player play the role of the vocal cords and the horn that of a (fixed-shape) vocal tract.

Avant-Garde Sounds

Programs and compilers[38] for more avant-garde sounds permitting the simultaneous synthesis of many *instruments*[39] were adapted to small on-line computers to permit the electronic composer to hear and modify the results of his inspiration almost instantly,[40] much like the conventional composer behind his analog equipment (piano). With wide availability of this new tool to exercise people's acoustic imagination and perception, recreational composing may come to occupy a significant portion of future generations' leisure time — which, no doubt, will be plentiful as a result of increasing automation.

Psychoacoustics

The success of on-line computing in speech research and sound synthesis led to the use of small computers for real-time experimentation in psychoacoustics.

As is well known, the collection of psychometric data is a time-consuming occupation, to say nothing of the difficulties due to boredom-induced shifting of perceptual thresholds and time-varying decision criteria. If each new stimulus or stimulus pair (or triplet) could be chosen in such a way that it extracts a maximum amount of information from the subject, a considerable speed-up in acquisition of psychometric data could be achieved. This is precisely what an on-line digital computer allows one to do. In determining, for example, the 50% point of a psychometric function, the computer can select that stimulus for presentation which is nearest to the expected mean as computed from past responses. The tracking of time-varying means[41] is likewise facilitated. Other parameters of psychometric functions can be similarly estimated.[42]

In some tests, the computer can generate the test stimuli in real time. In others, the stimuli have to be prerecorded. In either case, signals and noises of high reproducibility can be obtained.[43,44] With an on-line computer, it is not even necessary to prerecord all possible stimulus combinations. It suffices to record each test stimulus only once (or twice). A computer-controlled tape deck goes to the proper position on the stimulus tape for the presentation of the next stimulus. The computer knows the position of the tape from a shaft encoder coupled to the tape transport and a timing signal on a second tape track. This system has been successfully employed in a study of speech quality resulting from replacing the vocal excitation function by simplified waveforms and constraining the variability of successive fundamental periods.[45]

Multidimensional Scaling

One of the most remarkable advances made possible by computers is the multidimensional scaling of perceptual and other subjective data.[46,47] These scaling methods allow either linear or arbitrary monotonic transformations of the subject's data which may be in the form of subjective preferences or similarity judgments. Geometrical representations of the transformed data are calculated by the computer and the so constructed *subjective spaces* are plotted, two dimensions at a time by microfilm plotters or oscilloscopes. On-line computers in conjunction with three-dimensional display facilities allow the simultaneous representation of more than two dimensions and rotations of the display for more effective evaluation.

The perceptual aspects of a variety of acoustic stimuli including vowel sounds, tone-ringer signals and speakerphone systems have been explored by this powerful approach. In many cases a clear correlation between the physical parameters of the signal and its perceptual attributes could be established — a correlation that was obscure in the untransformed data.

Physiological Acoustics

The response of single neurons and neural nets[48] has been simuated off-line on digital computers. A model for the first-order auditory fibers has been shown[49] to match closely the resulting of physiological measurements on cats.

Our understanding of the peripheral hearing process was significantly advanced by a mathematical model of the basilar membrane.[50,51] Digital simulations of the model for a variety of acoustic stimuli were instrumental in interpreting psychoacoustic experiments on monaural and binaural hearing. As an ultimate tutorial refinement, three-dimensional motion pictures of waves traveling along the basilar membrane were made on the computer.[52]

Digital Simulation of Concert Hall Acoustics

Another complex problem which calls for model studies by simulation is the propagation of sound waves in reverberant enclosures. With the advent of large digital computers, classic scale-modeling techniques, using light rays and ultrasonic waves, were supplemented by digital simulation of concert hall acoustics.

This work originated with the digital simulation of *allpass* filters and *comb* filters to create natural-sounding artificial reverberation[53] and pseudostereophonic effects[54] without spectral distortion of the audio signal. This approach to adding three-dimensional realism and an illusion of space has been exploited to enhance computer-generated electronic sounds. Since the computer-simulated acoustics can be changed easily by program, illusions of auditoriums rapidly varying in size and reverberation characteristics are feasible.

Another astonishing percept made possible by computer processing of auditory signals is a kind of *super-stereophony*. Preprocessed sound signals, emanating from two closely spaced (±22.5°) loudspeakers in front of a listener, combine at his ears in a manner that closely resembles the sound pressure waves from a sound source at 90° (or some other, arbitrary, angle) off to the side.[55] The effect is astounding, particularly when the listener turns his head in an attempt to locate the (nonexistent) third sound source — at which point the unique perception vanishes because the computer program is based on the measured sound diffraction of a listener's head oriented in the *forward* direction.

The possibility of digitally stimulating realistic reverberation processes, together with the ability to add discrete arrivals (reflections off the stage, sidewalls and ceiling) with proper spectrum and directionality allows sound transmission in concert halls and auditoriums to be simulated on the computer.[56] Reverberation-free speech or music signals are fed into the computer which is programmed to add echoes and reverberation. The computer produces two or more audio output signals. These are radiated from loudspeakers in an anechoic chamber to create, at a listener's ears, sound fields which perceptually resemble those in an actual hall.

The two most important applications of this simulation method are: (1) auditory evaluation of new concert hall designs *before* construction, and (2) psychoacoustic studies of auditory perception in reverberant spaces. In the second application, the method can be used for instantaneous comparison of different reverberation processes acting on identical sound signals. The two reverberation processes can differ in any of a number of physical parameters (spectral composition of the direct sound, delay and direction of an early echo, direct-to-reverberant energy ratio, reverberation time as a function of frequency, shape of the decay curve, etc.). Needless to say, the degree of flexibility of computer simulation is impossible to achieve by listening in real concert halls. Owing to digital simulation, a better understanding of the complex relationship between the physical parameters of a reverberation process and its perceptual attributes has already been attained in a few cases.

Subjective Reverberation Time

One of the important applications of digital simulation of reverberation processes has been a study of subjective reverberation time.

A variety of nonexponential reverberation processes were simulated and compared with exponential decays in paired comparison tests.[57] Based on the results of these tests, a new criterion for objective reverberation time (best straight-line fit to the initial portion of the decay) was formulated which matches the subjective reverberation time surprisingly well (see Figure 1). How misleading the frequently used criterion, based on the decay between 5 dB and 35 dB, can be, is shown in Figure 1 for one extreme case with a subjective reverberation time of about 1.1 sec and an *objective* reverberation exceeding 2.5 sec!

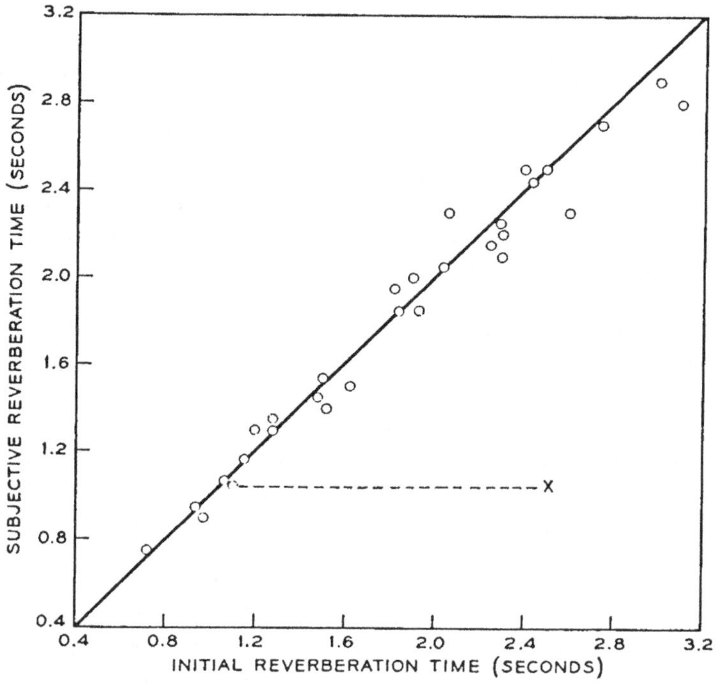

Figure 1. Ordinate: *subjective* reverberation time found in paired comparison tests of computer-made nonexponential reverberation processes with exponential decays. Abscissa: *physical* reverberation time based on a computed best straight-line fit to the initial 160-msec portion of the decay. The cross (×) shows the "objective" reverberation time for one extreme case based on an older but widely used definition of reverberation time.

Acoustic Measurements in Auditoriums

Such measurements used to be confined to what was possible by available, often inadequate, instruments. But our knowledge about human perception in reverberant sound fields is steadily increasing, and it is of utmost importance to measure what (and in a manner that corresponds to *how*) we perceive — using both appropriate test signals and subjectively meaningful evaluation procedures. Both of these tasks are greatly facilitated by digital computers.

The computer is ideally suited to act as a signal generator producing precisely tailored test signals to reveal acoustics deficiencies both of physical and perceptual significance. Such computer-generated test signals (*e.g.*, short tone-bursts with a third-octave bandwidth) are radiated from a loudspeaker on the stage and the resulting pressure responses are recorded at selected positions throughout the hall.[58] These responses are fed back into the computer for bandpass filtering (to minimize the influence of ambient noise on the results) and envelop detection. Integrals over selected time and frequency intervals and summations over various spatial positions are computed. The results of these computations are tabulated or plotted by means of a microfilm plotter attached to the computer. Subsidiary programs fit straight lines to decay curves, compute direct-to-reverberant energy ratios, *clarity*, and other quantities that are of subjective significance or that will aid in pinpointing physical defects. As new knowledge is accumulated, in part through comparison of different halls evaluated in this manner, programs can be modified or expanded to incorporate new insights into auditory perception in reverberant sound fields.

Computer-aided measurements in a large concert hall (Philharmonic Hall, Lincoln Center for the Performing Arts, New York)[59] revealed a considerable attenuation at low frequencies of the energy reflected from the suspended ceiling (*clouds*), see Figure 2. An unexpected lack of low frequencies in the direct sound has subsequently been explained as a *seat* effect.[60] The combination of these two effects led to a perceivable weakness of the low-frequency response on the main floor in Philharmonic Hall and a near inaudibility of the celli during tutti passages.[61]

Measurement of Reverberation Time

A new method[62] for obtaining the ensemble average of many decay curves and a correspondingly more accurate value of reverberation time was first applied in Philharmonic Hall using the computer method just described. In this method, the ensemble average decay is obtained by integrating the squared envelope of a *single* tone-burst response. The great variability of individual noise decay curves, which result from the randomness of the exciting noise signal, is avoided. As a result, significant double slopes and other perceptually important details of the sound decay are clearly revealed.

Dependence of Reverberation Time on Room Shape

Computers have led to both a more meaningful subjective criterion for reverberation time and a more accurate method of measuring it. But computers are also playing a part in refining the theory of reverberation and the dependence of reverberation time on room shape.

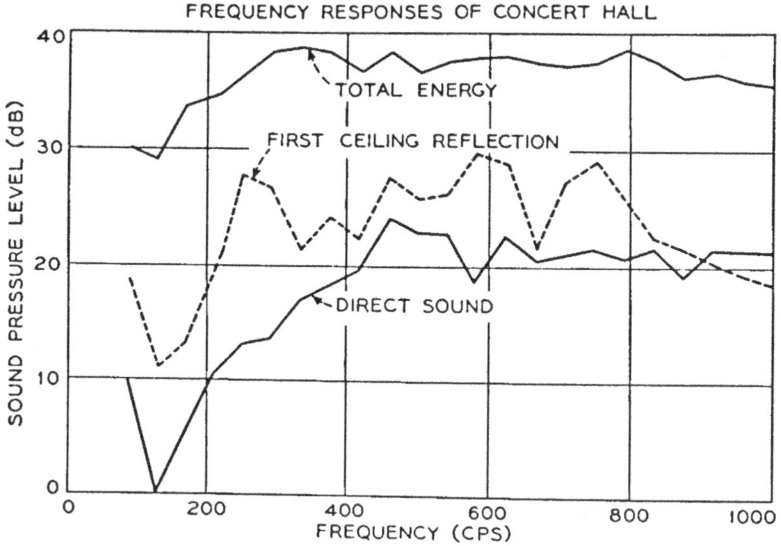

Figure 2. Frequency response curves obtained in Philharmonic Hall (New York) for a sound source on the stage (cello section) and a receiver position near the center of the main floor at ear level. Computer-generated test signals (short *tone bursts*) were used to excite the hall. The recorded sound pressure responses were fed back into the computer for filtering, envelope detection and integration over various time intervals corresponding to the total energy (top curve), the first ceiling reflection (*clouds*) and the direct sound (bottom curve). The observed lack of low frequencies in the direct sound and the ceiling reflection has been found responsible for the near inaudibility of the celli in tutti passages. The cause of these acoustical deficiencies has been traced to wave cancellation resulting from the periodic, lattice-like, structure of the cloud ceiling. Computer measurements in Philharmonic Hall with a redesigned, solid, ceiling have shown a perfect spectral balance of the energy reflected from the ceiling.

Existing formulas relating reverberation time to surface absorption depend on the total volume of the enclosure and the total surface area and absorption coefficient of the absorber. They do not account for the *shape* of the enclosure and the distribution of the absorber — contrary to experimental evidence. This problem has recently been attacked by ray tracing studies on a digital computer.[63] Figure 3 shows one of the results of this study, a comparison between values of reverberation time according to the widely used Eyring formula[64] and reverberation times found by ray tracing on the computer. The discrepancies exceed 50% (!) in 12 out of the 32 different room shapes examined. Similarly, large errors were found even for diffuse wall reflections and scattered placement of the absorbers. It appears that, until these discrepancies are more fully understood, the only way to derive a reliable formula for a given hall shape and absorber distribution is by ray tracing on the computer.

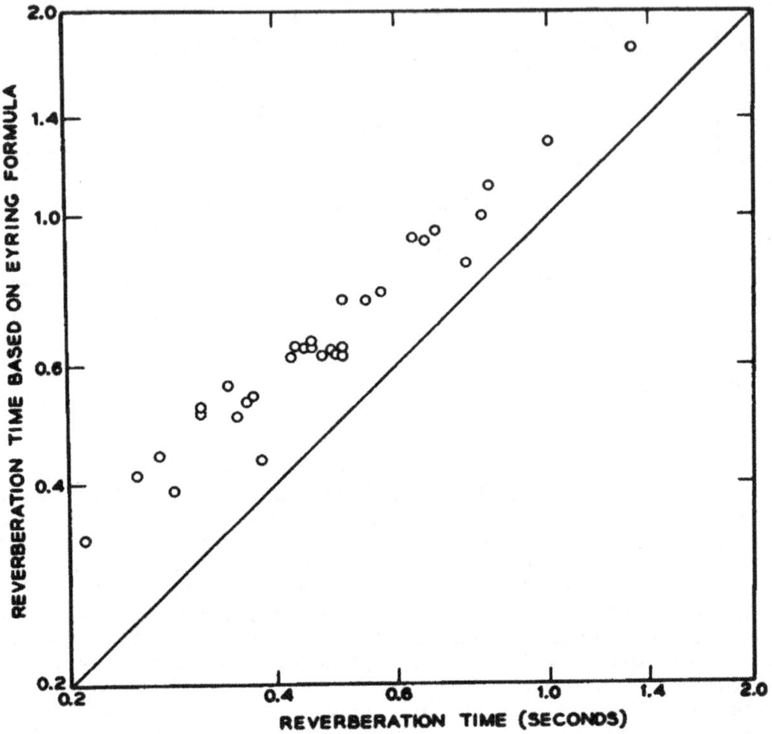

Figure 3. Ordinate: reverberation time according to the "Eyring" formula for 32 different room shapes. Abscissa: reverberation time found by ray tracing on the computer. The discrepancies depend on room shape and exceed 50% in 12 the of 32 cases.

Sound Propagation in the Ocean

Ray tracing and wavefront construction on the computer have also been used to study sound propagation in the ocean.

Slow-motion pictures of sound propagation in the deep ocean have been produced by a microfilm plotter in conjunction with a large computer programmed to repeatedly apply Snells' law of wave refraction.[65] Sensitive signal processing schemes, such as matched filtering and *volume focussing*[66] applied to the outputs of hydrophone arrays have likewise been studied by digital simulation,[67] permitting a rapid and reliable evaluation of their potential.

Random Wave Fields

Other wave propagation phenomenon which have been illuminated by computer simulation are the random wave fields resulting from wave interference in multipath propagation media. Some quantities of such wave fields can be computed analytically (for example, the expected density of interference maxima in space or frequency[68]) but others are beyond today's methods of analysis. An example of such a quantity is the expected level of the highest interference maximum in a given frequency range. This quantity, apart from being of interest in certain detection tasks, determines the stability of public address systems in auditoria.[69] To find this quantity, a computer was programmed to generate frequency response curves of reverberant rooms with appropriate statistical properties. The desired quantity was then determined by the computer using a Monte Carlo method.[70] Many other quantities pertaining to random wave fields have been calculated by Monte Carlo methods, such as correlation functions of spatial and frequency response curves[71] of random wave fields and the effect of frequency averaging on the fluctuations of frequency response curves.[72]

Image Reconstruction with Sound Waves

It is well known that the distribution of a field quantity obeying a linear wave equation inside a specified volume can be reconstructed from its values on the boundary. Since one or more field quantities change abruptly at the surface of an object whose impedance differs from that of the surrounding medium (*e.g.*, a rigid body in air or water, or an *air bubble* enclosed in a solid or liquid), the shape of the object can be computed from the (sound) field scattered by it and measured on some closed surface surrounding the object. This computation is a process not unlike wavefront reconstruction by holography. In fact, the holographic process can be viewed as an *analog* computer for wavefront reconstruction — with one important difference: With sound waves, one can record linear *complex amplitudes* of the scattered field, as opposed to merely some nonlinear function of *intensity* in the case of photographic film. Thus, for example, in the acoustical case, one can use a wide-spectrum or *impulsive* illumination of the object — as opposed to the essentially monochromatic illumination in optical holography — thereby adding another dimension to acoustic image reconstruction. To utilize the phase information available in the acoustic case, the computer is essential.

(a)

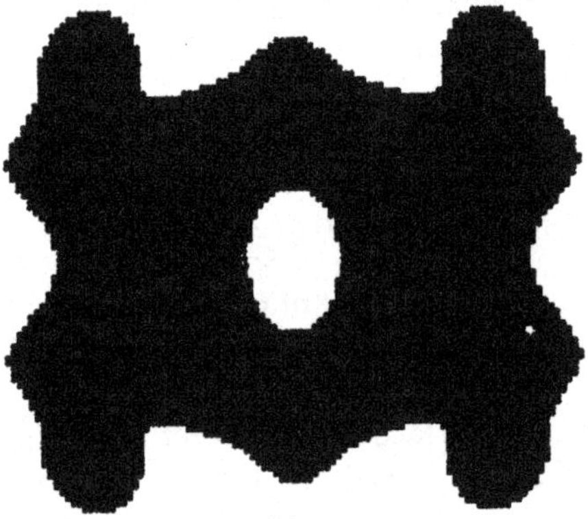

(b)

Figure 4. (a) Drawing of original object (letter H). (b) Magnitude of sound field scattered from object shown in Figure 4a at some distance behind the object relative to the insonifying source. The magnitude of the scattered field has been rounded to one (black) and zero (white). (c) Reproduction of microfilm output showing object reconstructed by the computer using only the scattered field (Figure 4b) and knowledge of the acoustic wavelength and the distances from the object to source and receivers.

Figure 4 illustrates the method of acoustic image reconstruction. The object is a letter " H " (Figure 4a). Figure 4b shows the magnitude of the scattered sound field at some distance from the object. Figure 4c shows the computer-reconstructed H, for the case of a sound wavelength equal to about 1/6 of the dimensions of the object.[73] Both the scattered sound field and the reconstructed image were computed as complex ampli-

(c)

Figure 4 (*continued*)

tudes. Figure 4 shows only the magnitude rounded to one (black) and zero (white). Potential applications of this method are nondestructive materials testing, geometrical shape determination of inaccessible objects and in medical diagnostics.

Conclusion

Digital computers, both large central installations and small "dedicated" laboratory models, have proven powerful research tools in many scientific disciplines. Selected applications of digital computers in acoustics research at Bell Laboratories illustrate how digital computers have deepened our understanding of physical processes of great complexity in terms of simple fundamental laws. Equally, our appreciation of perceptual phenomena and their relation to underlying physical principles and physiological structures has transcended the narrow bounds of the past. Thus, the digital computer, once viewed as but a large and expensive calculating machine, has forged a strong link between the two worlds in which man is rooted: the ideal world of symbolic representations and the real world of measurable events and quantifiable behavior.

References

1. P. M. Morse and R. H. Bolt, "Sound Waves in Rooms," *Rev. Mod. Phys.*, **16**, 69 (1944).
2. P. M. Morse, "Acoustics and Basic Physics," *J. Acoust. Soc. Am.*, **27**, 213 (1955).

3. P. M. Morse, *Vibration and Sound*, McGraw-Hill Book Company, Inc., New York (1948) 468 pp.
4. P. M. Morse and U. K. Ingard, *Theoretical Acoustics*, McGraw-Hill Book Company, Inc., New York (1968) 927 pp.
5. H. Dudley, *J. Acoust. Soc. Am.*, **11**, 169 (1939).
6. E. E. David, Jr., M. V. Mathews, and H. S. McDonald, *Proc. Natl. Electronics Conf.*, 766 (1958).
7. E. E. David, Jr., M. V. Mathews, and H. S. McDonald, *IEEE Western Joint Computer Conf.*, 354 (1959).
8. J. L. Kelly, C. Lochbaum, and V. A. Vyssotsky, *Bell System Tech. J.*, **40**, 669 (1961).
9. E. E. David, Jr., *Proc. 4th Intnl. Congr. on Acoust.*, **2**, 69 (1962).
10. M. R. Schroeder, in " The Human Use of Computing Machines, "*Proc. Bell Tel. Labs.*, Symposium, Murray Hill, N.J. (1966).
11. M. R. Schroeder, *Proc. IEEE*, **54**, 720 (1966).
12. M. R. Schroeder, B. F. Logan, and A. J. Prestigiacomo, *Proc. Stockholm Speech Comm. Seminar*, Stockholm (1962).
13. J. L. Flanagan and R. M. Golden, *Bell System Tech. J.*, **45**, 1493 (1966).
14. R. M. Golden, *J. Acoust. Soc. Am.*, **40**, 621 (1966).
15. M. R. Schroeder, J. L. Flanagan, and E. A. Lundry, *Proc. IEEE*, **55**, 396 (1967).
16. B. S. Atal and M. R. Schroeder, " Predictive Coding of Speech," *Conf. on Speech Comm. and Process*, Cambridge, Mass. (1967).
17. R. M. Golden, *Bell System Tech. J.*, **45**, 345 (1966).
18. M. V. Mathews, J. E. Miller, and E. E. David, Jr., *J. Acoust. Soc. Am.*, **33**, 179 (1961).
19. J. E. Miller and M. V. Mathews, *J. Acoust. Soc. Am.*, **35**, 1876(A) (1963).
20. M. V. Mathews, J. E. Miller, and E. E. David, Jr., *J. Acoust. Soc. Am.*, **35**, 1113(A) (1963).
21. A. M. Noll, *J. Acoust. Soc. Am.*, **36**, 296 (1964).
22. M. R. Schroeder, *J. Acoust. Soc. Am.*, **43**, 829 (1968).
23. M. M. Sondhi, *IEEE Trans. Audio Electroacoust.*, **16** (June 1968).
24. M. R. Schroeder and A. M. Noll, *Proc. 5th Intnl. Congr. Acoust.*, Liège (1965).
25. J. W. Cooley and J. W. Tukey, *Math. of Comp.*, **19**, 297 (1965).
26. See: Special Issue on Fast Fourier Transform, *IEEE Trans. Audio Electroacoust.*, **15**, 43–117 (1967).
27. M. R. Schroeder, " Images from Computers and Microfilm Plotters," Comm. of Assoc. for Computing Machinery (to be published).
28. A. M. Noll, *Computers and Automation*, **14**, 32 (1965).
29. J. C. Risset, M. V. Mathews, J. E. Miller, J. R. Pierce, and J. Tenney, *J. Acoust Soc. Am.*, **38**, 912(A) (1965).
30. P. B. Denes and M. V. Mathews, *J. Acoust. Soc. Am.*, **32**, 1450 (1960).
31. S. Pruzansky and M. V. Mathews, *J. Acoust. Soc. Am.*, **36**, 2041 (1964).
32. P. Mermelstein, *J. Acoust. Soc. Am.*, **39**, 1219 (1966).
33. M. R. Schroeder, *J. Acoust. Soc. Am.*, **41**, 1002 (1967).
34. J. L. Kelly, Jr., and L. J. Gerstman, *J. Acoust Soc. Am.*, **33**, 835(A) (1961).

35. P. B. Denes, *Proc. 5th Intnl. Congr. Acoustics.*, Liège (1965).
36. C. H. Coker and O. Fujimura, *J. Acoust. Soc. Am.*, **40**, 1271(A) (1966).
37. J. L. Flanagan and L. L. Landgraf, *IEEE Trans. Audio Electroacoust.*, **16**, 57 (1968).
38. M. V. Mathews, *Bell System Tech. J.*, **40**, 677 (1961).
39. M. V. Mathews, *Science*, **142**, 553 (1963).
40. M. V. Mathews, *J. Acoust. Soc. Am.*, **39**, 1245(A) (1966).
41. H. Levitt and D. E. Bock, *J. Acoust. Soc. Am.*, **42**, 911 (1967).
42. J. L. Hall, " Maximum-Likelihood Sequential Procedure for Estimation of Psychometric Functions," 75th Meeting Acoust. Soc. Am., Ottawa (1968). (Abstract to be publ. in *J. Acoust. Soc. Am.*)
43. B. S. Atal, M. R. Schroeder, and K. H. Kuttruff, *Proc. 4th Intnl. Congr. Acoustics*, Copenhagen (1962).
44. S. M. Pfafflin and M. V. Mathews, *J. Acoust. Soc. Am.*, **39**, 340 (1966).
45. A. E. Rosenberg, "Effect of Pitch Averaging on the Quality of Natural Vowels," *75th Meeting Acoust. Soc. Am.*, Ottawa (1968). (Abstract to be published in *J. Acoust. Soc. Am.*)
46. R. N. Shepard, *Human Factors*, **5**, 33 (1963).
47. J. B. Kruskal, *Psychometrika*, **29**, 1 (1964).
48. L. D. Harmon, J. Levinson, and W. A. van Bergeijk, *Kybernetik*, **1**, 89 (1961); *J. Acoust. Soc. Am.*, **35**, 1924 (1963).
49. R. C. Lummis and E. A. Lundry, *J. Acoust. Soc. Am.*, **42**, 1207(A) (1967).
50. J. L. Flanagan, *Bell System Tech. J.*, **39**, 1163 (1960).
51. J. L. Flanagan, *Bell System Tech. J.*, **41**, 959 (1962).
52. R. C. Lummis, A. M. Noll, and M. M. Sondhi, *J. Acoust. Soc. Am.*, **39**, 1252(A) (1966).
53. M. R. Schroeder, *J. Audio Eng. Soc.*, **10**, 219 (1962).
54. M. R. Schroeder, *J. Acoust. Soc. Am.*, **33**, 1061 (1961).
55. M. R. Schroeder, B. S. Atal, and C. Bird, *Proc. 4th Intnl. Congr. Acoustics*, Copenhagen (1962).
56. M. R. Schroeder and B. S. Atal, *IEEE Intnl. Conv. Record* (1963).
57. B. S. Atal, M. R. Schroeder, and G. M. Sessler, *Proc. 5th Intnl. Congr. Acoustics*, Liège (1965).
58. B. S. Atal, M. R. Schroeder, G. M. Sessler, and J. E. West, *J. Acoust. Soc. Am.*, **40**, 428 (1966).
59. M. R. Schroeder, B. S. Atal, G. M. Sessler, and J. E. West, *J. Acoust. Soc. Am.*, **40**, 434 (1966).
60. G. M. Sessler and J. E. West, *J. Acoust. Soc. Am.*, **36**, 1725 (1964).
61. G. Szell (personal communication).
62. M. R. Schroeder, *J. Acoust. Soc. Am.*, **37**, 409 (1965).
63. B. S. Atal and M. R. Schroeder, *J. Acoust. Soc. Am.*, **41**, 1598(A) (1967).
64. A. D. Fokker, *Physica*, **4**, 262 (1924).
65. M. M. Sondhi (unpublished work).
66. M. R. Schroeder (unpublished work).
67. J. L. Flanagan, L. Landgraf, and D. J. MacLean, *J. Acoust. Soc. Am.*, **42**, 1165(A) (1967).
68. M. Schröder, *Acustica*, **4**, 594 (Beiheft 2) (1954).

69. M. R. Schroeder, *J. Acoust. Soc. Am.*, **36**, 1718 (1964).
70. M. R. Schroeder and K. H. Kuttruff, *J. Acoust. Soc. Am.*, **34**, 76 (1962).
71. M. R. Schroeder, *J. Acoust. Soc. Am.*, **34**, 1819 (1962).
72. M. R. Schroeder, B. S. Atal, G. M. Sessler, and J. E. West, *Proc. 3rd Acoustics Conf.*, Budapest (1964).
73. M. M. Sondhi and M. R. Schroeder (work in progress).

J. E. White
Marathon Oil Company
Denver Research Center

10 Computed Waves in a Nonlinear Solid

Analyses of seismic waves traveling in the earth and measurements on vibrating rock samples have established that earth materials lose energy in a characteristic fashion. Some of the experimental data are: When a sinusoidal stress cycle is applied to a small rock sample, the fractional energy lost per cycle is independent of the frequency; when a slender rod is driven near a resonant frequency, the factor Q (designating sharpness of resonance) is independent of frequency; for traveling waves, each frequency component suffers an exponential decrease with distance of travel, the exponent itself being proportional to frequency; and the same loss behavior is noted over a tremendous range of vibration amplitude. There is no reason to suppose that a single loss mechanism is dominant in all earth materials, and in fact experiments on some samples do not fit the above pattern. However, hundreds of observations on a wide range of rock types, covering a frequency range of nine decades, have led more than one author to conclude that seismic waves are subject to attenuation which varies as the first power of frequency, and therefore that a loss mechanism must be operative which yields this kind of exponential attenuation coefficient.

It has been suggested on a purely theoretical basis that the loss mechanism must be nonlinear. The intrinsically nonlinear phenomenon of sliding between solid surfaces in contact (Coulomb friction) is almost certainly involved in the distortion of granular materials. In any case, attention has been directed toward possible nonlinear stress-strain relations which would be consistent with the observed properties of typical rocks. The tools of linear theory cannot be expected to apply exactly, but perhaps superposition and causality apply approximately to a slightly nonlinear solid. Fourier analysis of observed transient waves has led to self-consistent results, suggesting that superposition does apply. These experiments also indicate that the observed attenuation is accompanied by the small variation of phase velocity with frequency which is demanded by the causality condition.

Herein are reported some calculations on a simple model of a particular nonlinear solid which was selected in relation to the observed properties of earth materials. The model consists of mass-elements connected by nonlinear springs. For ease of visualization, each mass element is considered to be one cubic centimeter of continuous solid. Then each mass is numerically equal to the bulk mass density; force in each spring is numerically equal to stress; and displacement difference between adjacent masses is equal to strain. The selected stress-strain relation is shown in Figure 1. As stress (p_{xx}) increases from zero, strain

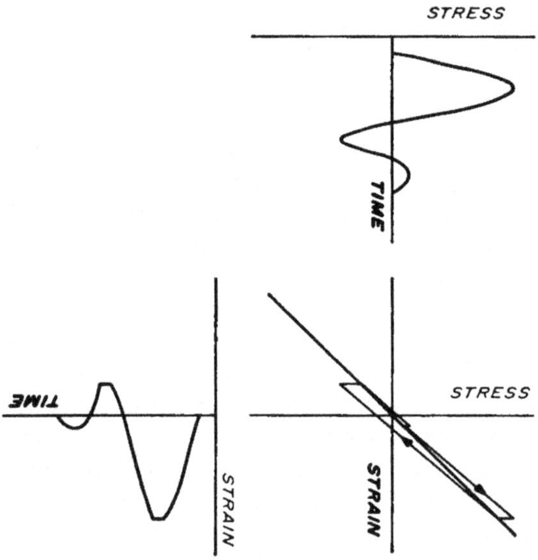

Figure 1. Stress-strain diagram for the nonlinear medium.

ε_{xx} increases linearly according to the relation

$$p_{xx} = (1 + q)M\varepsilon_{xx} \tag{1}$$

where M is an elastic constant and q is a dimensionless loss parameter. As stress begins to decrease, strain remains constant until stress drops to the *elastic line*, i.e., the line relating stress and strain if q were zero. One may consider that the solid is *locked in place* by solid friction. As stress decreases further, strain decreases along a line whose slope is $(1+q)M$. When stress reverses again, strain remains constant until the elastic line is intersected, after which strain increases. At each reversal, the stress-strain locus intersects the elastic line and follows the line with the greater slope. From the past and present displacements of two adjacent mass-elements, the present force in the connecting spring is computed. Equating the difference between the two forces acting on a given mass-element to the acceleration of the element yields displacement at the next future time step. This then determines the forces at one step in the future, permitting a further projection of displacements, and so on. In the process, displacements of every element and forces in every spring are computed at every time step.

Calculations were made for a solid having the properties of a competent sandstone:

$$\text{Density } \rho = 2.5 \text{ gm/cm}^3$$
$$\text{Modulus } M = 4 \times 10^{11} \text{ dyne/cm}^2$$
$$\text{Parameter } q = 0.02$$

Time increment was 5×10^{-7} sec.

In one type of calculation, stress at one end of the medium was required to vary as a simple transient function of time. Computed strains at two distances from the applied stress are shown in Figure 2. It was

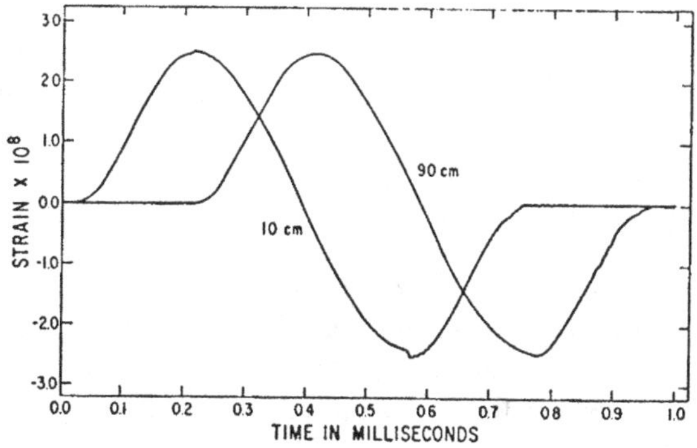

Figure 2. Strain wave forms at two distances in the nonlinear medium.

assumed that the strain can be expressed as a Fourier integral, and that the strain is a wave traveling in the positive direction with a complex propagation constant γ:

$$\varepsilon_{xx}(x, t) = \frac{1}{2\pi} \int_{-\infty}^{\infty} E(0, \omega) e^{-\gamma x} e^{i\omega t} \, d\omega \tag{2}$$

The propagation constant $\gamma = a + i\omega/c$, where attenuation constant a and phase velocity c may vary with frequency. Fourier analyses of the wave forms in Figure 2 gave amplitude and phase, from which attenuation and phase velocity were calculated for the strongest frequency components. The open circles of Figure 3 show attenuation values from

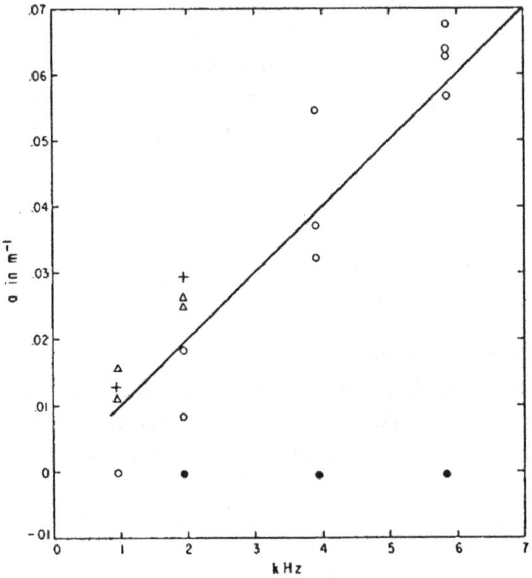

Figure 3. Attenuation values obtained from the three types of calculations.

several pairs of waveforms treated in this fashion. The solid circles show the substantially zero attenuation when the loss parameter q was set to zero.

A second type of calculation concerned the stress-strain relation in a short sample. A clamped-free bar was simulated by 10 mass elements, driven at one end and clamped at the other. On the basis that this models a continuous medium of length L and propagation constant γ,

$$\begin{aligned} U(-L, \omega) &= P(-L, \omega) \, \gamma \tanh(\gamma L)/\rho\omega^2 \\ &\doteq P(-L, \omega)\gamma^2 L/\rho\omega^2 \end{aligned}$$

Since

$$\gamma^2 = (-\omega^2/c^2)(1 + a^2c^2/\omega^2 - i2ac/\omega)$$
$$\doteq (-\omega^2/c^2)\exp(-i2ac/\omega)$$
$$U(-L, \omega)/L \doteq (-P(-L, \omega)/\rho c^2)\exp(-i2ac/\omega)$$

when

$$P_{xx}(-L, t) = P_0 \sin \omega_0 t$$
$$U(-L, t)/L = (-P_0/\rho c^2)\sin(\omega_0 t - 2ac/\omega) \quad (3)$$

The expectation is that when the applied stress is a sine wave beginning at $t = 0$, the displacement should be a transient which eventually settles down to the sine wave expressed in Equation 3. By means of the numerical model, the displacement for several cycles was computed. The motion during a time coinciding with one of the later cycles of the applied stress was taken to be representative of the steady-state displacement. The phase angles between stress and average strain are given in Table 1. The number in parentheses gives the cycle of the applied force during which displacement was Fourier analyzed. Phase angles were converted to attenuation through the relation $a = (\pi f\theta/c)$, and the resulting values are plotted as triangles in Figure 3. Data are also given in Table 1 for

Table 1. Stress-Strain in Short Rod

Frequency	Phase Angle	Equivalent a
0.98 kHz	0.0205 (2nd)	0.0158 m^{-1}
	0.0144 (4th)	0.0111
1.95 kHz	0.0163 (2nd)	0.0250
	0.0173 (7th & 8th)	0.0265
0.98 } *	0.0139 (2nd)	0.0107
1.95 }	0.0244 (3rd & 4th)	0.0374
0.98 } *	0.0126 (4th)	0.0097
1.95 }	0.0198 (7th & 8th)	0.0303

* Both frequencies applied at the same time.

two calculations when the applied force consisted of a fundamental and its first multiple, of equal amplitudes. These sketchy data suggest that superposition applies only approximately.

A third calculation concerns the fractional energy loss per cycle, in a short length of the lossy medium. For the same 10-cm sample, energy supplied by the source was calculated for several cycles of oscillation.

The ratio of energy lost per cycle ΔW to elastic energy stored W appears in Table 2. These were converted to equivalent attenuation by means of the relation $a = (f/2c)(\Delta W/W)$ and plotted as crosses in Figure 3.

Table 2. Energy Loss in Short Rod

Frequency	$(\Delta W/W)$	Equivalent a
0.98 kHz	0.104	0.0127 m^{-1}
1.95	0.121	0.0295

The straight line in Figure 3 is proposed as a plausible interpretation of the data from these three sources. When attenuation is present, the causality condition requires that phase velocity vary with frequency. When the attenuation coefficient is directly proportional to frequency within a band of frequencies, the phase velocity is a constant plus a term varying as the logarithm of frequency. The solid line in Figure 4 shows the variation which is consistent with the straight line in Figure 3. Open circles in Figure 4 show velocities computed from phase shifts

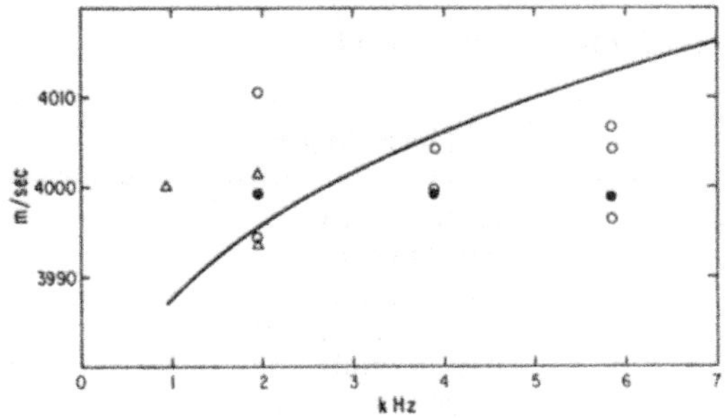

Figure 4. Computed phase velocities for the nonlinear medium.

between pairs of pulses. The triangles show velocities obtained from the ratio of stress to strain in the short rod discussed above. (The solid circles are phase velocities from pulses in the lossless medium, substantially constant and very near the expected value of 4000 m/sec.) Although the data are scattered, a constant value of 4000 m/sec fits the data better than does the logarithmic curve. This suggests that a closer look at the application of the causality principle to this nonlinear medium is in order.

A fourth calculation demonstrates the behavior when a transient vibration travels through a medium which has been prestressed. A bar 120 cm in length is assumed to have been preloaded by increasing stress from zero to a little more than 10^4 dyne/cm^2, producing a uniform strain of 2.5×10^{-8}. Beginning at $t = 0$, a transient stress is added to the static stress. Figure 5 shows the passage of this transient at a point 10 cm

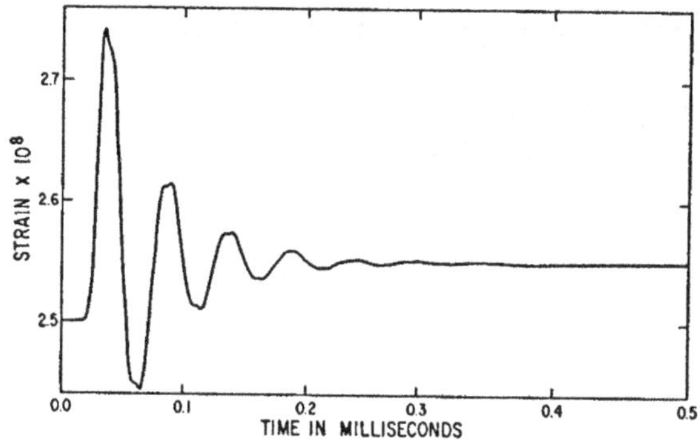

Figure 5. Transient strain in the prestressed nonlinear medium.

from the source. Note that the static strain after passage of the wave is greater than the initial strain. The final combination of stress and strain lies on the elastic line discussed in relation to Figure 1. This feature of the nonlinear medium duplicates experience with strain seismometers on the earth, in that stepchanges in strain are often associated with earthquakes. The known large displacements near the epicenter must produce step changes in strain at considerable distances, and the observed steps may generally be caused by this shift. However, the results presented here suggest that step changes may also be caused by stresses existing near the observation point prior to the passage of a wave.

L. J. Chu and Jerome Glaser
Massachusetts Institute of Technology

11 Synthesis of Radiation Patterns from Circular and Elliptical Apertures

Introduction

This paper deals with the problem of finding a current distribution on a circular or elliptical aperture which produces a given far field radiation pattern. The related synthesis problem for a finite linear aperture has been solved by several authors.[1,2] For the linear aperture, current distributions have been found which optimize the relationship between beam width and side-lobe level of the radiation pattern.[3] Several solutions to the circular problem, which consist of expanding the current distribution in a Fourier-Bessel series, have been published.[4,5,6]

In this paper, a new method of circular and elliptical aperture synthesis based on the finite linear aperture synthesis is formulated. It is shown that the far field pattern of a symmetrical circular aperture current distribution is equal to the pattern of a finite linear aperture current distribution in a cut parallel to the linear aperture when the length of the linear aperture is equal to the diameter of the circular aperture and when the current on the linear aperture is related to that on the circular aperture by an integral equation. This integral equation is inverted, so that the equivalent circular current distribution is expressed in terms of the linear current distribution by another integro-differential equation. In this way, the circular synthesis problem is made equivalent to the linear synthesis problem.

The procedure for synthesis is as follows: The desired radiation pattern is first synthesized by a finite linear aperture current distribution. Then, by means of the integral equation derived in this paper, the equivalent circular aperture current distribution is obtained.

The elliptical aperture synthesis problem is formulated in a similar way. The pattern in a cut parallel to the major axis of the ellipse is shown to be equal to the pattern of a linear aperture in a cut parallel to the linear aperture when the length of the linear aperture is equal to the major axis of the ellipse and when current on the linear aperture is related to that on the ellipse by an integral equation. The integral equation is then inverted so that the equivalent current distribution on the ellipse is expressed in terms of the current on the linear aperture. The pattern from such a current distribution on the ellipse is not circularly symmetric but in all cuts is simply related to that in the major axis cut.

The synthesis method obtained is applied to the lambda function patterns, first used by Chu, and to the lambda functions of complex argument, which optimize the relationship between beam width and side-lobe level.

1. Linear Aperture Synthesis

In this section, a review of the Fourier transform relation between current distribution and pattern for a linear aperture will be given as a starting point for the circular and elliptical synthesis problem. The current distributions which produce lambda function and modified lambda function patterns will be given.

The far field vector potential will be taken as the "pattern" of a current distribution, because it is simply related to the currents, and because the fields can be obtained from it by differentiation. The far field vector potential for a sinusoidally time varying current distribution on the x axis of length $2a$ is given by

$$\underline{\bar{A}} = \frac{\mu_0 e^{-jkr}}{4\pi r} \int_{-a}^{a} \underline{\bar{I}}(x) e^{jkx \cos \theta_x} \, dx \tag{1}$$

where $\underline{\bar{A}}$ is the complex amplitude of the far field vector potential, $\underline{\bar{I}}$ is the complex vector current in amperes, r is the distance from the origin to the far field point, k is the wave number, $\cos \theta_x$ is the direction cosine between the r vector and the x axis. Here $\underline{\bar{I}}$ is presumably produced by many small dipoles distributed along x with arbitrary orientation. For simplicity, only x directed dipoles will be considered.

After normalizing x to a and defining the new angle variable u

$$v = \frac{x}{a} \tag{2}$$

$$u = ka \cos \theta_x \tag{3}$$

the x component of $\underline{\bar{A}}$ can be written as

$$\underline{A}_x = \frac{\mu_0 a}{4\pi r} e^{-jkr} F(u) \tag{4}$$

where

$$F(u) = \int_{-1}^{1} \underline{I}_x(v) e^{juv} \, dv \tag{5}$$

$F(u)$ determines the angular variation of $\underline{A}$ on a sphere of constant r and hence, is known as the "pattern function." It is an analytic function of u and Equation 5 indicates that it is the Fourier transform of $\underline{I}_x(v)$. Therefore, $\underline{I}_x(v)$ is equal to the inverse Fourier transform of $F(u)$.

$$\underline{I}_x(v) = \int_{-\infty}^{\infty} F(u) e^{-juv} \frac{du}{2\pi} \tag{6}$$

$F(u)$ is usually specified for u in the range $-ka$ to ka, since $\cos \theta_x$ lies in the range -1 to 1; however, $F(u)$ can be analytically continued over the entire real u line, since it is analytic. Therefore, Equation 6 constitutes the solution to the aperture synthesis problem; namely, given a realizable analytic $F(u)$, by Equation 6, $\underline{I}_x(v)$ is uniquely determined.

A one-parameter family of realizable $F(u)$ functions introduced by Chu are the lambda functions given by

$$\Lambda_p(u) = \frac{\Gamma(p+1) J_p(u)}{\left(\frac{u}{2}\right)^p} \tag{7}$$

$$\Lambda_p(0) = 1$$

The lambda functions are essentially Bessel functions of order p, $J_p(u)$ with the zero at $u = 0$ removed (p is restricted to be greater than $-1/2$).

The current distributions which produce the lambda function patterns are given by

$$\underline{I}_x(v, p) = \frac{(1 - v^2)^{p-1/2} \Gamma(p+1)}{\sqrt{\pi}\, \Gamma(p + 1/2)} \qquad |v| < 1$$

$$= 0 \qquad |v| > 1 \tag{8}$$

The $p = 1/2$ case corresponds to the uniformly illuminated aperture. The $p = -1/2$ and $p = 1/2$ case were considered by Taylor, who noted that the functions would still remain realizable if the argument u was replaced by $\sqrt{u^2 - c^2}$ where c is a real number. This modification has the advantageous effect of reducing the side-lobe level. The same substitution may be applied to the lambda functions in general, to obtain the two parameter family of realizable pattern functions

$$\Lambda_p(\sqrt{u^2 - c^2}) = \Gamma(p + 1) \frac{J_p(\sqrt{u^2 - c^2})}{\left(\dfrac{\sqrt{u^2 - c^2}}{2}\right)^p} \tag{9}$$

The current distributions which produce the modified lambda function patterns is given by

$$\underline{I}_x(v, 0, c) = \frac{\Gamma(p + 1)}{\sqrt{\pi}} \left(2 \frac{\sqrt{1 - v^2}}{c}\right)^{p - 1/2} I_{p-1/2}(c\sqrt{1 - v^2}) \tag{10}$$

where $I_{p-1/2}$ is a modified Bessel function. For $c \geq 0$, the p must be greater than or equal to 1/2 in order for the current to be finite everywhere on the aperture.

2. Circular Aperture Synthesis. In this section, the symmetric circular aperture synthesis problem will be made equivalent to the linear aperture synthesis problem discussed in the previous section.

The far field vector potential for a circular current distribution lying in the $x - y$ plane of radius b is given by

$$\underline{\bar{A}} = \frac{\mu_0 e^{-jkr}}{4\pi r} \int_0^{2\pi} \int_0^b \rho \, d\rho \, d\psi \underline{\bar{K}}(\rho, \psi) e^{jk(x \cos \theta_x + y \cos \theta_y)}$$

where

$$\begin{aligned} x &= \rho \cos \psi \\ y &= \rho \sin \psi \end{aligned} \tag{11}$$

where $\cos \theta_x$ and $\cos \theta_y$ are the direction cosines of the far field point with respect to the x and y axes respectively; $\underline{\bar{K}}$ is the complex vector surface current in amperes per meter, and ρ, ψ are cylindrical coordinates defined above. If we consider only y directed currents and assume that the current distribution is only a function of radius ρ, the x

component of vector potential will be a circular symmetric function in the $x - y$ plane. Therefore, all cuts in the $x - y$ plane will be identical. The $x - z$ cut is given by

$$\underline{A}_x(x - z \text{ cut}) = \frac{\mu_0 e^{-jkr}}{4\pi r} \int_0^{2\pi} \int_0^b \rho d\rho \, d\psi \underline{K}_x(\rho) e^{jkx \cos \theta_x} \tag{12}$$

The integration is now split into strips and can be written as follows

$$\underline{A}_x(x - z \text{ cut}) = \frac{\mu_0 e^{-jkr}}{4\pi r} \int_{-b}^{b} dx [e^{jkx \cos(\theta_x)}] \int_{-\sqrt{b^2 - x^2}}^{+\sqrt{b^2 - x^2}} \underline{K}_x(\rho) \, dy \tag{13}$$

After defining new variables, as in the linear case,

$$v' = \frac{x}{b} \tag{14}$$

$$u' = kb \cos \theta_x \tag{15}$$

Equation 13 can be written as

$$\underline{A}_x(x - z) = \frac{\mu_0 b e^{-jkr}}{4\pi r} \int_{-1}^{1} dv' e^{ju'v'} \int_{-b\sqrt{1 - v'^2}}^{b\sqrt{1 - v'^2}} \underline{K}_x(\rho) \, dy \tag{16}$$

If we compare Equation 16 with Equation 4, it is clear that the circular aperture will have the same $x - z$ cut pattern as that of the linear aperture discussed in Section 1, if

1. The linear aperture length is equal to the diameter of the circle $2a = 2b$;
2. The linear aperture current is equal to the following integral of the circular aperture current:

$$\underline{I}_x(v') = \int_{-b\sqrt{1 - v'^2}}^{b\sqrt{1 - v'^2}} \underline{K}_x(\sqrt{b^2 v'^2 + y^2} \, dy \tag{17}$$

Equation 17 is clear physically because it states that all the currents in the strip of constant x on the circle add up linearly with the same phase retardation $e^{+jkx \cos \theta}$ in contributing to the far field in the $x - z$ cut.

Equation 17 can be written in terms of radius and linear dimension by a change of variables.

$$\underline{I}_x(x) = 2 \int_x^a \frac{\rho d\rho \underline{K}_x(\rho)}{\sqrt{\rho^2 - x^2}} \tag{18}$$

It can be shown that Equation 18 can be inverted to yield the following inverse relation

$$\underline{K}_x(\rho) = -\frac{1}{\pi\rho}\frac{d}{d\rho}\int_\rho^a \frac{x dx \underline{I}_x(x)}{\sqrt{x^2-\rho^2}} \tag{19}$$

(Note: Equation 19 will be derived in a paper to be published at a later date.) Equations 18 and 19 constitute a transform pair and provide the basis of the circular aperture synthesis scheme.

It is now possible to determine the circular current distributions which produce the lambda function patterns from the linear current distributions which produce the lambda function patterns, given in Section 1.

The circular currents which produce a lambda function pattern are given by Table 1.

Table 1 Circular Currents

Pattern	Circular Current $K_x(\rho, p, c)$
$\Lambda_p(u)$	$\frac{p}{\pi a}\left((1-\frac{\rho^2}{a^2}\right)^{p-1} \quad p>0$
$\Lambda_p(\sqrt{u^2-c^2})$	$\frac{\Gamma(p+1)}{\pi a}\left(\frac{2\sqrt{1-\frac{\rho^2}{a^2}}}{c}\right)^{p-1} I_{p-1}\left(c\sqrt{1-\frac{\rho^2}{a^2}}\right)$

(The special case $p = 0$ is produced by an impulse of current distributed around the periphery of the circle.)

For $c \geq 0$, the p must be greater than or equal to 1 in order for the current to be finite over the aperture. An interesting relationship between the circular and linear currents which produce the lambda function patterns is

$$\underline{K}_x(\rho, p, c) = \frac{\underline{I}_x\left(\frac{\rho}{a}, p-\frac{1}{2}, c\right)\Gamma(p+1)}{\sqrt{\pi}\, a\Gamma\left(p+\right)\frac{1}{2}} \tag{20}$$

3. Elliptical Aperture Synthesis

In this section, it shall be shown that there exists a one-to-one relation between the current distribution on an elliptical aperture and that on a linear aperture such that the patterns resulting from both are identical in one cut. The current on the elliptical aperture K, will be expressed in terms of the currents on *similar* ellipses defined by the parameter R, where

$$\begin{aligned} x &= R\cos\phi \\ y &= \left(\frac{b}{a}\right) R\sin\phi \end{aligned} \qquad 0 < 0 < 2\pi \quad 0 < R < a \tag{21}$$

The far field vector potential of an elliptical aperture as defined above is given by

$$\underline{A} = \frac{\mu_0}{4\pi r} e^{-jkr} \int_0^{2\pi} \int_0^a R\, dR\, d\phi \frac{b}{a} \underline{K}(R) e^{jk(x\cos\theta_x + y\cos\theta_y)} \tag{22}$$

The currents are again assumed to flow in the x direction only. The vector potential therefore only has an x component. The vector potential for the $x - z$ cut is then given by

$$\underline{A}_x(x - z \text{ cut}) = \frac{\mu_0}{4\pi r} e^{-jkr} \int_0^{2\pi} \int_0^a \underline{K}_x(R) e^{jkx\cos\theta_x} R\, dR\, d\phi \frac{b}{a} \tag{23}$$

Splitting the integration into strips as in the circular case and defining the new variables

$$v' = \frac{x}{a} \tag{24}$$

$$u' = ka\cos\theta_x \tag{25}$$

Equation 23 can be written as

$$\underline{A}_x(x - z \text{ cut}) = \frac{\mu_0 a}{4\pi r} e^{-jkr} \int_{-1}^{1} dv' e^{ju'v'} \int_{-b\sqrt{1-v'^2}}^{b\sqrt{1-v'^2}} \underline{K}_x(R)\, dy \tag{26}$$

If we compare Equation 26 with Equation 4, it is clear that the elliptical aperture will have the same pattern as the linear aperture in the $x - z$ cut if

1. The linear aperture length $2a$ is equal to the major axis of the ellipse;
2. The linear aperture current is equal to the following integral of the elliptical aperture current.

$$\underline{I}_x(x) = \int_{-b\sqrt{1-(x/a)^2}}^{b\sqrt{1-(x/a)^2}} \underline{K}_x\left(\sqrt{x^2 + \left(\frac{ay}{b}\right)^2}\right) dy \tag{27}$$

Equation 27 can be written in terms of the variables x and R

$$\underline{I}_x(x) = \frac{2b}{a} \int_x^a \frac{\underline{K}_x(R) R \, dR}{\sqrt{R^2 - x^2}} \tag{28}$$

Equation 28 can be inverted to yield the following inverse relation:

$$\underline{K}_x(R) = -\frac{a}{b} \frac{1}{\pi R} \frac{d}{dR} \int_R^a \frac{\underline{I}_x(x) x \, dx}{\sqrt{x^2 - R^2}} \tag{29}$$

This completes the synthesis problem for the $x - z$ cut.

It is important to determine the nature of the pattern in all cuts when K is synthesized according to the $x - z$ cut. It can be shown that if the pattern in the $x - z$ cut is $A_0(u_x)$ where $u_x = ka \cos \theta_x$, then the pattern in any cut is given by

$$A = A_0(\sqrt{u_x^2 + u_y^2}) \qquad \text{where } u_y = kb \cos \theta_y \tag{30}$$

So that for example, the $y - z$ cut pattern will be given by

$$A(y - z \text{ cut}) = A_0(kb \cos \theta_y) \tag{31}$$

The functional shape of the pattern in the $y - z$ cut is identical to that in the $x - z$ cut but the nulls and side-lobes are shifted.

Conclusions

The symmetrical circular and elliptical aperture synthesis problems have, therefore, been solved by reducing them to an equivalent linear aperture synthesis problem. The method was applied to the synthesis of the modified lambda function patterns, and it was shown that the circular current distribution which produces the lambda function pattern of order p is of the same functional form as the linear current distribution which produces the lambda function pattern of order $p - 1/2$. In a paper to be published, the details of inverting the integral equations relating the equivalent linear aperture current and the circular or elliptical aperture current will be presented. Numerical curves and optimization of the two-parameter lambda function patterns with regard to gain and side-lobe level will be presented elsewhere.

References

1. C. L. Dolph, "A Current Distribution for Broadside Arrays Which Optimizes the Relationship Between Beam Width and Side-lobe Level," *IRE Proc.*, Vol. 34, 6, June 1946, pp. 335, 342.

2. T. T. Taylor, " Design of Line Source Antennas for Narrow Beamwidths and Low Side-lobes," *PGAP*, Vol. **AP-3**, 1, Jan. 1955, pp. 16–28.
3. E. J. Powers, " Utilization of the Lambda Functions in the Analysis and Synthesis of Monopulse Antenna Difference Patterns," *Trans. IEEE, Antennas and Propagation*, *PGAP*, Vol. **AP-15**, 6, Nov. 1967, pp. 771–777.
4. T. T. Taylor, " Design of Circular Apertures for Narrow Beamwidths and Low Side-Lobes," *Trans. IEEE.*, *Antenna and Propagation*, Vol. **AP-8**, Jan. 1960, pp. 17–22.
5. A. Ishimaru and G. Held, " Analysis and Synthesis of Radiation Patterns from Circular Apertures," *Can. J. Phys.*, Vol. **38**, 1, Jan. 1960, pp. 78–99.

Peter J. Westervelt
Brown University

12 Acoustics and General Relativity*†

1. Introduction

Obtaining the energy, or especially the momentum and angular momentum, radiated by extended gravitational sources can be quite a tedious operation,[1,2,3] as has recently been demonstrated by F. Cooperstock, a former student of mine. The present motivation for such calculations stems from the possibility that gravitational radiation may account for dissipation[4,5] and transfer[6] of energy in the domain of relativistic astrophysics.

I will show that, within the context of linearized relativity theory, all physical quantities normally calculated with the aid of the second rank field tensor can be far more easily obtained by means of an equivalent scalar theory. That this approach works in a special case, namely the gravitational energy radiated by an extended Einstein-Rosen oscillator or by a pulse of light, has already been demonstrated.[7]

The *scalarization* of a normally tensorial theory is accomplished by introducing isotropic tensor sources which, by their very nature, cannot alter the physical content of the theory. Introduction of isotropic sources leads to a radical computational simplification, and to a formal analogy with the linearized theory of acoustics, hence the title of this paper.

The following section introduces the technique of scalarizing; in Section 3 the method is applied to an extended oscillator; in Section 4 the method is applied to a spinning rod.

* This paper is dedicated to Professor Philip M. Morse. His summer course on General Relativity given in 1950 at M.I.T. inspired my interest in the present topic.

† Supported in part by the U.S. Atomic Energy Commission (Report No. NYO-2262TA-192).

2. *Scalarization*

In the notation of Landau and Lifshitz[8] adopted in this paper, the square of the line element ds^2 is related to ${}^0g_{ik}$ and h_{ik}, the unperturbed and perturbation metrics respectively. Thus

$$ds^2 = -g_{ik}\,dx^i\,dx^k \qquad \text{where } g_{ik} = {}^0g_{ik} + h_{ik} \qquad {}^0g_{\alpha\beta} = \delta_{\alpha\beta}$$
$${}^0g_{00} = -1 \qquad {}^0g_{0\alpha} = 0 \tag{1}$$

Recall that the radius vector of three space is x^α and that $x^0 = ct$. It is convenient to express the perturbations in terms of quantity ψ_i^k, thus:

$$h_i^k = \psi_i^k - \tfrac{1}{2}\delta_i^k\psi \qquad \text{where } \psi = \psi_i^i \tag{2}$$

for then ψ_i^k satisfies the inhomogeneous wave equation

$$\Box\psi_i^k = -16\pi c^{-4}\,G\tau_i^k \tag{3}$$

with τ_i^k, the energy momentum tensor of matter and nongravitational radiation serving as sources. In Equation 3, G is the gravitational constant and c the velocity of light.

Conservation of energy and momentum, in the Einstein weak-field approximation adopted in this paper, is expressed by

$$\frac{\partial\tau_i^k}{\partial x^k} = 0 \tag{4}$$

which implies the following auxiliary conditions:

$$\frac{\partial\psi_i^k}{\partial x^k} = 0 \tag{5}$$

The retarded potential solution is

$$\psi^{ik} = 4c^{-4}G\int r^{-1}[\tau^{ik}]_{x^0-r}\,dv \tag{6}$$

Here $r = [(x^1-y^1)^2 + (x^2-y^2)^2 + (x^3-y^3)^2]^{1/2}$, and $dv = dy^1\,dy^2\,dy^3$, in which y^α denote the coordinates of the source while the x^α denote the coordinates of the field point.

For a locally plane progressive wave the energy momentum pseudo-tensor,

$$ct_i^k = (64\pi G)^{-1}c^4\left[2\psi_{,i}^{\ lm}\psi_{lm}^{\ ,k} - \psi_{,i}\psi^{,k} + \delta_i^k\left(\frac{1}{2}\psi_{,l}\psi^{,l} - \psi_{,n}^{\ lm}\psi_{lm}^{\ ,n}\right)\right]$$

becomes

$$ct_i^k = (64\pi G)^{-1}c^4[2\psi_{,i}^{\ lm}\psi_{lm}^{\ ,k} - \psi_{,i}\psi^{,k}] \tag{7}$$

This is because $\psi_{,i}^{\ lm}\psi_{lm}^{\ ,i}$ and $\psi_{,i}\psi^{,i}$ are zero, which is a consequence of the fact that $\Box\psi^{lm}\psi_{lm} = \Box\psi_{lm} = \Box\psi^2 = \Box\psi = 0$.

We confine ourselves to propagation along the x^1 axis in which case it is easy to show that Equation 7 becomes either

$$ct_i^{\,k} = (64\pi G)^{-1}c^4[2\psi_{22,i}\psi_{22}{}^{,k} + 2\psi_{33,i}\psi_{33}{}^{,k} + 4\psi_{23,i}\psi_{23}{}^{,k} - \psi_{,i}\psi^{,k}] \tag{8}$$

or

$$ct_i^{\,k} = (64\pi G)^{-1}c^4[\psi_{,i}\psi^{,k} + 2(2\psi_{23,i}\psi_{23}{}^{,k} - \psi_{22,i}\psi_{33}{}^{,k} - \psi_{33,i}\psi_{22}{}^{,k})] \tag{9}$$

At this stage, we can introduce an isotropic field

$$\zeta_{\alpha\beta} = -\psi_{22}\,\delta_{\alpha\beta}, \quad \zeta_{00} = -\psi_{22}$$

which when added to $\psi_l^{\,m}$ yields a new field $\Psi_l^{\,m} = \psi_l^{\,m} + \zeta_l^{\,m}$. The components of $t_i^{\,k}$ are obviously invariant with respect to this transformation. Thus Equation 9 becomes

$$ct_i^{\,k} = (64\pi G)^{-1}c^4[\Psi_{,i}\Psi^{,k} + 4\Psi_{23,i}\Psi_{23}{}^{,k}] \tag{10}$$

This result greatly simplifies the calculation of angular momentum radiation rates **N** since

$$\mathbf{N} = -(64\pi G)^{-1}c^4 \oint d\mathbf{S}\cdot[\nabla\Psi(\mathbf{r}\times\nabla\Psi) + 4\nabla\Psi_{23}(\mathbf{r}\times\nabla\Psi_{23})] \tag{11}$$

If we are willing to restrict ourselves to a calculation of the average rate of loss of power, or of linear momentum, a scalar theory can be constructed as follows. From Equation 9, we obtain the Poynting vector ct^{01}:

$$ct^{01} = c^3(64\pi G)^{-1}[\dot\psi^2 + 4(\dot\psi_{23}{}^2 - \dot\psi_{22}\dot\psi_{23})] \tag{12}$$

Introduce the isotropic field $g(r, t)\delta_{\alpha\beta}$ such that the following two conditions, in which $\langle\ \rangle$ denotes time averaging, are satisfied,

$$\langle g\dot\psi\rangle = 0, \tag{13}$$

and

$$\langle g^2\rangle = \langle\dot\psi_{23}{}^2 - \dot\psi_{22}\dot\psi_{33}\rangle \tag{14}$$

It is always possible to satisfy these conditions for either transient or steady-state radiation. However, I will demonstrate this for the latter case only. Now in terms of $\Psi = \dot\psi + 2g$, the Poynting vector **P** becomes

$$\mathbf{P} = -(64\pi G)^{-1}c^4\,\dot\Psi\nabla\Psi \tag{15}$$

and the total power radiated is

$$\oint \mathbf{P}\cdot d\mathbf{S} = (64\pi G)^{-1}c^3\oint \dot\Psi^2\, dS \tag{16}$$

The rate of momentum radiation is minus the reaction force **F** with **F** given by

$$\mathbf{F} = -c^{-1} \oint \mathbf{P}\, dS \qquad (17)$$

At this point an immediate analogy with acoustical problems results if the following definitions for the pressure p, velocity $\mathbf{v}$, source strength density[9] q, characteristic density of the vacuum μ_0, and the mass density of gravitational source $\bar{\mu}$ are defined as follows:

$$\begin{aligned} p &= -(16\pi G)^{-1}c^2\ddot{\Psi} & \mathbf{v} &= -\frac{1}{4}c^2 \nabla \int \Psi\, dt \\ \Box\Psi &= -16\pi c^{-4}G\tau & \tau &= -\bar{\mu}c^2 \\ \bar{\mu} &= \mu + \sigma & \psi &= 16\pi c^{-4}G\mu c^2 \end{aligned} \qquad (18)$$

For simple harmonic radiation with circular frequency ω,

$$\begin{aligned} p &= (16\pi G)^{-1}\omega^2 c^2 \Psi \qquad \mu_0 = (4\pi G)^{-1}\omega^2 \\ -q &= \mu_0{}^{-1}\bar{\mu} = 4\pi G\omega^{-2}\bar{\mu} \end{aligned} \qquad (19)$$

For a plane wave $v = \frac{1}{4} c\Psi$, and the characteristic impedance $z_0 = (4\pi G)^{-1}\omega^2 c$.

If we use the above acoustical analogies and methods previously developed by me[10,11,12] and Dr. Maidanik[13] to calculate acoustical fluxes, it is possible to recast the expressions for the total power radiated, Equation 16, and the force, Equation 17, into integrals over sources, thus

$$\oint \mathbf{P} \cdot d\mathbf{S} = -\frac{1}{4} \int \dot{\mu} c^2 \Psi\, dV \qquad (20)$$

and

$$\begin{aligned} \langle \mathbf{F} \rangle &= -\int \langle \mu_0 q \mathbf{v} \rangle\, dV \\ &= -\frac{1}{4} \int \left\langle \dot{\mu} c^2 \nabla \int \psi\, dt \right\rangle dV \\ &= +\frac{1}{4} \int \langle \bar{\mu} c^2 \nabla\Psi \rangle\, dV \end{aligned} \qquad (21)$$

Further, in certain special cases (one of which is the spinning rod) when the rate of radiation of angular momentum is correctly given by

$$\mathbf{N} = -(64\pi G)^{-1}c^4 \oint d\mathbf{S} \cdot [\nabla\Psi(\mathbf{r} \times \nabla\Psi)] \qquad (22)$$

N may also be found from the following integral over sources:

$$\langle \mathbf{N} \rangle = -\frac{1}{4} \int \langle \bar{\mu} c \mathbf{r} \times \nabla \Psi \rangle dV \tag{23}$$

In connection with these two formulas, Equations 22 and 23, for the rate of angular momentum radiated by scalar waves, the recent incorrect statement[14] has been made that a scalar field cannot carry off angular momentum.

The routine for calculating Ψ will now be outlined. First the space-space components of the time derivative of ψ_{ij} are expressed as a sum of bivectors. Thus

$$\dot{\psi}_{\alpha\beta} = \sum_i a_\alpha{}^i a_\beta{}^i f^i(r, t) \tag{24}$$

in which the $\mathbf{a}^i$ are unit vectors. It is simple to show that

$$\dot{\psi}_{22}\dot{\psi}_{33} - \dot{\psi}_{23}{}^2 = \frac{1}{2} \sum_{i \neq j} \sum_j [\mathbf{n} \cdot \mathbf{a}^i \times \mathbf{a}^j]^2 f^i f^j \tag{25}$$

where $\mathbf{n} = r^{-1}\mathbf{r}$

We wish ultimately to obtain

$$\Psi_{\alpha\beta} = \dot{\psi}_{\alpha\beta} + g(r, t)\,\delta_{\alpha\beta} \tag{26}$$

with g subject to the conditions given in Equations 13 and 14.

Suppose we treat the harmonic case in which

$$\dot{\psi} = F(r)e^{-i\omega t} \tag{27}$$

and

$$g = G(r)e^{-i\omega t}$$

The conditions $\langle g\dot{\psi} \rangle = 0$ requires that

$$\text{Re } GF^* = 0. \tag{28}$$

If we further let $F = d(r)e^{i\alpha(r)}$ and $G = h(r)e^{i\beta(r)}$, Equation 28 requires that

$$\beta(r) = \alpha(r) \pm \frac{\pi}{2} \tag{29}$$

The condition $\langle g^2 \rangle = \langle \dot{\psi}_{22}{}^2 - \dot{\psi}_{22}\dot{\psi}_{22} \rangle$ requires that

$$GG^* = h^2 = \sum_{i \neq j} \sum_j \langle [\mathbf{n} \cdot \mathbf{a}^i \times \mathbf{a}^j]^2 f^i f^j \rangle \tag{30}$$

from which we find

$$h = \pm \left[\sum_{i \neq j} \sum_j \langle [\mathbf{n} \cdot \mathbf{a}^i \times \mathbf{a}^j]^2 f^i f^j \rangle \right]^{1/2} \tag{31}$$

Finally, combining the above results, the time derivative of the scalar field Ψ, turns out to be

$$\Psi = \psi \pm 2i\left[\sum_{i \neq j}\sum_{j} \langle[\mathbf{n} \cdot \mathbf{a}^i \times \mathbf{a}^j]^2 f^i f^j\rangle\right]^{1/2} e^{-i\omega t + i\alpha(r)} \tag{32}$$

A particularly simple result occurs in the event $\psi_{\alpha\beta} = a_\alpha a_\beta f(r, t)$, since now $\Psi = \psi$. This is so for the extended Einstein-Rosen oscillator which is discussed in the following section.

3. The Extended Einstein-Rosen Oscillator

An extended Einstein-Rosen oscillator is a one-dimensional oscillator, long with respect to λ_g, the gravitational wavelength, which reduces in all of its characteristics to the familiar Einstein-Rosen point quadrupole when l, its length, is very much smaller than λ_g. I have previously calculated[15,7] the power radiated by such an oscillator. I now wish to repeat the calculation by an alternative method, and I wish also to compute the rate of momentum radiation.

This oscillator is constructed by placing a heavy laser at the origin, $x = y = z = 0$, and pointing the collimated laser beam at an absorber located at $x = l$, $y = z = 0$. The laser is supposed to be 100% modulated at the circular frequency ω, and the maximum transverse dimension of the laser beam is assumed very small with respect to λ_g. We ignore all contributions oscillating with optical frequencies, since it can be shown (but we won't go into it here) that such high-frequency terms do not couple into the gravitational field.

The mass density distribution for such an oscillator can be found by a simple application of energy and momentum conservation, or by taking the Fourier transform of the $\tau_s{}^{00}$ (for both source and sink!) in Reference 16. The source density turns out to be

$$\begin{aligned}\mu = P_E(i\omega c^2)^{-1}\{\delta(x) + c(i\omega)^{-1}\delta'(x) \\ - [\delta(x - l) + c(i\omega)^{-1}\delta'(x - l)]e^{ikl}\}\delta(y)\delta(z)e^{-i\omega t}\end{aligned} \tag{33}$$

In the above, P_E is the average power emitted by the laser beam and $\delta' = d\delta(x)/dx$. The asymptotic field, owing to the source at the origin only, is

$$\psi = 4GP_E c^{-4} r^{-1}(1 + \cos\theta)e^{-i\omega t + ikr} \tag{34}$$

in which r is the field point, $k = c^{-1}\omega$, and $\cos\theta = r^{-1}x$. The field along the axis, $y = z = 0$, is

$$\psi = -4GP_E(k^2c^5)^{-1}\left\{\frac{e^{2ikl-ikx}}{(l-x)^2} + \frac{(1-2ikx)}{x^2}e^{ikx}\right\}e^{-i\omega t} \tag{35}$$

When $l \gg \lambda_g$, no interference between source and sink occurs so the power radiated is twice that given by Equation 16 in which Ψ is replaced by $\dot{\psi}$ from Equation 34. This results in a radiated gravitational power P_G equal to

$$P_G = \frac{4}{3}Gc^{-5}P_E{}^2 \qquad \text{when } l \gg \lambda_g \tag{36}$$

When l is the same order of magnitude or less than λ_g, the work done by the source on the sink and *vice versa* must be added on to Equation 36. This addition is given by Equation 20 in which $\bar{\mu}$ is replaced by μ from Equation 33, and Ψ is replaced by $\dot{\psi}$ from Equation 35, and the divergent terms are ignored since they are correctly accounted for in Equation 36. The interference terms are easily found to be

$$\frac{1}{4}\int \langle \mu c^2 \dot{\psi} \rangle \, dv = Gc^{-5}(kl)^{-3}[\sin 2kl - 2kl] \tag{37}$$

The complete answer is given by the sum of Equations 36 and 37, *viz*:

$$P_G = \frac{4}{3}Gc^{-5}P_E{}^2\left[1 + \frac{3}{4}(kl)^{-3}(\sin 2kl - 2kl)\right] \tag{38}$$

valid for any l.

In the limit when $l \ll \lambda_g$, Equation 38 becomes

$$\begin{aligned} P_G &= \frac{4}{5}Gc^{-5}P_E{}^2(kl)^2 \\ &= \frac{1}{15}GI_0{}^2c^{-5}\omega^2 \end{aligned} \tag{39}$$

In the above, I_0 is the amplitude of the moment of inertia of the oscillator, a familiar result.[17]

The linear momentum radiated by our extended oscillator is easily found in a similar manner as was the energy. When $l \gg \lambda_g$, the momentum radiated is minus twice that given by Equation 17 thus

$$\begin{aligned} \mathbf{F} &= -(32G)^{-1}c^2 \int \langle \dot{\psi}^2 \rangle \, d\mathbf{S} \\ &= -\frac{2}{3}Gc^{-6}P_E{}^2\mathbf{i} \end{aligned} \tag{40}$$

where $\mathbf{i}$ is the unit vector directed along the x axis.

When l is the same order of magnitude as λ_g, the force exerted by the source on the sink and *vice versa* must be added on to Equation 40. This addition is given by Equation 21 in which $\bar{\mu}$ is replaced by μ from Equation 33, Ψ is replaced by ψ from Equation 35 and, as before, divergent self-interactions are ignored. Thus the interference terms are found to be

$$\frac{1}{4}\int \langle \mu c^2 \, \nabla\psi \rangle \, dv$$

$$= -Gc^{-6}P_E{}^2(kl)^{-4}[3\cos 2kl - 3 + kl\sin 2kl + 4(kl)^2]\mathbf{i} \quad (41)$$

The correct result is the sum of Equations 40 and 41 thus

$$\mathbf{F} = -\frac{2}{3}Gc^{-6}P_E{}^2\left\{1 - \frac{3}{2}(kl)^{-4}[3\cos 2kl - 3 + kl\sin 2kl + 4(kl)^2]\right\}\mathbf{i} \quad (42)$$

In the long wavelength limit,

$$\mathbf{F} = -\frac{2}{315}Gc^{-6}P_E{}^2(kl)^4\mathbf{i} \quad (43)$$

The ratio of the radiation reaction force to the power radiated is

$$-\frac{1}{42}c^{-1}(kl)^2 \qquad \text{when } \lambda_g \gg l \qquad \text{or}$$

$$-\frac{1}{2}c^{-1} \qquad \text{when } \lambda_g \ll l$$

We look at angular momentum radiation in the final section.

4. *The Spinning Rod*

In this section, I will apply the more extended formalism of Section 2 to a calculation of the radiation from a spinning rod.

Consider a rod with moment of inertia I spinning with angular velocity ω in the $x - y$ plane. The angle made by the rod and the y axis is $\omega t \pm \pi/2$. The volume integral of the flux of momentum density tensor for the whirling rod is easily shown to be

$$\int \tau_{\alpha\beta}\, dV = I\omega^2(b_\alpha{}^1 b_\beta{}^1 - b_\beta{}^2 b_2{}^\alpha) \quad (44)$$

in terms of the vectors

$$\mathbf{b}^1 = (\mathbf{i} + i\mathbf{j})e^{-i\omega t}$$

and

$$\mathbf{b}^2 = (-i\mathbf{i} + \mathbf{j})e^{-i\omega t}$$

In order to be able to express the field components in the manner of Equation 24, we adopt the following set of four unit vectors.

$$\mathbf{a}^1 = \frac{1}{\sqrt{2}}(\mathbf{i}+\mathbf{j}) \qquad \mathbf{a}^2 = \frac{1}{\sqrt{2}}(\mathbf{i}-\mathbf{j})$$

$$\mathbf{a}^3 = \mathbf{i} \qquad \mathbf{a}^4 = \mathbf{j} \tag{45}$$

The space-like components of the field tensor are obtained from

$$\psi_{\alpha\beta} = 4Gc^{-4}r^{-1}\left[\int \tau_{\alpha\beta}\, dV\right]_{t-r/c}$$

Thus, in terms of the set of vectors, Equation 45, we find

$$\dot{\psi}_{\alpha\beta} = 8Gc^{-4}I\omega^3 r^{-1}[a_\alpha{}^1 a_\beta{}^1 - a_\alpha{}^2 a_\beta{}^2 + i(a_\alpha{}^3 a_\beta{}^3 - a_\alpha{}^4 a_\beta{}^4)]e^{-2i\omega t+2ikr} \tag{46}$$

It is obvious that $\dot{\psi}_\alpha{}^\alpha = 0$; therefore, since $\psi_{11} = \psi_{00}$, it follows that

$$\dot{\psi} = \dot{\psi}_\alpha{}^\alpha + \dot{\psi}_0{}^0 = -\dot{\psi}_{\alpha\beta}n_\alpha n_\beta$$

thus

$$\dot{\psi} = -8Gc^{-4}I\omega^3 r^{-1}[(\mathbf{n}\cdot\mathbf{a}^1)^2 - (\mathbf{n}\cdot\mathbf{a}^2)^2 + i(\mathbf{n}\cdot\mathbf{a}^3)^2 - i(\mathbf{n}\cdot\mathbf{a}^4)^2]e^{-2i\omega t+2ikr} \tag{47}$$

Now introducing θ and ϕ, the polar and azimuthal angles, respectively, defined by $\cos\theta = \mathbf{n}\cdot k$ and $\cos\phi = \mathbf{n}\cdot\mathbf{j}$, Equation 47 becomes

$$\dot{\psi} = 8iGc^{-4}I\omega^3 r^{-1}\sin^2\theta e^{-2i\omega t-2i\phi+2ikr} \tag{48}$$

In passing, note that Equation 48 could be obtained directly from the formula

$$\dot{\psi} = -\frac{2}{3}Gc^{-4}r^{-1}\dddot{D}_{\alpha\beta}n_\alpha n_\beta]_{t-r/c} \tag{49}$$

in which $D_{\alpha\beta} = \int \rho(3x_\alpha x_\beta - r^2\delta_{\alpha\beta})\,dv$ is the mass quadrupole moment of the rod, with ρ the rod's density.

From Equation 29, obtain

$$\alpha = \beta \pm \frac{\pi}{2} = -2\phi + kr \pm \pi \tag{50}$$

and from Equation 31, obtain

$$h = \pm 8Gc^{-4}I\omega^3 r^{-1}[(\mathbf{n}\cdot\mathbf{a}^1\times\mathbf{a}^2)^2 + (\mathbf{n}\cdot\mathbf{a}^3\times\mathbf{a}^4)^2]^{1/2}$$

$$= \pm 16\sqrt{2}Gc^{-4}I\omega^3 r^{-1}\cos\theta \tag{51}$$

If we combine Equation 32, 48, 50, and 51, the complete scalar wave function for the spinning rod is found to be

$$\Psi = 8Gc^{-4}I\omega^3r^{-1}[i \sin^2 \theta + 2\sqrt{2} \cos \theta]e^{-2i\omega t - 2i\phi + 2ikr} \tag{52}$$

The energy flux, obtained by introducing Equation 52 into Equation 15 and taking the time average, is

$$\mathbf{P} = \frac{1}{2} Gc^{-5}I^2\omega^6r^{-2}(\sin^4 \theta + 8 \cos^2 \theta)\mathbf{n} \tag{53}$$

The power radiated is

$$\oint \mathbf{P} \cdot d\mathbf{S} = \frac{32}{5} Gc^{-5}I^2\omega^6 \tag{54}$$

a well known result.[17]

The rate of angular momentum radiation calculated from Equation 22 is easily found to be

$$\mathbf{N} = \frac{32}{5} Gc^{-5}I^2\omega^5\mathbf{k}$$

in agreement with the far more complicated calculation of F. I. Cooperstock and D. J. Booth.[3]

I wish to conclude with a few comments about the related acoustical problem of a rod spinning in a compressible fluid. For this problem the sound pressure turns out to be

$$\begin{aligned} p &= \frac{c^2\omega^2\psi}{16\pi G} \\ &= -4I\omega^4\pi^{-1}c^{-2}r^{-1} \sin^2 \theta e^{-2i\omega t - 2i\phi + 2ikr} \end{aligned} \tag{55}$$

in which we now understand c to be the velocity of sound. The acoustic power radiated is

$$(\rho c)^{-1} \int \langle p^2 \rangle \, dS = \frac{2(2\omega)^8 I^2}{15\pi\rho c^5} \tag{56}$$

The moment of inertia I appearing both in Equation 55 and in Equation 56 is that which the rod would have were it to be constructed out of matter with the same density ρ as the fluid. A whirling aspherical blob of turbulence would be an example of such a source. Notice that M. J. Lighthill's[18] famous law of dependence of radiated power on the eighth power of the velocity is correctly given. The rate at which angular momentum is carried away by this acoustic wave is, of course, simply ω^{-1} times the emitted acoustic power.

References

1. F. I. Cooperstock, *Phys. Rev.*, **163**, 1368 (1967).
2. F. I. Cooperstock, *Phys. Rev.*, **165**, 1424 (1968).
3. F. I. Cooperstock and D. J. Booth, *Angular Momentum Flux for a Gravitational Field*, to be published.
4. B. B. Braginskii, *Usp. Fiz. Nauk*, **86**, 433 (1965) [English transl.: *Sov. Phys. — Usp.*, **8**, 513 (1966)].
5. C. W. Chin, *Phys. Rev.*, **139**, 13761 (1965).
6. W. A. Fowler, *Rev. Mod. Phys.*, **36**, 545 (1964).
7. P. J. Westervelt, *JETP Lett.*, **4**, 225 (1966).
8. L. D. Landau and E. M. Lifshitz, *The Classical Theory of Fields*, Addison-Wesley Publishing Company, Inc., Reading, Mass. 2nd ed., revised (1962).
9. P. M. Morse, *Vibration and Sound*, p.313, McGraw-Hill Book Company, Inc., New York, N.Y., 2nd ed. (1948).
10. P. J. Westervelt, *J. Acoust. Soc. Am.*, **23**, 312 (1951).
11. P. J. Westervelt, *J. Acoust.. Soc Am.*, **29**, 26 (1957).
12. H. Olsen, H. Wergeland, and P. J. Westervelt, *J. Acoust. Soc. Am.*, **30**, 633 (1958).
13. G. Maidanik, *J. Acoust. Soc. Am.*, **30**, 620 (1958).
14. R. E. Morganstern, *Phys. Rev.*, **163**, 1357 (1967).
15. M. E. Rose, Ed. *Proceedings of the Eastern Theoretical Physics Conference*, Gordon and Breach, Science Publishers, Inc., N.Y.-London (1963).
16. P. J. Westervelt, *Acta Phys. Polon.*, **27**, 831 (1965).
17. J. Weber, *General Relativity and Gravitational Wave*, p.97, Interscience Publishers, Inc., New York (1961). See p.94 for a simple derivation, originally due to Einstein, of t_i^k, the energy momentum pseudotensor.
18. M. J. Lighthill, *Proc. Roy. Soc.* (*London*), **A211**, 564 (1952).

L. I. Schiff
Stanford University

13 Nongeodesic Motion*

1. Introduction

The motion of a test particle that is free of nongravitational forces is described to good approximation by a geodesic path in the general theory of relativity. The description need not be precise, however, even for a test particle of infinitesimal mass. It was recognized by Mathisson[1] more than thirty years ago that a spinning test particle follows a nongeodesic path, the deviation from geodesic motion being proportional to the angular momentum per unit mass of the particle,

$$\mathbf{s} = \mathbf{S}/M \tag{1}$$

where $\mathbf{S}$ is the spin angular momentum and M is the rest mass of the particle.

The primary objective of the present paper is the calculation of the differential acceleration between spinning and nonspinning test particles that are in the same approximately Newtonian orbit about a spherically symmetric gravitating body. The hope is that the measurement of such a differential acceleration can be incorporated in the experiment now under way[2] to measure the precession of the spin axis of a gyroscope[3] that is in a satellite orbiting the earth.[4] The secondary objective is the calculation of the effect of the possible existence of a nucleonic *heavy-particle charge*, arising from the heavy-particle gauge transformation proposed by Lee and Yang,[5] on such a differential acceleration measurement.[6]

* Research sponsored by the Air Force Office of Scientific Research, Office of Aerospace Research, U.S. Air Force, under AFOSR Contract No. AF49(638)1389.

Existing Eötvös-type experiments[7,8] set a limit on the nucleonic gauge effect, since they consist of comparisons of the ratio of gravitational to inertial mass for substances of different atomic mass number. They do not provide any information on the nongeodesic motion of spinning particles, since the objects compared thus far have had negligible angular momenta. We shall see that the expected spin effect is extremely small even under the most favorable experimental conditions. Nevertheless, the inclusion of the differential acceleration measurement in the satellite-borne gyroscope experiment is believed to be worth while, since an unexpected spin effect might be present, and also since an improvement in the Eötvös-Dicke experiment by several orders of magnitude may be possible.

2. *Nongeodesic Motion of a Gyroscope*

As with the earlier calculation of the precession of the spin axis,[3] we find it convenient to start from the covariant equations of motion derived by Papapetrou[9] using the method of Fock,[10] although other approaches are possible.[11,12] Corinaldesi and Papapetrou[13] have expressed these equations in terms of a rectangular coordinate system in which the central gravitating body is at rest at the origin. This implies that the ambiguity associated with the existence of six rather than three independent components of the spin angular momentum tensor $S^{\alpha\beta}$ ($\alpha, \beta = 1, 2, 3, 4$) is to be removed by imposing the supplementary condition $S^{i4} = 0$ ($i = 1, 2, 3$). The choice of this rather than the Pirani[11] supplementary condition $S^{\alpha\beta}u_\beta = 0$, where u^α is the four-velocity of the gyroscope, was discussed in detail in Reference 3. The argument given there can be extended to show that the result of the present calculation is unaffected by the choice of the supplementary condition, or by the use of the standard or isotropic form of the Schwarzschild metric.

We choose units in which the Newtonian gravitational constant G and the speed of light c are equal to unity, and denote the mass of the central body by m. The quantities $\mathbf{r}$, $\mathbf{v} = d\mathbf{r}/dt$, and $\mathbf{j} = \mathbf{r} \times \mathbf{v}$ denote the coordinate, velocity, and orbital angular momentum per unit mass, respectively, of the gyroscope. Equation (4.6) of Reference 13 then becomes, to first order in m and $\mathbf{v}$:

$$\frac{d^2\mathbf{r}}{dt^2} = \frac{-m\mathbf{r}}{r^3} + \mathbf{a}_S \tag{2}$$

$$\mathbf{a}_S = \left(\frac{3m}{r^5}\right)[\mathbf{j}(\mathbf{r} \cdot \mathbf{s}) + (\mathbf{r} \times \mathbf{s})(\mathbf{r} \cdot \mathbf{v})] \tag{3}$$

The first term on the right side of Equation 2 is the usual Newtonian acceleration. The spin-dependent acceleration $\mathbf{a}_S$ is proportional to the spin angular momentum per unit mass, Equation 1, of the gyroscope, as expected.

Before numerical estimates are obtained from Equation 3, it is interesting to derive the expression for $\mathbf{a}_S$ that corresponds to the most general static, spherically symmetric metric:[14]

$$ds^2 = \left[1 - \left(\frac{2\alpha m}{r}\right) + \left(\frac{2\beta m^2}{r^2}\right) + \cdots\right] dt^2$$
$$- \left[1 + \left(\frac{2\gamma m}{r}\right) + \cdots\right](dx^2 + dy^2 + dz^2) \qquad (4)$$

The numerical coefficients α, β, γ, ... are all equal to unity in general relativity theory, and we always assume that the equations of motion of matter and electromagnetic radiation are given in terms of this metric by Einstein's equations. Further, with the conventional definition of the gravitational constant G, we must choose $\alpha = 1$ in order for planetary orbits to agree with Newtonian theory in lowest order. The Brans-Dicke theory,[15] which modifies general relativity by including a scalar field as well as Einstein's tensor field, can be cast in this form by a suitable choice of the coefficients. With sufficient accuracy to treat first-order effects, it can be shown[15,16] that

$$\alpha = 1 \qquad \gamma = \frac{\omega + 1}{\omega + 2} \qquad (5)$$

The Brans-Dicke parameter ω is essentially the ratio of the coupling strength associated with the tensor field to that of the scalar field, and is infinite in the Einstein theory. Recent observations of solar oblateness have been accounted for by assuming a value for ω in the neighborhood[17] of 6, although this interpretation is by no means established.

The expression for $\mathbf{a}_S$ that corresponds to the metric, Equation 4, with $\alpha = 1$ is obtained by a straightforward recalculation of the work of Reference 13. The result is

$$\mathbf{a}_S = \left(\frac{m}{r^5}\right)[(1 + 2\gamma)\mathbf{j}(\mathbf{r} \cdot \mathbf{s}) + (2 + \gamma)(\mathbf{r} \times \mathbf{s})(\mathbf{r} \cdot \mathbf{v}) - (1 - \gamma)\mathbf{r}(\mathbf{j} \cdot \mathbf{s})] \qquad (6)$$

It is evident that this reduces to Equation 3 when $\gamma = 1$, as of course it must.

The nucleonic gauge effect[5] can be taken into account by replacing Equation 2 by

$$\frac{d^2\mathbf{r}}{dt^2} = \frac{-m\mathbf{r}}{r^3} + \mathbf{a}_S + \mathbf{a}_N \tag{7}$$

$$\mathbf{a}_N = \frac{-km\mathbf{r}}{r^3} \qquad k = \frac{\eta^2\varepsilon}{GM_N{}^2} \tag{8}$$

Here η is the unknown heavy-particle charge of a nucleon and M_N is its average mass. The dimensionless number ε is proportional to the difference between the ratio of mass number to atomic weight of the gyroscope and of its housing, since only the differential acceleration can be measured. Since $\mathbf{a}_S$ and $\mathbf{a}_N$ are both defined to be the acceleration of the quartz gyroscope with respect to its nonrotating copper housing,[2] a simple computation shows that $\varepsilon = 5.7 \times 10^{-4}$.

3. Numerical Results

Circular Orbits. For an unperturbed orbit that is a circle, it is easily seen that the average differential acceleration over a period is zero, so that there is no secular change in velocity or position of the gyroscope with respect to its housing. We choose the orbit to have radius R and to lie in the xy-plane, and $\mathbf{s}$ to lie in the xz-plane. It then follows from Equations 6, 7, and 8 that

$$\begin{aligned} a_{Sx} + a_{Nx} &= -\left(\frac{m}{R^2}\right)[(1-\gamma)\omega_0 s_z + k]\cos\omega_0 t \\ a_{Sy} + a_{Ny} &= -\left(\frac{m}{R^2}\right)[(1-\gamma)\omega_0 s_z + k]\sin\omega_0 t \\ a_{Sz} + a_{Nz} &= \left(\frac{m}{R^2}\right)(1+2\gamma)\omega_0 s_x \cos\omega_0 t \end{aligned} \tag{9}$$

where ω_0 is the orbital angular velocity.

As a numerical example, we consider the third of Equations 9, and restore its dimensions by multiplying the right side by G/c^2. With $\mathbf{s}$ along the x-axis, we can replace s_x by $(2/5)R_g{}^2\omega_g$ for a homogeneous spherical gyroscope of radius R_g and angular velocity ω_g. Finally, we choose $\gamma = 1$ and replace Gm/R^2 by the acceleration of gravity g at the orbit to obtain

$$a_{Sz} + a_{Nz} = \left(\frac{6}{5}\right) g \left(\frac{R_g{}^2\omega_g\omega_0}{c^2}\right) \cos\omega_0 t \tag{10}$$

In the experiment now[2] in progress, $R_g = 0.75$ inch and $\omega_g = 2000$ rad/sec. For a near-earth orbit, the amplitude of the periodic acceleration in Equation 10 is 0.9×10^{-17} cm/sec^2.

As another example, we consider the first of Equations 9, and again choose $\gamma = 1$ so that the amplitude is gk. Now k cannot exceed the uncertainty in the Eötvös-Dicke experiment,[8] which is about 10^{-11}. We thus obtain an upper limit of roughly 10^{-8} cm/sec^2 for the amplitude of the periodic acceleration arising from the nucleonic gauge effect. It is apparent that there is an enormous range for improvement of the Eötvös-Dicke experiment before the expected spin acceleration (Equation 10) can be detected.

Elliptical Orbits. When the unperturbed orbit is an ellipse, there is a periodic acceleration similar to Equation 9. More interestingly, there is also a secular change in velocity, and hence also in position, of the gyroscope with respect to its housing, which arises from $\mathbf{a}_S$ but *not* from $\mathbf{a}_N$. We again choose the orbit to be in the xy-plane, with aphelion on the positive x-axis, semi-major axis a, and eccentricity e; the $\mathbf{s}$ may have any orientation. A straightforward calculation based on Equation 6 then shows that the time integral of the differential acceleration over a period is

$$\oint a_{Sx}\, dt = -\frac{3\pi m\gamma e s_z}{a^2(1-e^2)^2} \qquad \oint a_{Sy}\, dt = 0 \qquad \oint a_{Sz}\, dt = -\frac{3\pi m\gamma e s_x}{a^2(1-e^2)^2} \tag{11}$$

A more important quantity from the experimental point of view is the time average of the secular acceleration, which is obtained by multiplying Equation 11 by $\omega_0/2\pi$, where $\omega_0 = (m/a^3)^{1/2}$. The optimum orbit is that which maximizes this average secular acceleration for a fixed perihelion distance $a(1-e) = R$. The maximum value of the quantity

$$\frac{e}{a^2(1-e^2)^2 a^{3/2}} \propto \frac{e(1-e)^{3/2}}{(1+e)^2}$$

is attained when $e = 0.275$. We want the closest distance of approach of the satellite to the earth to be as small as possible, so that the perihelion distance for the elliptical orbit would be chosen equal to the radius of the circular orbit considered earlier. Then if $g = Gm/R^2$ is the acceleration of gravity at perihelion, the optimum secular acceleration is equal to the amplitude of Equation 10 multiplied by

$$\frac{e}{2(1+e)^2} = 0.085$$

Thus the largest secular acceleration expected with the presently planned gyroscope is about 0.8×10^{-18} cm/sec^2. This corresponds to a displacement of the gyroscope with respect to its housing of $4 \times 10^{-4}\ Y^2$ cm in Y years, provided that the orientation of the orbit does not change appreciably in that time; this is discouragingly small.

It is a pleasure to thank Dr. Phillip K. Chapman for suggesting that a differential acceleration measurement might be included in the gyroscope experiment, and Professor William M. Fairbank for discussion of the experimental possibilities.

References

1. M. Mathisson, *Acta Phys. Polon.*, **6**, 163 (1937).
2. C. W. F. Everitt and W. M. Fairbank, *Proc. Tenth International Conf. on Low Temperature Physics*, Moscow (August 1966).
3. L. I. Schiff, *Proc. Nat. Acad. Sci.*, **46**, 871 (1960).
4. The inclusion of such a measurement was suggested by P. K. Chapman (private communication), who is working on the derivation of this and related matters from the field point of view of W. E. Thirring, *Ann. Phys.* (N.Y.), **16**, 96 (1961).
5. T. D. Lee and C. N. Yang, *Phys. Rev.*, **98**, 1501 (1955).
6. The extension of the satellite-borne gyroscope experiment to include a search for the nucleonic gauge effect was suggested some time ago by C. N. Yang (private communication).
7. R. V. Eötvös, D. Pekár, and E. Fekete, *Ann. Physik*, **68**, 11 (1922).
8. P. G. Roll, R. Krotkov, and R. H. Dicke, *Ann. Phys.* (N.Y.), **26**, 442 (1964).
9. A. Papapetrou, *Proc. Roy. Soc.* (London), **A209**, 248 (1951).
10. V. A. Fock, *Jour. Phys. USSR*, **1**, 81 (1939).
11. F. A. E. Pirani, *Acta. Phys. Polon.*, **15**, 389 (1956).
12. A. Schild, in *Relativity Theory and Astrophysics* (1. Relativity and Cosmology), J. Ehlers, ed., Amer. Math. Soc. (1967), p. 59.
13. E. Corinaldesi and A. Papapetrou, *Proc. Roy. Soc.* (London), **A209**, 259 (1951).
14. L. I. Schiff, in *Relativity Theory and Astrophysics* (1. Relativity and Cosmology), J. Ehlers, ed., Amer. Math. Soc. (1967), p. 106.
15. C. Brans and R. H. Dicke, *Phys. Rev.*, **124**, 925 (1961).
16. D. R. Brill, *Zeits. Naturforschung*, **22a**, 1336 (1967); R. F. O'Connell, *Phys. Rev. Letters.*, **20**, 69 (1968).
17. R. H. Dicke and H. M. Goldenberg, *Phys. Rev. Letters*, **18**, 313 (1967).

Hüseyin Yilmaz
Communications Machines, Inc.

14 Statistical Theory of Fields

I would like to dedicate this article to a great teacher and scientist, and a great human being — Professor Philip M. Morse — wishing him a long, happy life.

1. Introduction

The purpose of this paper is to formulate a new approach to quantum field theory. We deviate from the conventional approach only through the imposition of certain statistical constraints on the field operators, but the formal apparatus of the theory as it depends on these operators remains unaltered. Consequently the forms of the equations of motion, Lagrangian, Feynman graphs etc. are the same. Differences begin to surface only when we try to evaluate the Feynman graphs, because the new theory has modified commutators and propagators. Among other things all the serious divergences will be shown to disappear.

The conditions that we impose shall, of course, entail some new physical assumptions. However, it will be shown in this introduction that they emerge quite naturally from considerations of a relativistic nature. Since these relativistic considerations are also necessary in constructing the theory in a manifestly covariant manner we shall present the essential ideas and their relativistic origins in some detail. Let $\phi_{k_0}(x)$ be a permissible plane-wave solution of the field equations belonging to a four-momentum eigenvalue k_0. If, from this, we should reach a new momentum k by a Lorentz transformation, we must obtain a new eigenfunction, $\phi_k(x)$. Suppressing tensor indices, etc., for simplicity, we may write†

† Note that here the Lorentz transformations act on the momentum variables, rather than on the coordinates.

$$\phi_k(x) = L_{kk_0}\phi_{k_0}(x) \tag{1}$$

Since such solutions can be superposed to give more general solutions, we may also write

$$\phi(x) = \sum_k \alpha_k \phi_k(x) = \sum_k \alpha_k L_{kk_0}\phi_{k_0}(x) \tag{2}$$

We shall assume that in Equation 2 all the permissible Lorentz frames appearing in the conventional theory are included. For example if we write the Lorentz group as $L(\Lambda k + \ell)$ where Λ and ℓ are the homogeneous and the inhomogeneous parts, Λ transforms k_0 into k and ℓ contributes a phase, etc. It is clear that the set of Ω transformations defined by

$$\Omega = \sum_k \alpha_k L_{kk_0} \tag{3}$$

acts as a certain generalization of Lorentz transformations. However such a generalization can be legitimate only if it can be reduced to ordinary Lorentz transformations in the limiting circumstances. For example if all the α- coefficients except a particular $\alpha_{k'}$ become zero, this particular $\alpha_{k'}$ must have unit magnitude. We shall write the general condition of reducibility as

$$\Omega^*\Omega = \sum_{kk'} \alpha_k{}^*\alpha_{k'} L^T_{kk_0} L_{k'k_0} = 1 \tag{4}$$

which, by virtue of $L^T_{kk_0} L_{k'k_0} = L_{k_0k} L_{k'k_0} = \delta_{kk'}$, implies that

$$\sum_k |\alpha_k|^2 = 1 \tag{5}$$

With this condition Equation 2 admits two equivalent interpretations[1]: (a) The generalized frame $\Omega = \sum_k \alpha_k L_{kk_0}$ is made up of all permissible Lorentz frames L_{kk_0} with a probability amplitude α_k being assigned to each L_{kk_0}. (b) The generalized field $\phi(x)$ is made up of all permissible operators $\phi_k(x)$ a probability amplitude α_k being assigned to each $\phi_k(x)$. Note that Equation 5 expresses the normalizability of every permissible field $\phi(x)$ although it does not contain information as to how the normalization itself shall be achieved. In other words the normalizability condition, Equation 5, has to do with the geometric structure of the world as is evident from Equation 3 rather than the dynamics of the individual fields.

To explore further the nature of the restrictions imposed by the above considerations, we may write Equation 5 in the familiar, infinite volume limit[2]

$$\frac{V}{(2\pi)^3}\int d^3k|\alpha_k|^2 = 1 \tag{6}$$

which translates the condition, Equation 5, into

$$|\alpha_k|^2 \leq k^{-3-|\varepsilon|} \qquad \text{for } k \to \infty \tag{7}$$

$$|\alpha_k|^2 \leq k^{-3+|\varepsilon|} \qquad \text{for } k \to 0 \tag{8}$$

where ε is a small nonzero quantity. We shall see in Section 2 that α_k coefficients appear explicitly in the propagators and affect the numerical values of the integrals over intermediate states. The behavior at $k \to \infty$ will be shown to suppress all ultraviolet divergences.

Another, perhaps less obvious consequence of the normalizability condition is that it imposes certain restrictions on the variation $\delta\phi(x)$ of the field operator in a Lagrangian treatment. To see this remember that the Ω transformations are generalized Lorentz transformations and as such the condition, Equation 5, is kinematical in nature rather than dynamical. The essence of a variational treatment, on the other hand, is that one ignores the dynamical constraints and considers all possible variations of the field in their absence while obeying the kinematical constraints.[3] Therefore if $\phi(x)$ satisfies any such conditions, then $\phi'(x) = \phi(x) + \delta\phi(x)$ must also. But we shall find in the next section that the normalization of field operators is to be implemented through their commutators (or anti-commutators). By equating the commutator (anticommutator) $[\phi(x), \phi(x')]_\pm$ and $[\phi'(x), \phi'(x')]_\pm$ and then integrating over $dV \to d\sigma_\mu$ we find

$$\int_\sigma f^\mu(x')[\phi(x), \delta\phi(x')]_\pm \, d\sigma_\mu = 0 \tag{9}$$

where $f^\mu(x)$ is a nonzero vector and " $-$ " and " $+$ " indicate commutator and anticommutator, respectively. As a consequence of normalizability this equation simply says that $\phi(x)$ and $\delta\phi(x)$ are orthogonal in the operator space of $\phi(x)$. Note carefully, however, that Ω transformations do not determine the vector $f^\mu(x)$ nor do they contain any information as to why the normalization is to be achieved through commutators. In Section 2 we shall see that this information comes from the action principle.

2. *The Action Principle*

With these preliminaries in mind, let us now give a closer look at the conventional action principle. For the purpose of economy we shall consider here only the relevant points and for detail the reader is referred to the excellent exposition by Jauch and Rohrlich.[4] Consider as in Reference 4 an action integral

$$W = \int L(\phi, \partial_\mu \phi)\, d^4x \tag{10}$$

and a unitary transformation $U^{-1} = U^*$ such that

$$\phi(x) = U\phi(x_0)U^{-1} \tag{11}$$

If U is an infinitesimal unitary transformation, we may write

$$U = 1 + iF, \qquad F = F^*$$

where F is an infinitesimal quantity called the generator of the infinitesimal transformation U. Equation 11 now translates into

$$\delta\phi(x) = \phi(x_0) - \phi(x) = i[F, \phi(x)] \tag{12}$$

The action principle states that the generator F and the action W are related as

$$\delta W = F(\sigma) - F(\sigma_0) = W_{\sigma\sigma_0} = \int_{\sigma_0}^{\sigma} L\, d^4x \tag{13}$$

where σ and σ_0 are two space-like surfaces. Following Reference 4, one has

$$F(\sigma) = \int_\sigma [(\pi^\mu \partial^\nu \phi - g^{\mu\nu} L)\delta x_\nu - \pi^\mu \delta\phi]\, d\sigma_\mu \tag{14}$$

with $\pi^\mu = \partial L/\partial_\mu \phi$. From Equation 14, it follows that the variation in F owing to $\delta\phi$ arises from the $\pi^\mu\ \delta\phi$ term alone. Introducing this variation of F (owing to $\delta\phi$) back into Equation 12, we find

$$\delta\phi(x) = i\left[\phi(x), \int_\sigma \pi^\mu \delta\phi\right] d\sigma_\mu \tag{15}$$

which is Equation (1.104) of Reference 4 with tensor indices suppressed. By using the mathematical identity as given in Reference 4, Equation (1.105)

$$[A, BC] = [A, B]_\pm C \mp B[A, C]_\pm$$

where "$-$" and "$+$" denote commutator and anticommutator, respectively, we obtain

$$\delta\phi(x) = i \int_\sigma [\phi(x), \pi^\mu(x')]_\pm\, \delta\phi(x')\, d\sigma_\mu \mp i \int_\sigma \pi^\mu(x')[\phi(x), \delta\phi(x')]_\pm\, d\sigma_\mu \tag{16}$$

At this point the conventional action principle leads to a difficulty, because although it implies that the normalization itself will have to be achieved through the commutation properties (the first integral of

Equation 17 is in the form of a normalization), no information regarding the second integral is contained in it. As mentioned in the previous section our Ω transformations provide this information. In fact the second integral vanishes. We obtain the commutator relation

$$\delta\phi(x) = i \int_\sigma [\phi(x), \pi^\mu(x')]_\pm \, \delta\phi(x') \, d\sigma_\mu \tag{17}$$

Since in the conventional theory the second integral of Equation 16 is "assumed" to vanish anyway one might conclude that from here on our theory becomes identical to the conventional theory. This, however, is not the case. The conventional "assumption" does not provide any further information, whereas the invariance of the Lagrangian under the generalized transformation implies further the existence and the normalizability of the α_k coefficients. This is very essential, because without a clear separation between the kinematical constraints and the dynamically allowable variations, there is no meaning to the idea of a variational principle, especially for operators. Such an action principle does not exist.[5] Thus although the two theories shall have identical forms in terms of $\phi(x)$ the differences will become apparent as soon as we expand $\phi(x)$ into some eigenoperators or try to construct propagators from the normalization conditions, Equation 17. The essential difference lies in the fact that the conventional theory equates the commutator (anticommutator) in Equation 17 to a $\delta^\mu(x - x')$ function, whereas we are prevented from doing this because of the α_k coefficients. However, if we let the Ω transformation tend to a particular Lorentz transformation in Equation 17, we obtain the familiar

$$[\pi_k{}^\mu(x), \phi_{k'}(x')]_\pm = i\delta^\mu_{kk'}(x - x') \tag{18}$$

Consequently, the present theory provides the commutation relations

$$[\pi^\mu(x), \phi(x')]_\pm = i\delta^\mu(x, x') = i \sum_k |\alpha_k|^2 \delta^\mu_{kk'}(x - x') \tag{19}$$

With the help of this new $\delta(x, x')$ function, we can now construct the propagators. Thus, in the absence of derivative coupling (*e.g.*, electrodynamics),

$$\Delta(x, x') = \frac{+1}{(2\pi)^4} \int_c d^4k |\alpha_k|^2 \frac{e^{+ik(x-x')}}{+k^2 + m^2} \tag{20}$$

$$S(x, x') = \frac{+1}{(2\pi)^4} \int_c d^4p |\alpha_p|^2 \frac{e^{+ip(x-x')}}{+i\gamma p + m} \tag{21}$$

where the contours c in the complex k_0 plane are to be taken the same way as in the conventional theory. These are now the functions to be used in the evaluation of Feynman graphs. It can be verified that they satisfy equations similar to the conventional ones, namely,

$$\Delta(x, x') = i[\phi(x), \phi(x')]_- \qquad (\partial_\mu \partial^\mu + m^2)\, \Delta(x, x') = \delta(x, x') \quad (22)$$

$$S(x, x') = i[\bar{\psi}(x), \psi(x')]_+ \qquad S(x, x') = (\gamma^\mu \partial_\mu - m)\, \Delta(x, x') \quad (23)$$

and so on. We note that although the conventional functions are recoverable from these by setting $|a_k|^2 = |\alpha_p|^2 = 1$, the conventional theory as a whole is not recoverable because of the normalizibility conditions. Such a limit makes the integrals meaningless.[6] Accordingly, the impression created by the similarity of equations that the present theory might be a generalization of the conventional one is not true. On the other hand, our theory of Ω transformations as presented so far has a serious drawback. This is because it leaves the α_k coefficients arbitrary within the conditions of normalizability. Therefore, unless a way is found to determine the whole set of α_k, the above formulas are useless. In Section 3 we shall remedy this situation.

3. Statistical Interpretation

The problem associated with the arbitrariness of the α_k coefficients is very serious, because within the normalizability conditions one can choose them in such a way that any experimentally verified result of quantum electrodynamics will come out wrong. Conversely, by adjusting the infinitely many α_k parameters in certain ways practically any claimed result, right or wrong, can be reproduced. Clearly, a new statement that determines the whole set of α_k consistently with the few phenomenological quantities entering into the Lagrangian is needed. For this purpose, we put forward the hypothesis that "the α_k coefficients are statistical amplitudes through which the observed characteristics of the elementary particles are recovered." We shall implement this statement in terms of an entropy principle.

Let us consider again $|\alpha_k|^2$ as a probability and define a density operator ρ such that $\rho_{kk'} = \alpha_k \alpha_{k'}^*$,

$$\rho = \alpha\alpha^* \qquad \text{trace}\, \rho = \sum_k |\alpha_k|^2 = 1 \qquad (24)$$

Let us also consider the observed quantities of elementary particles, such as energy E, momentum P_i, spin $s_{\mu\nu}$, etc., as averages over the quantities ω_k, k_i, $s_{\mu\nu}$, etc.

$$E = \sum \alpha_k{}^* \omega_k \alpha_k \qquad P_i = \sum_k \alpha_k{}^* k_i \alpha_k \qquad \sigma_{\mu\nu} = \sum \alpha_k{}^* s_{\mu\nu} \alpha_k \quad (25)$$

We define furthermore, an entropy operator $S_{op} = -\ln \rho$ and construct the average entropy as

$$S = -\sum_k |\alpha_k|^2 \ln |\alpha_k|^2 \quad (26)$$

These formulas can also be written as $E = \text{trace}(\rho\omega)$, $S = -\text{trace}\,(\rho \ln \rho)$, $\sigma_{\mu\nu} = \text{trace}\,(\rho s^{\mu\nu})$, etc. Since entropy can be interpreted as the "lack of information," the statistical hypothesis may now be stated as follows: "Subject to the known values of energy, momentum, spin ... etc., the entropy S of the field is as large as possible." The technique of the Lagrangian multipliers then leads to the distribution

$$\rho = |A|^2 e^{-2\zeta_\mu k^\mu - \lambda_{\mu\nu} s^{\mu\nu} - \cdots} \quad (27)$$

which is Boltzmann in character because α_k coefficients are ordinary numbers.† By splitting ρ into α_k, $\alpha_k{}^*$ for the purpose of introducing the phase variables, we may write this result as

$$\alpha_k = Ae^{-(\zeta_\mu + i x_\mu)k^\mu - 1/2(\lambda_{\mu\nu} - i\theta_{\mu\nu})s^{\mu\nu} - \cdots} \quad (28)$$

and combine the spatial variables in Equations 20 and 21 with $\alpha_\kappa{}^* \alpha_{\kappa'}$. In this manner the whole set of α_κ can be determined and high momentum limit $k \to \infty$ exponentially decays. It is clear that this will leave all the low-energy consequences of the conventional quantum electrodynamics practically intact and modify the high-energy region in a drastic manner. In particular, all the ultraviolet-divergent integrals will now converge. Although not absolutely necessary, a renormalization will also be possible. For the purpose of completeness, we emphasize again that the renormalization here is possible if and only if $|\alpha_k|^2$ satisfies the normalizability condition. (For example, if $\zeta_0 > 0$.) If $|\alpha_k|^2 = \text{const}$, the theory does not exist and the renormalization becomes pointless.

The form of Equation 28 is suggestive of complex Lorentz transformations. It is in fact possible to reach the present theory equivalently through the following steps: Consider the fields entering the conventional Lagrangian with complex arguments. The field equations obtained

† Note: The Boltzmann character of Equation 27 should not be confused with the statistics of the collections of real particles which obey either Bose-Einstein or Fermi-Dirac statistics. The latter are dynamical in origin. This is clear from the fact that if $\phi(x)$ were not an operator, the Equation 27 would still be true, due to the normalizability condition and the statistical interpretation. The operator nature of $\phi(x)$ is required by the Action principle.

therefrom will have double the solutions as before with decaying and growing exponentials appended to each. Discard the ones with growing exponentials since they become catastrophic for $k \to \infty$. Use the decaying exponentials as "statistical" parameters through which observed characteristics of single particle states are recovered. The result will be the same as the present theory. In this process the vanishing of the second integral of Equation 16 will not be easy to see, but it can be reiterated through the key word "statistical," above. Although it taxes intuition rather heavily, this scheme is workable and it may sometimes be useful as a shortcut. Another interesting observation is that the real part of the exponent in Equation 27 is (up to an additive constant) the entropy operator. Therefore this section could also be conceived to emerge from an extended variational principle in which action is generalized into a complex quantity $\Gamma + iW$ where Γ is half of the entropy operator. This at least formal unification of the two major extremal principles of physics appears, to us, quite satisfying.

We conclude this section by emphasizing that although the present theory is able to remove the ultraviolet divergences the infrared divergence still survives. However, some chronic headaches of field theory, such as the ghost states and Haag's theorem, no longer pose problems. (The LSZ requirement that vertex part vanish for $k^2 \to 0$ is met and the singular $\delta(x - x')$ used by Haag's theorem is rejected.[7] As to the infrared divergence itself (which is quite harmless here as is in the conventional theory), we may remark that it arises from the infinite volume $V \to \infty$ limit of the normalizability conditions. If the universe is somehow finite, the problem does not exist. Here a combination of general relativity and zero point energy (which is conjectured to be quite physical in the next section) can remove the difficulty consistently by providing a finite (although perhaps unbounded) universe. The point we are trying to make is that the removal of the infrared divergence through impositions of a dynamical nature (such as a non vanishing photon mass) would be quite inappropriate since the problem really arises from kinematical considerations. At either limit, then, the normalizability conditions point toward the inadequacy of the special relativistic kinematics represented by the ordinary Lorentz transformations.

4. *Quantum electrodynamics, Test Possibilities*

Let us now consider more specifically the quantum electrodynamics. We have the field operators

$$A_\mu(x) = N^{\frac{1}{2}} \sum_k e^{-\zeta k} A_{\mu k}(x) \tag{29}$$

$$\psi(x) = N^{\frac{1}{2}} \sum_p e^{-\zeta p} \psi_p(x) \tag{30}$$

and the corresponding propagators

$$D(x, x') = \frac{N}{(2\pi)^4} \int_c d^4k e^{-2\zeta k} \frac{e^{ik(x-x')}}{k^2} \tag{31}$$

$$S(x, x') = \frac{N'}{(2\pi)^4} \int_c d^4p e^{-2\zeta p} \frac{e^{ip(x-x')}}{i\gamma p + M} \tag{32}$$

For $\zeta \to 0$ these expressions reduce to the usual wave-function renormalizations with $N \to Z_3^{-1}$, $N' \to Z_2^{-1}$. Thus although the limit $\zeta \to 0$ is not legitimate (due to Equation 5) our theory, when supplemented with a mass counter-term, recovers the renormalized quantum electrodynamics as a limit. A less objectionable procedure is to keep ζ finite but let $k \to 0$. Then the conventional theory may be regarded as a legitimate low-energy approximation to the present theory.

If the statistical theory of fields presented in this paper is in anyway valid, the effect of the ζ_μ parameters ought to be present somewhere in the experiment. The most natural place to look for such effects is, of course, the high-energy region. At extreme high energies, the theory implies that cross sections for production and scattering processes diminish owing to the exponentially decaying nature of the α coefficients. In fact this behavior is fairly simple to evaluate, because our statistical factor acts there as a relativistic cut-off for energies $E \gtrsim \zeta_0^{-1}$ (in the rest frame, the space part ζ_i disappears because of the momentum averaging). Present high-energy experiments on the photoproduction of μ-mesons, wide-angle scattering, electron proton scattering, and the electroproduction of μ-mesons, taken together, indicate that ζ_0 should be comparable or smaller than 5.10^{-15} cm. Experiments on the g-factor of μ-mesons give 2.10^{-15} cm. or less. In the latter case, a discrepancy of about twice the estimated error seems to be present.[8] No indication as to the actual value of ζ_0 is anywhere available. Other accurately known results such as the Lamb shift, anomolous magnetic moment, etc., are essentially low-energy phenomena that are insensitive to the value of ζ_0^{-1} as long as it is sufficiently high.

Under these circumstances, we must turn to indirect clues depending on the structure of the theory rather than the actual value of ζ_0. For example in the usual theory, the zero point energy E_{vac} diverges and is discarded from the theory for that reason. Since in our theory, this is now finite, we are in a better position to interpret it physically. Therefore, if the zero-point energy should experimentally be verifiable, it may be counted as a good mark for the new theory. Experimental proof of the existence of vacuum energy is in principle possible by measuring the forces it presumably sets up in polarizable media and between two extremely close sheets of neutral conductors as it was pointed out by Casimir and later reported to be present upon observation.[9] The existence of vacuum energy is very important, because it is independent of the interaction term and hence kinematical in origin. In fact there are reasons to suggest that ξ_0^{-1} might be equal to that energy. Other effects that might be associated with the presence of the vacuum energy, such as possible low-energy contribution to some atomic effect, a systematic shift of $|\alpha_k|^2\omega_k/2$ in the Planck distribution, etc., might be informative to study. Possible origin of $\pi^{\pm}$, π^0 mass difference as an electromagnetic self energy,[10] phenomena having to do with cosmic ray showers, the interiors of stars, size of the universe etc., would again be welcome considerations.

As to the nature of ζ_0, we offer the following heuristic argument: If such a quantity exists, it cannot come from the mass, charge, or spin values of individual particles. Because according to the suggestive interpretation of Section 3, the inverse of this quantity is essentially the average rest energy of the ensemble of quanta making up the field $\phi(x)$. Now from the equipartition of energy which surprisingly applies here (statistics is formally Boltzmann), this is likely to be the same for all fields. In other words ζ_0 ought to be considered, at least provisionally, as a general kinematical property of nature.

References

1. H. Yilmaz, "On the Theory of S-Matrix," p. 9, Report of Arthur D. Little, Inc. (1963) (Unpublished).
2. G. Barton, *Introduction to Advanced Field Theory*, p. 3, Interscience Publishers, Inc., New York (1963).
3. H. Yilmaz, *Introduction to the Theory of Relativity*, p. 48, Blaisdel Publishing Co., New York (1965).
4. J. M. Jauch and R. Rohrlich, *Theory of Photons and Electrons*, pp. 22, 23, Addison-Wesley Publishing Co., Inc., Reading, Mass. (1955).
5. R. Rohrlich, *Classical Charged Particles*, p. 238, Addison-Wesley Publishing Co., Inc., Reading, Mass. (1965).
6. L. Streit, in *Quantum Electrodynamics*, Paul Urban, ed., p. 14, Springer-Verlag, Inc., New York (1965).
7. G. Barton, *loc. cit.*, pp. 71, 156.
8. R. Weinstein, Private communication. See also J. D. Björken, in "Quantum Electrodynamics," Reference 6.

9. H. B. G. Casimir, "Proc. Nederlands aka Wetenschappen," Amsterdam, **60**, 793 (1948). Sparnaay, M. Phisica, **24**, 751 (1958).
10. G. Källén, in *Quantum Field Theory*, Proceedings of the Twelfth Conference on Physics (Solvay Conference) R. Stoops, ed., p. 150, Interscience Publishers, Inc., New York (1961). See also articles by W. Heitler and R. P. Feynman in the same volume.

W. Shockley
Bell Telephone Laboratories and Stanford University
K. K. Thornber
Stanford University

15 The "Hidden Momentum" Equivalent to Magnetic Charges for a Bound-State Dirac Electron*§

Recent publications[1,2] review the linear momentum $\mathbf{G}_h = \boldsymbol{\mu} \times \mathbf{E}/c$ hidden in a stationary, electrically neutral current loop of magnetic moment $\boldsymbol{\mu}$ in an electric field $\mathbf{E}$.‡ $\mathbf{G}_h$ is very small: [A proton of mass M and momentum G_h due to $\mu = e\hbar/2mc$ (a Bohr magneton) in a field of $E = e/a_0{}^2 = e^5m^2/\hbar^4 \approx 5 \times 10^9$ V/cm has energy $(m/4M)(e^2/2a_0)$

* This essay first appeared in *Phys. Letters*, **27A**, No. 8, 534–535 (1968).
§ Supported in part by the Advanced Projects Research Agency through the Center for Materials Research at Stanford University, and by the Joint Services Electronics Program, Office of Naval Research, Contract No. Nonr-225(83).

‡ The earliest publication predicting the physical reality of the magnetic dual of the Lorentz force appears to be Penfield[4]; however this reference did not stress the conclusion that this new term should have directly measurable, new physical consequences as did Costa De Beauregard[5] and later but independently Shockley and James[3].

Production of $\mathbf{G}_h$ by relativistic mass changes due to kinetic energy was discussed in 1963 lecture notes by Haus and Penfield but not definitively published[6] until 1967, and was apparently rediscovered later by Coleman and Van Vleck.[7] The mass-change model and a mechanical stress model were shown equivalent to a magnetic charge model by Shockley and James.[3] Additional papers treat other models,[1] hidden momentum in a Dirac electron wave packet,[8] and applications to the relativistic energy momentum tensor.[2]

$\times (\alpha = e^2/\hbar c)^4 \approx 5 \times 10^{-12}$ eV. R. P. James' very sensitive measurements of smaller momenta in ferrites appear, as discussed in Reference 2, adequate to reject the related Minkowski $\mathbf{D} \times \mathbf{B}$ momentum formula.] $\mathbf{G}_h$ arises naturally from Maxwell's equations applied to a Dirac electron of charge $q = -|e|$ bound to a stationary non-quantum nucleus with potential energy $q\phi_n = -q^2/|\mathbf{r}|$ in the presence of slowly varying classical "driving" fields $\mathbf{E}_d$ and $\mathbf{H}_d$ derived from potentials ϕ_d and $\mathbf{A}_d$. A consequence of $\mathbf{G}_h$ is that the rate of change of momentum $\mathbf{G}_b$ of bodily motion of the atom in its rest frame is

$$\begin{aligned}\dot{\mathbf{G}}_b &= (\boldsymbol{\eta} \cdot \nabla)\mathbf{E}_d + (\boldsymbol{\mu} \cdot \nabla)\mathbf{H}_d + (\dot{\boldsymbol{\eta}} \times \mathbf{H}_d + \mathbf{E}_d \times \dot{\boldsymbol{\mu}}(1/c) \\ &= (\boldsymbol{\eta} \cdot \nabla)\mathbf{E}_d + (\dot{\boldsymbol{\eta}} \times \mathbf{H}_d/c) + \nabla(\boldsymbol{\mu} \cdot \mathbf{H}_d) - \dot{\mathbf{G}}_h \qquad (1)\end{aligned}$$

where $\boldsymbol{\eta}$ and $\boldsymbol{\mu}$ are the expectation values for electric and magnetic dipole moments for the atom. These moments change as the electron's wave function varies adiabatically with ϕ_d and $\mathbf{A}_d$ in an eigenstate of energy $\mathscr{E}(t)$. The first expression in Equation 1 is exactly that expected for $\boldsymbol{\eta}$ and $\boldsymbol{\mu}$ produced by electric and magnetic charges without any contributions from amperian current loops and the second expression equals the first from Maxwell's equations. Dirac's equation gives for the electron's kinetic momentum

$$\begin{aligned}\langle \mathbf{p} - (q/c)\mathbf{A} \rangle &= \langle [\mathscr{H} c\boldsymbol{\alpha} + c\boldsymbol{\alpha}\mathscr{H} - 2q\phi c\boldsymbol{\alpha}]/2c^2 \rangle \\ &= \int [(\mathscr{E} - q\phi)/c^2]\psi^\dagger\, c\boldsymbol{\alpha}\psi \, dV \\ &= (-1/c^2) \int \phi \mathbf{J}_e \, dV \equiv \mathbf{G}_{ha} + \mathbf{G}_h \qquad (2)\end{aligned}$$

in which $\phi = \phi_n + \phi_d$ and $\mathbf{A} = \mathbf{A}_d$, the "hidden momentum" $\mathbf{G}_h \equiv -\int \phi_d \mathbf{J}_e \, dV/c^2 = -\int \mathbf{r}(\mathbf{E}_d \cdot \mathbf{J}_e) \, dV/c^2 = \boldsymbol{\mu} \times \mathbf{E}_d/c^2$, $\mathbf{J}_e \equiv q\psi^\dagger c\boldsymbol{\alpha}\psi$, $\boldsymbol{\mu} = (\frac{1}{2}c) \int \mathbf{r} \times \mathbf{J}_e \, dV$, and $\mathbf{G}_{ha} \equiv -\int \phi_n \mathbf{J}_e \, dV/c^2$. Eq. (2) correctly gives the relativistic internal mass transport needed[3] to compensate for the divergence of Poynting's vector. The quantity $(\mathscr{E} - q\phi)/c^2 = (mc^2 + T)/c^2$ is to be interpreted as the rest mass plus kinetic-energy T mass that is carried by the probability current density $\mathbf{J}_e/q$ to produce the "hidden momentum" of the electron in its eigenstate.

To establish Equation 1, we assume initially that zero force acts on the stationary nucleus: $\mathbf{F}_n \equiv \mathbf{F}_{nd} + \mathbf{F}_{ne} = -q(\mathbf{E}_{do} + \mathbf{E}_{no}) = 0$, where subscript zero means evaluated at the nucleus ($\mathbf{r} = 0$) and $\mathbf{J}_e$ and $\rho_e \equiv q\psi^\dagger\psi$ produce $\mathbf{E}_e$, $\mathbf{H}_e$. Dirac's equation also gives

$$\dot{\mathbf{G}}_{ha} + \dot{\mathbf{G}}_h \equiv (d/dt)\langle \mathbf{p} - (q/c)\mathbf{A}\rangle$$
$$= \int [\rho_e(\mathbf{E}_d + \mathbf{E}_n) + (\mathbf{J}_e \times \mathbf{H}_d/c)]\, dV \equiv \mathbf{F}_{ed} + \mathbf{F}_{en} \quad (3)$$

where $\mathbf{H}_n = 0$ does not appear since the nucleus is stationary, and $\mathbf{F}_{ed}$ and $\mathbf{F}_{en}$ are the respective parts of the integral of force density. Vanishing Maxwell stress tensor integrals over a large sphere and Gauss's theorem lead to

$$0 = \mathbf{F}_{en} + \mathbf{F}_{ne} + \dot{\mathbf{G}}_{p(ne)} = \mathbf{F}_{en} - \mathbf{F}_{nd} - \dot{\mathbf{G}}_{ha} \quad (4)$$

where $\mathbf{F}_n = 0$ gives $\mathbf{F}_{ne} = -\mathbf{F}_{nd}$ and $\mathbf{G}_{p(ne)} \equiv \int (\mathbf{E}_n \times \mathbf{H}_e/4\pi c)\, dV = -\mathbf{G}_{ha}$ follows from manipulations equivalent to the Shockley–James[3] conclusion that total "flow" momentum vanishes for a stationary (i.e., adiabatic limit) system.

Combining eqs. (3) and (4) leads to $\dot{\mathbf{G}}_h = \mathbf{F}_{ed} + \mathbf{F}_{nd} \equiv \mathbf{F}_{ad}$ signifying that the $\mathbf{F}_n = 0$ condition requires that the "driving force" $\mathbf{F}_{ad}$ on the atom is absorbed by the hidden momentum $\mathbf{G}_h$. Evidently any excess of $\mathbf{F}_{ad}$ over $\dot{\mathbf{G}}_h$ will result in rate of change in bodily momentum given by $\dot{\mathbf{G}}_b = \mathbf{F}_{ad} - \dot{\mathbf{G}}_h$. Equation 1 results from evaluating $\mathbf{F}_{ad}$ by adding $\mathbf{F}_{ed}$ of eq. (3) to $\mathbf{F}_{nd} = -q\mathbf{E}_{d0} = \int \rho_n \mathbf{E}_d\, dV$. For macroscopic driving fields, $\mathbf{E}_d$ and $\mathbf{H}_d$ may be linearly approximated by $\mathbf{E}_{d0} + (\mathbf{r}\cdot\nabla)\mathbf{E}_d$ and $\mathbf{H}_{d0} + (\mathbf{r}\cdot\nabla)\mathbf{H}_d$ for $\mathbf{F}_{ed}$ of Equation 3; the result is the second form of Equation 1. [$\boldsymbol{\eta} \equiv \int (\rho_e + \rho_n)\mathbf{r}\, dV = \int \rho_e \mathbf{r}\, dV$ since ρ_n gives $-q$ at $\mathbf{r} = 0$. In $\int \mathbf{J}_e \times [\mathbf{H}_{d0} + (\mathbf{r}\cdot\nabla)\mathbf{H}_d]\, dV$, the relationships $\dot{\boldsymbol{\eta}} = \int \mathbf{J}_e\, dV$ and $0 = \int (xJ_{ey} + yJ_{ex})\, dV$, etc. are used.]

The authors acknowledge encouragement from colleagues at Bell Telephone Laboratories and Stanford University.

References

1. H. A. Haus and P. Penfield, Jr., *Phys. Letters*, **26A**, 412 (1968).
2. W. Shockley, *Proc. Nat. Acad. Sciences*, **60**, 807 (1968).
3. W. Shockley and R. P. James, *Phys. Rev. Letters*, **18**, 876 (1967); *Science*, **156** 542 (1967).
4. P. Penfield, Jr., *IEEE Convention Record*, Part 7, pp. 152–154, March 21–25 (1966).
5. O. Costa De Beauregard, *Phys. Letters*, **24A**, 177 (1967).
6. P. Penfield, Jr., and H. A. Haus, *Electrodynamics of Moving Media*, M.I.T. Press, Cambridge, Mass. (1967).
7. S. Coleman and J. H. Van Vleck, *Phys. Rev.*, Aug. 25, 1968.
8. W. Shockley, *Phys. Rev. Letters*, **20**, 343 (1968).

Pearl R. Lichtenstein
Rensselaer Polytechnic Institute

16 Diffusion Calculations and Solar Proton Spectra

The description of solar proton events in terms of a simple diffusion process[1,2] has had considerable success. Specific models have been fitted in detail to satellite observations, for example by Krimigis[3] and Rinehart.[4] But while in some respects the models fit the data well, many departures from the theory have not been explained and cast doubt on the applicability of the diffusion model. In this paper, we investigate the effects of treating the spectral energy distribution of the solar protons, rather than dealing only with monoenergetic particles. We will show that inclusion of the spectrum can account for some of the apparent disagreements with the diffusion relations.

We select as a specific model for our calculations the following simple conditions:

1. **An instantaneous burst of solar protons is assumed to occur on the sun at time $t = 0$, with a differential spectral energy distribution given by**

$$\frac{dN}{dK} = N_0 K^{-\gamma} \qquad \text{(power law)}$$

or

$$\frac{dN}{dK} = N_0' e^{-K/K'} \qquad \text{(exponential)}$$

where N is the number of particles, K the kinetic energy, γ the spectral index, and K' the e-folding energy.

2. These protons diffuse by a random-walk process through a medium having spherical symmetry. The diffusion coefficient is taken to be a linear function of particle velocity $\beta = v/c$ and distance r with constant coefficient m. The parameter m has the dimension of velocity and is expressed in astronomical units/hour,

$$D = m\beta r.$$

3. No losses occur. Particle energies remain unchanged in the scattering process.
4. All calculations are for $r = 1$ A.U.

Under these conditions the intensity J in the monoenergetic case may be written[4]

$$J \approx \frac{1}{(m\beta t)^3} e^{-1/m\beta t}$$

Thus $\log(Jt^3)$ should be proportional to $1/t$. Further, the intensity is maximum at a time t_{max} which varies as $1/\beta$, so that the lower the energy the later the maximum. At long times after the burst, the intensity at a fixed location decays as $1/t^3$.

In extending the solution to include the spectral energy distribution, we limit our consideration to energies below about 100 MeV, where nonrelativistic relations between velocity and kinetic energy are sufficiently accurate. The extension to relativistic energies is straightforward; the limitation is made partly for simplicity, but mainly because the effects of the spectrum are large only at low energies.

Under these assumptions, the intensity of protons in a given energy range K_1 to K_2, or velocity range β_1 to β_2, may be expressed in the form

$$J_\gamma(mt) \approx \frac{1}{(mt)^3} \int_{\beta_1}^{\beta_2} \beta^{-2\gamma-1} e^{-(1/\beta mt)}\, d\beta$$

or

$$J_{\beta'}(mt) \approx \frac{1}{(mt)^3} \int_{\beta_1}^{\beta_2} \beta^{-1} e^{-\beta^2/\beta'^2} e^{-(1/\beta mt)}\, d\beta$$

These expressions have been integrated numerically for a series of values of the parameter γ or β' and over a number of energy ranges in the 1 to 100 MeV interval. Notice that time enters these expressions only in the combination (mt). In any specific case, the value of m is chosen to make the intensity maximum occur at the desired time. The results of the integration are presented in the form of graphs or tables of log J vs log mt. Since only relative values of J have been calculated, the coordinates of both axes are arbitrary to within an additive constant. Calculation of absolute values of J would require knowledge of the actual number of protons emitted. With any particular value of γ or β', however, the relative intensities in a pair of energy ranges, for example, are fixed, as well as the ratios of the times of maximum.

In order to illustrate the effect of the integration over energy, we have calculated intensity *vs* time curves for two energy ranges designated as *low* and *high*. The *low* range is taken to cover approximately 1.5 to 20 MeV or β from 0.06 to 0.20; the *high* range corresponds to K about 40 to 100 MeV, or β, 0.30 to 0.42. The values of β corresponding to the mean energies are 0.15 and 0.35, respectively. These ranges have been selected for comparison with observations such as those made with the satellite Injun I.

The results are presented in Figure 1, where parts a, b, and c represent power-law spectra of index $\gamma = 2.5$, 4.0, and 5.5, respectively, and part d

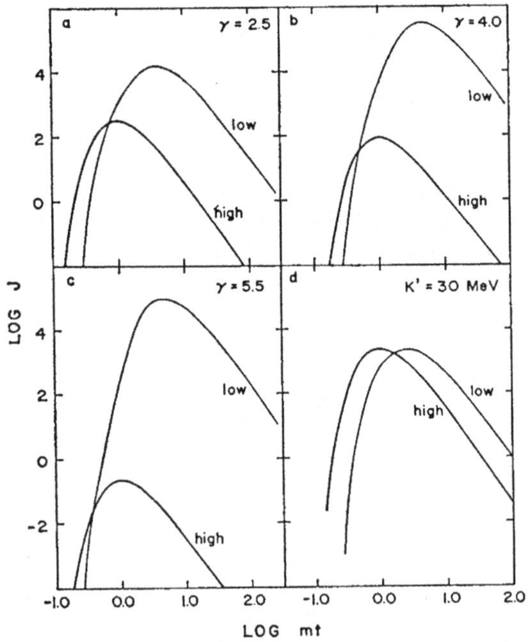

Figure 1. Calculated intensity *vs* time curves: Parts a, b, c are for the power-law spectrum, index γ. Part d is for exponential spectrum, *e*-folding energy K'. *High* energy, $\beta = 0.30$ to 0.42. *Low* energy, $\beta = 0.06$ to 0.20.

is for the case of an exponential spectrum of *e*-folding energy 30 MeV. We note that while on these log-log plots the shapes of the curves differ only slightly from case to case, the ratio of maximum intensities and the times of maximum change significantly. Notice also that a good approximation to the total intensity in the energy range 1.5 to 100 MeV is obtained by summing the two curves in each case. These resultant curves differ quite noticeably, especially in the breadth of the maximum, which decreases as the spectrum steepens.

As a further illustration of the effects of including the particle spectrum, Columns 1 and 2 of Table 1 compare the ratio of *low* to *high*

Table 1. Comparison of Maximum Intensities and Times of Maximum Calculated with Mean Energies and Spectral Distributions

Power Law	$(J_{low}/J_{high})_{max}$		$\Delta \log t_{max}$	
γ	Mono-energetic	Inte-grated	Mono-energetic	Inte-grated
2.5	30	48	0.37	0.60
4.0	380	3.6×10^3	0.37	0.65
5.5	4.8×10^3	4.0×10^5	0.37	0.70
Exponential K', MeV				
30	2.1	0.9	0.37	0.37
20	5.1	2.1	0.37	0.40

energy maximum intensities as calculated in the monoenergetic approximation and exactly, by numerical integration. Columns 3 and 4 give a similar comparison of the ratio of the times of maximum of the low- and high-energy components. The error in the time of maximum increases as the spectrum softens, since the effective mean energy deviates more and more from the normal value.

Consider how we would interpret a fairly soft solar proton event, of spectral index 4.0, observed just outside the earth's atmosphere by a satellite carrying two detectors corresponding to our *low* and *high* energy ranges. The record obtained would be that of Figure 1b. Let us proceed to analyze these curves on the basis of the mean energy monoenergetic results. First, from the ratio of maximum intensities, 3.6×10^3, we conclude that the spectral index of these solar protons is about 5, a considerable overestimate. Next, we plot log t_{max} vs log $\bar{\beta}$, and find the slope to be -1.8, rather than -1.0, as expected. (This result should be compared with Webber[5], Figure 27-10, where such curves for a number of events are plotted. Many of these have slopes near -2.)

Last, we look for agreement with the diffusion model by plotting $\log(Jt^3)$ vs $1/t$ (Figure 2). The numerical value $m = 0.15$ has been

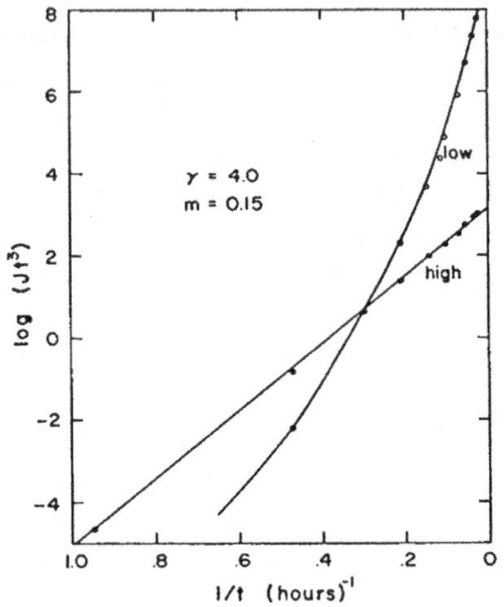

Figure 2. Log(Jt^3) versus $1/t$, calculated with power-law spectrum of index 4.0.

chosen arbitrarily; this choice determines the scale factor but not the shape of the curves. Note that while the high-energy values lie along a straight line, the low-energy points deviate strongly from linearity. Such a deviation from linearity has been pointed out by Krimigis and Van Allen[6] (Figure 11) for the low energy intensities of the February 5, 1965 event as observed by Mariner 4.

Let us finally compare our theoretical results with a real event, that of July 18, 1961, which was extensively observed by Injun I as reported by Maehlum and O'Brien[7] and Pieper[8] *et al.* In this event, the observed ratio of maximum intensities of the low- to the high-energy components was about 1.3, while $\Delta \log t_{max}$ was about 0.3. Clearly none of the power-law cases of Table 1 can fit these observations; a spectral index much smaller than 2.5 would be indicated by the small intensity ratio. An approximate fit is possible, however, with an exponential spectrum of e-folding energy between 20 and 30 MeV, as seen from Table I and Figure 1d. From the observed time of maximum of the high energy protons, at 7 to 8 hours after the initiating event, we find $\log m \simeq -0.87$, or $m \simeq 1.3$. A replot of the high-energy intensity *vs* time curve of Figure 4 in Maehlum and O'Brien[7], results in a good fit except for the earliest point, where the intensity is appreciably larger than the computed value. On the other hand, the observed low-energy curve is considerably narrower than that of Figure 1d, suggesting that the true spectrum in the low energy region was much steeper than the assumed exponential.

Conclusions

It has been demonstrated that neglect of the spectral energy distribution of solar protons can lead to serious discrepancies between observed and predicted diffusion results, for even the simplest models. These discrepancies are most marked at low energies and when the sensitivity range of the detector exceeds a few MeV. Before more complicated models are adopted, therefore, allowance must be made for the effects of the proton spectrum. When such corrections are included, improved values of solar burst and diffusion parameters are obtained.

Acknowledgments

I wish to thank Dr. R. Stoeckly for his assistance in the computations. This work was supported by NSF Grant GA 353.

References

1. Eugene N. Parker, *Interplanetary Dynamical Processes*, Interscience Publishers, Inc., New York (1963).
2. G. C. Reid, "A Diffusive Model for the Initial Phase of a Solar Proton Event," *J. Geophys. Res.*, **69**, 2659–2667 (1964).
3. S. M. Krimigis, "Interplanetary Diffusion Model for the Time Behavior of Intensity in a Solar Cosmic Ray Event," *J. Geophys. Res.*, **70**, 2943–2960 (1965).
4. Marion C. Rinehart, "Diffusion of Solar Protons," *J. Geophys. Res.*, **72**, 3459–3470 (1967).
5. William R. Webber, "A Review of Solar Cosmic Ray Events," AAS-NASA Symposium on the Physics of Solar Flares, *NASA SP-50*, 215–255 (1963).
6. S. M. Krimigis and J. A. Van Allen, "Observations of the February 5–12, 1965 Solar Particle Event with Mariner 4 and Injun 4," *J. Geophys. Res.*, **72**, 4471–4486 (1967).
7. B. Maehlum and B. J. O'Brien, "Solar Cosmic Rays of July 1961 and Their Ionospheric Effects," *J. Geophys. Res.*, **67**, 3269–3279 (1962).
8. G. F. Pieper, A. J. Zmuda, C. O. Bostrom, and B. J. O'Brien, "Solar Protons and Magnetic Storms in July 1961," *J. Geophys. Res.*, **67**, 4959–4981 (1962).

Herman Feshbach
Massachusetts Institute of Technology

17 Overlapping Potentials and Multiple Scattering Theory*

1. Introduction

In this note we shall be concerned with the interpretation of recent experiments involving the collision of protons ranging in energy from 600 MeV to 20 GeV with complex nuclei.[1] It is expected that these and similar future investigations will probe the small-scale properties of nuclear wave functions. The theoretical discussion has been based on impact parameter multiple scattering theory. Considerable success[2] has been obtained so that it becomes of some importance to examine the approximations involved. In this note we examine some of these approximations paying most attention to the assumption that the scattering from a given nucleon of the target nucleus is independent of the position of the other target nucleons.

It is probably unnecessary to add that these considerations may be useful for multiple scattering theory in a wide range of physical systems in addition to the nucleus.

2. The Approximation

We shall begin with a review of the fundamental assumptions which are made in the semiclassical high-energy approximation, limiting the discussion for the most part to the approximation for the elastic scattering amplitude.

* This work is supported in part through funds provided by the Atomic Energy Commission under Contract AT (30-1) 2098.

Assumption 1. It is assumed that the target nucleons making up the target nucleus are stationary during the passage of the incident nucleon through the nucleus. This assumption is based on the large value for the ratio of the periods of target nuclear motion to the transit time for the projectile. It is valid if there are no close collisions between the incident and target nucleon for then the state of motion of the target nucleon is not substantially changed in the course of the collision. We shall not discuss this approximation. Suffice it to say it is most accurate for small-angle scattering of the incident nucleon by the target nucleon. The critical angle is of the order of $(1/kr_0)$ where k is the wave number for the incident nucleon and r_0 is the average distance between the target nucleus nucleons.

Once Assumption 1 is made it becomes possible to consider the progress of the projectile through the nucleus as a set of multiple scatterings from the target nucleons. The resulting amplitude is then averaged over the possible positions of the latter. It is at this point that the wave function for the target system makes its appearance. Assumptions are made with regard to both the scattering by the individual nucleons and by the ensemble.

Assumption 2. It is assumed that the scattering amplitude generated by a fixed nucleon can be written:

$$f(\mathbf{q}, k) = \frac{ik}{2\pi} \int e^{i\mathbf{q}\cdot\boldsymbol{\rho}}\Gamma(\boldsymbol{\rho}, k)\, d\boldsymbol{\rho} \tag{1}$$

Here $\boldsymbol{\rho}$ is a vector perpendicular to the initial direction $\mathbf{k}_i$ of incident beam while

$$\mathbf{q} \equiv \mathbf{k}_f - \mathbf{k}_i \tag{2}$$

where the scattering angle θ is defined by

$$\cos\theta = \hat{\mathbf{k}}_f \cdot \hat{\mathbf{k}}_f \tag{3}$$

For simplicity we shall omit the spin and isospin dependence of $f(\mathbf{q}, k)$. The important elements to be noticed in this representation are (1) the absence of any dependence on the longitudinal momentum $\mathbf{q}\cdot\hat{\mathbf{k}}$ and (2) the absence of any dependence on the vectors $\mathbf{k}_f$ and $\mathbf{k}_i$ in the factor Γ. Starting with Equation 1, a further approximation is made in using the Fourier integral theorem to obtain

$$\Gamma(\boldsymbol{\rho}, k) = \frac{1}{2\pi ik} \int e^{-i\mathbf{q}\cdot\boldsymbol{\rho}} f(\mathbf{q}, k)\, d\mathbf{q}_T \tag{4}$$

where $\mathbf{q}_T$ is the component of $\mathbf{q}$ perpendicular to $\mathbf{k}_i$.

Equation 4 assumes knowledge of the scattering amplitude for nonphysical values of θ since k is fixed. However, if f drops off rapidly with increasing angle, inversion Equation 4 is well defined. The conditions in this case are generally satisfied when Equation 1 is valid.

We finally note that unitarity in the high-energy limit requires that Γ take on the form

$$\Gamma = 1 - e^{i\chi} \tag{5}$$

where χ is real in the absence of absorption, and χ is interpreted as the phase shift for a ray with impact parameter ρ.

Assumption 3. We now consider the nucleon-nucleus scattering. It is assumed that the scattering amplitude operator F, for this system, can also be written in the form of Equation 1 so that

$$F = \frac{ik}{2\pi} \int e^{i\mathbf{q}\cdot\boldsymbol{\rho}}[1 - e^{i\chi(\boldsymbol{\rho},k)}]\, d\boldsymbol{\rho} \tag{6}$$

It is assumed that $\chi(\boldsymbol{\rho}, k)$ is composed *additively* of the phase shifts generated by each nucleon in the nucleus:

$$\chi(\boldsymbol{\rho}, k) = \sum_i \chi(\boldsymbol{\rho} - \boldsymbol{\rho}_i, k) \equiv \sum \chi_i \tag{7}$$

where $\boldsymbol{\rho}_i$ gives the transverse coordinates of the ith nucleon. The operator F is a function of $\boldsymbol{\rho}_i$. Its expectation value with respect to the target nuclear ground state yields the scattering amplitude.

From Equations 4 and 5, we find

$$e^{i\chi_i} = 1 - \Gamma_i \tag{8}$$

$$= 1 - \frac{1}{2\pi ik} \int e^{-i\mathbf{q}\cdot\boldsymbol{\rho}} f(\mathbf{q}, k; \boldsymbol{\rho}_i)\, d\mathbf{q} \tag{9}$$

The final expression for the scattering amplitude F is

$$F(\mathbf{q}k; \boldsymbol{\rho}_1 \cdots \boldsymbol{\rho}_A) = \frac{ik}{2\pi} \int e^{i\mathbf{q}\cdot\boldsymbol{\rho}}[1 - \Pi(1 - \Gamma_i)]\, d\boldsymbol{\rho} \tag{10}$$

Note that F can be expanded in a series of one, two, etc., body operators:

$$F = F_1 + F_2 + \cdots \tag{11}$$

where

$$F_1 = \frac{ik}{2\pi} \sum_i \int e^{i\mathbf{q}\cdot\boldsymbol{\rho}} \Gamma_i\, d\boldsymbol{\rho} \tag{12}$$

$$F_2 = -\frac{ik}{2\pi} \sum_{i>j} \int e^{i\mathbf{q}\cdot\boldsymbol{\rho}} \Gamma_i \Gamma_j \, d\boldsymbol{\rho} \tag{13}$$

Upon taking expectation values of F_1 with respect to the ground state target nucleon wave function, we find that the consequent scattering depends only upon the target nucleon density distribution. The term originating from F_2, the two particle operator, will also depend upon the correlations in the ground state wave functions.

This concludes our review.

3. *Remarks on the Validity of Assumptions 2 and 3*

The various assumptions made in Section 2 cannot be generally correct. It is hoped they are valid in a particular range of the experimental variables; namely, for high-energy incident nucleons and for small angle scattering.

Let us first consider Approximation 2. The exact expression for the scattering amplitude is

$$f(\mathbf{q}, k) = -\frac{1}{4\pi} \int e^{-i\mathbf{k}_f \cdot \mathbf{r}} U(\mathbf{r}) \psi^{(+)}(\mathbf{k}_i, \mathbf{r}) \, d\mathbf{r} \tag{14}$$

where

$$U = \frac{2m}{\hbar^2} V$$

We see that the only dependence on $\mathbf{k}_f$ is contained in the exponential factor. Approximate Expression 1 is correct if $\exp(ik_{fz} z)$ can be replaced by a function of k only, for example, e^{ikz}. Since k_{fz} is $k \cos \theta$ it follows that Equation 1 is valid if

$$k(1 - \cos \theta) b \ll 1 \tag{15}$$

where b measures the nuclear force range. If we make the small angle, approximation Equation 15 becomes

$$\theta < \sqrt{\frac{2}{kb}} \tag{16}$$

This is, of course, a sufficient condition only.

It will be useful for our later considerations to obtain an explicit but necessarily approximate expression for $\Gamma(\boldsymbol{\rho}, k)$ and χ. Toward this end we employ the semiclassical approximation:[3]

$$e^{-ik_i z} \psi^{(+)}(\mathbf{k}_i, \mathbf{r}) \cong \exp i \left\{ \int_{-\infty}^{z} \left[\sqrt{k^2 - \frac{2m}{\hbar^2} V(\boldsymbol{\rho}, z')} - k \right] dz' \right\} \tag{17}$$

From Equation 17 it follows that

$$\Gamma(\rho, k) = \frac{i}{2k}\int_{-\infty}^{\infty} dz\, V(\rho, z) \exp i\int_{-\infty}^{z}\left(\sqrt{k^2 - \frac{2m}{\hbar^2}V} - k\right) dz' \quad (18)$$

To obtain a useful expression for χ of Equation 5, it has been customary to expand the square root in Expression 18 retaining only the linear term in V. Expansion in the exponent is always a dangerous procedure.[4] We therefore adopt another method in which the approximation is made to the multiplicative factor V in Equation 18 which we necessarily assume, as is implicit in the derivation of Equation 17, does not vary too rapidly. For this purpose we introduce the differential phase function

$$\Phi(\rho, z) \equiv \sqrt{k^2 - \frac{2m}{\hbar^2}V(\rho, z)} - k \quad (19)$$

In terms of Φ, V becomes

$$\frac{2m}{\hbar^2}V = -\Phi(2k + \Phi) \quad (20)$$

We now insert Equation 19 into the exponent of Equation 18 and replace the V factor multiplying the exponential function in Equation 18 by

$$\frac{2m}{\hbar^2}V \simeq -2k\Phi \quad (21)$$

This approximation involves errors of the same order as those involved in the derivation of Equation 17.

Equation 18 becomes

$$\Gamma(\rho, k) \simeq -i\int_{-\infty}^{\infty} dz\, \Phi(\rho, z) \exp\left[i\int_{-\infty}^{\infty} \Phi(\rho, z')\, dz'\right] \quad (22)$$

$$= 1 - e^{i\chi(\rho, k)}$$

$$\chi(\rho, k) = \int_{-\infty}^{\infty} \Phi(\rho, z)\, dz \quad (23)$$

We can now turn to Approximation 3 and in particular to the linearity assumption, Equation 7. The assertion implicit in Equation 7 is that the phase shift suffered by the incident particle in passing the ith target nucleon is independent of the positions of the other target nucleons. This approximation is clearly not valid when the target nucleons are close for then the scattering of the incident particle is simultaneously under the influence of the forces exerted by each of the target nucleons. An alternative description refers to the possibility that in the first scattering the incident particle goes off the energy shell requiring a second scattering to return. Such processes are not included in Equation 10. To include such effects in terms similar to that of Equation 10 it is clearly necessary to include amplitudes which describe scattering from two particles. If three particle clusters are of interest amplitudes for scattering from three target nucleons are important, and so on. In what follows, we shall formulate such a theory basing it upon the semiclassical approximations Equations 18 and 22.

Formally the exact expression for $\chi(\boldsymbol{\rho}, k)$ is obtained by generalizing Equation 7 so that it contains not only the one-body terms χ_i but also two-body, three-body, ... A body terms, *viz*:

$$\chi = \sum_i \chi(\boldsymbol{\rho} - \boldsymbol{\rho}_i, k) + \sum_{i>j} w_{ij}(\boldsymbol{\rho} - \boldsymbol{\rho}_i, \boldsymbol{\rho} - \boldsymbol{\rho}_j) + \cdots \tag{24}$$

For simplicity, we shall restrict our considerations to systems in which the first two terms are important.

We shall now relate w_{ij} to the effects obtained by scattering from two target nucleons. Such a scattering amplitude according to Equation 24 is given by

$$f^{(2)}(\mathbf{q}, k; \boldsymbol{\rho}_1, \boldsymbol{\rho}_2) = \frac{ik}{2\pi} \int e^{i\mathbf{q}\cdot\boldsymbol{\rho}}[1 - (1 - \Gamma_1)(1 - \Gamma_2)e^{iw_{12}}]\, d\boldsymbol{\rho} \tag{25}$$

We also introduce the more approximate $\tilde{f}^{(2)}$ in which the two body term w_{12} is put equal to zero:

$$\tilde{f}^{(2)}(\mathbf{q}, k; \boldsymbol{\rho}_1\boldsymbol{\rho}_2) = \frac{ik}{2\pi} \int e^{i\mathbf{q}\cdot\boldsymbol{\rho}}[1 - (1 - \Gamma_1)(1 - \Gamma_2)]\, d\boldsymbol{\rho} \tag{26}$$

Then

$$\Delta f_{12} \equiv f^{(2)} - \tilde{f}^{(2)} = \frac{ik}{2\pi} \int e^{i\mathbf{q}\cdot\boldsymbol{\rho}}(1 - \Gamma_1)(1 - \Gamma_2)(1 - e^{iw_{12}})\, d\boldsymbol{\rho} \tag{27}$$

Upon approximate inversion,

$$(1 - \Gamma_1)(1 - \Gamma_2)(1 - e^{iw_{12}}) = \frac{1}{2\pi ik} \int e^{-i\mathbf{q}\cdot\boldsymbol{\rho}}\, \Delta f_{12}\, d\mathbf{q} \tag{28}$$

Hence,

$$e^{i\chi} = \prod_{\ell} (1 - \Gamma_{\ell}) \prod_{i>j} \left\{1 - \frac{1}{2\pi ik} \frac{\int e^{-i\mathbf{q}\cdot\boldsymbol{\rho}}\, \Delta f_{ij}\, d\mathbf{q}}{(1 - \Gamma_i)(1 - \Gamma_j)}\right\} \tag{29}$$

The scattering amplitude is thus

$$F = \frac{ik}{2\pi} \int e^{i\mathbf{q}\cdot\boldsymbol{\rho}} \times \left\{1 - \prod_{\ell} (1 - \Gamma_{\ell}) \prod_{i>j} \left[1 - \frac{1}{2\pi ik} \frac{\int e^{-i\mathbf{q}'\cdot\boldsymbol{\rho}} \Delta f_{ij}(\mathbf{q}', k; \boldsymbol{\rho}_i, \boldsymbol{\rho}_j)\, d\mathbf{q}'}{(1 - \Gamma_i)(1 - \Gamma_j)}\right]\right\} d\mathbf{q} \tag{30}$$

Equation 29 reduces to Equation 10 if Δf is put equal to zero. The one- and two-body parts of this improved formula for F are

$$F_1 = \frac{ik}{2\pi} \sum_i \int e^{i\mathbf{q}\cdot\boldsymbol{\rho}} \Gamma_i \, d\boldsymbol{\rho}$$
$$F_2 = \frac{-ik}{2\pi} \sum_{i>j} \int e^{i\mathbf{q}\cdot\boldsymbol{\rho}} (-\Gamma_i \Gamma_j + H_{ij})\, d\boldsymbol{\rho} \tag{31}$$

where

$$H_{ij} \equiv 1 - e^{iw_{ij}} \tag{32}$$

We see that F_1 is unchanged from Equation 12 while F_2 differs from Equation 13 by the nonfactorable H_{ij}. To first order in the two-body operators

$$-\Gamma_i \Gamma_j + H_{ij} \simeq \chi_i \chi_j - iw_{ij} \tag{33}$$

This approximation is useful only if $\chi_i \chi_j$ and w_{ij} are individually small.

Formula 29 is useful to the extent that the two-body scattering amplitude $f^{(2)}$ and of course the one-body amplitude f are known. The one body amplitude can be determined experimentally. But for the nuclear case the only two-body system available as a target is the deuteron so that the two-body scattering amplitude must be calculated theoretically. This does not seem to be completely out of the question because the two scattering particles are assumed to be fixed. The experimental situation may not be as unfavorable in other fields of physics, particularly when the scatterers are macroscopic.

An approximate evaluation is possible when the semiclassical approximation Equation 17 is used. The differential phase function for the case of scattering from a set of fixed scattering centers is, according to Equation 19,

$$\Phi(\boldsymbol{\rho}, z) = \sqrt{k^2 - \frac{2m}{\hbar^2} \sum V_i} - k \tag{34}$$

where V_i is the potential of the incident particle in the field of the ith target nucleon. Let

$$\Phi_i = \sqrt{k^2 - \frac{2m}{\hbar^2} V_i} - k \tag{35}$$

Hence

$$\frac{2m}{\hbar^2} V_i = -\Phi_i(2k + \Phi_i)$$

If we substitute in Equation 34 and expand the square root to second order in Φ_i, we obtain

$$\Phi = \sum \Phi_i - \frac{1}{k} \sum_{i>j} \Phi_i \Phi_j + \cdots \tag{36}$$

We now evaluate $\Gamma(\rho, k)$ of Equation 17,

$$\Gamma(\rho, k) = \frac{i}{2k} \int_{-\infty}^{\infty} dz \sum V_i e^{i\int_{-\infty}^{z} \Phi(\rho, z')\, dz'}$$

Again to order $\sum V_i/E$ in the amplitude

$$\Gamma \equiv 1 - e^{i\chi}$$

$$\chi = \sum \chi_i + \sum_{i>j} w_{ij}$$

where

$$\chi_i = \int_{-\infty}^{\infty} \Phi_i \, dz \tag{37}$$

$$w_{ij} = \frac{1}{k} \int_{-\infty}^{\infty} \Phi_i \Phi_j \, dz \tag{38}$$

The integrand for the approximate amplitude for F_2, Equation 31, as given by Equation 33, is

$$\chi_i \chi_j - i w_{ij} = \int_{-\infty}^{\infty} \Phi_i \, dz \int_{-\infty}^{\infty} \Phi_j \, dz - \frac{i}{k} \int dz \, \Phi_i \Phi_j \tag{39}$$

The elastic scattering amplitude arising from the two-body term F_2 is $\langle e^{i\mathbf{P}\cdot\mathbf{R}}\Psi_0, F_2 \Psi_0 \rangle$ where Ψ_0 is the ground state wave function for the target nucleus, $\mathbf{P}$ is the momentum of the recoiling final target nucleus, and $\mathbf{R}$ is the coordinate of its center of mass.

It will suffice for the point we wish to make here to evaluate F_2 for a simple analytic potential. Take

$$\frac{2m}{\hbar^2} V_i = U e^{-\alpha^2(\mathbf{r}-\mathbf{r}_i)^2} \quad \text{and} \quad \Phi_i \approx -\frac{1}{2k}\frac{2m}{\hbar^2} V_i \tag{40}$$

where $\mathbf{r}$ is the coordinate of the incident nucleon and $\mathbf{r}_i$ that of the ith target nucleon. The parameter $(1/\alpha)$ is a measure of the range of the potential while U gives its strength. Then

$$\begin{aligned} 4k^2[\chi_i\chi_j - iw_{ij}] &= U^2 e^{-\alpha^2[(\rho-\rho_i)^2+(\rho-\rho_j)^2]} \\ &\quad \times \left[\int_{-\infty}^{\infty} e^{-\alpha^2(z-z_i)^2}\, dz \int_{-\infty}^{\infty} e^{-\alpha^2(z-z_j)^2}\, dz \right. \\ &\quad \left. -\frac{i}{k}\int_{-\infty}^{\infty} e^{-\alpha^2[(z-z_i)^2+(z-z_j)^2]}\, dz\right] \\ &= \frac{\pi}{\alpha^2} U^2 e^{-\alpha^2[(\rho-\rho_i)^2+(\rho-\rho_j)^2]} \cdot \left[1 - \frac{i}{\sqrt{2\pi}}\frac{\alpha}{k} e^{-\alpha^2(z_i-z_j)^2/2}\right] \end{aligned} \tag{41}$$

Finally,

$$\begin{aligned} 4k^2 F_2 &= -\frac{i\pi}{4}\frac{kU^2}{\alpha^4} e^{-q^2/8\alpha^2} \sum_{i>j} [e^{i\mathbf{q}\cdot(\rho_i+\rho_j)/2-\alpha^2(\rho_i-\rho_j)^2/2}] \\ &\quad \cdot \left[1 - \frac{i}{\sqrt{2\pi}}\frac{\alpha}{k} e^{-\alpha^2(z_i-z_j)^2/2}\right] \end{aligned} \tag{42}$$

The effect of the overlapping potentials is given by the z dependent term in the second bracket. It is identical in form to the correlation term which is present in the factor $|\Psi_0|^2$ which appears in the scattering amplitude $\langle e^{i\mathbf{P}\cdot\mathbf{R}}\Psi_0, F_2\Psi_0\rangle$. The quantity $|\Psi_0|^2$ will have the typical factor $[1 + C(\mathbf{r}_i - \mathbf{r}_j)]$ where C is the correlation function. It is clear that before C can be extracted from the data it is necessary to account for the effect of the overlapping potentials. A rough estimate of the ratio of the effect of the overlap of the scattering potentials to that of the correlation C which is obtained by inserting rough forms for C and Ψ_0 shows it to be of the order of $(1/k\ell)$ with

$$\ell \cong \frac{\sqrt{2\pi}}{3}\frac{\alpha^4}{\gamma(\alpha^2 + 2\gamma^2)^2} \tag{43}$$

Here γ^{-1} is approximately the length at which C is zero. If the incident particle is a nucleon, we take $\gamma \simeq \alpha$ (it is probably somewhat larger) $\ell \simeq \sqrt{2\pi}/(27\alpha) \sim (1/10\alpha)$. The length ℓ is thus of the order of $(1/5)\ 10^{-13}$ cm. We may thus expect this correction arising from the overlap of the potentials to be comparable with the correlation term for incident nucleon energies of the order of 500 MeV and to decrease like $1/\sqrt{E}$ from that energy. For other incident projectiles, we can obtain other values for ℓ. If $\gamma \ll \alpha$, for example, for pions, then $\ell \sim \gamma^{-1}$ and the energy range over which the independence approximation is valid is considerably larger than in the nucleon case. If $\gamma \gg \alpha$, *e.g.*, electron scattering, $\ell \simeq \alpha^4/4\gamma^5$ and the independence approximation would lead to a serious error in the determination of the correlation function.

3. *Summary*

A formulation of multiple scattering theory is proposed which makes use of (a) the impact parameter approximation and (b) the expansion of the phase shift function in one-, two-, and many-body operators.[4] Explicit results are given for the case in which the expansion is truncated after two terms, retaining the one- and two-body terms. Evaluation is obtained by appropriate use of the semiclassical physical optical approximation. It is used to show that the effect of the overlap in the incident nucleon-target nucleon potentials can be important for the determination of correlations in the target nuclear wave function by elastic scattering. The use of experiments which are more sensitive to correlation effects is thus recommended.

References

1. H. Palevsky, J. L. Friedes, R. J. Sutter, G. W. Bennett, G. J. Igo, W. D. Simpson, G. C. Phillips, D. M. Corley, N. S. Wall, R. L. Stearns, and B. Gottschalk, *Phys. Rev. Letters*, **18**, 1200 (1967). Also, E. T. Boschitz, W. K. Roberts, J. S. Vincent, K. Gotow, P. C. Gugelot, C. F. Perdrisat, and L. W. Swenson, *Phys. Rev. Letters*. Also, G. Belletini, G. Cocconi, A. N. Diddens, E. Lillethum, G. Matthiae, J. P. Scanlon, and A. M. Wetherell, *Nucl. Phys.*, **79**, 609 (1966).
2. R. J. Glauber, in *High Energy Physics and Nuclear Structure*, G. Alexander, ed., North Holland Publishing Co., Amsterdam (1967). Also, W. Czyz and L. Lesniak, *In High Energy Physics and Nuclear Structure*, G. Alexander, ed., North Holland Publishing Co., Amsterdam (1967). Also, R. H. Bassel and C. Wilkin, Brookhaven National Laboratory, preprint *BNL* 12430, Dec. 1967.

3. R. J. Glauber, *High Energy Collision Theory,* Lectures in Theoretical Physics, Vol. **1**, W. E. Brittan and L. G. Dunham, eds., Interscience Publishers, Inc., New York (1959). Also, H. Feshbach, *Semi-Classical Approximation for High Energy Scattering and Reactions in Proceedings of the International School of Physics*, Course XXXVIII, T. E. O. Ericson, ed., Academic Press, Inc., New York (1967).
4. D. J. Vezzetti and J. B. Keller, *J. Math. Phys.*, **8**, 1861 (1967). In this paper an exact expansion involving " pure " n particle scattering functions is obtained for the calculation of the refractive index for waves in a polarizable medium.

J. E. Young and M. Rich
Los Alamos Scientific Laboratory

18 Single π-Production in π-Nucleon Collisions*†

We study the process $\pi + N \rightarrow 2\pi + N$ within the framework of simple models of the sort proposed by Amado, *et al.*,[1,2] and Lovelace.[3] Such models, as we shall elucidate, are meant to describe transitions between states of a three-body system. The models assume that bound states and resonances of the two-body subsystems occurring dominate the three-body transition amplitudes. They also exhibit manifest two-, and three-body unitarity.

The work discussed here is restricted to the approximation of the static model. It is expected that the structure of the equations so obtained will persist in a fully covariant, relativistic description. No static-model calculations are planned nor are any in progress. These are not interesting, *e.g.*, the Dalitz plot degenerates into a line. Relativistic calculations are being undertaken and these will ultimately be reported.

Our analysis is restricted to energies below 1 BeV. At such energies the model proposed may well have at least first-order validity. The upper limit in energy excludes ρ-dominant production of the final pions.

The compositeness assumption (dominance of two-body bound states and resonances) noted above leads to a parametrization of the production amplitude that is constrained to be consistent with known two-body systematics. At the first step it forces us to regard the initial

* A preliminary version of this work was submitted to the Second International Conference on High Energy Physics and Nuclear Structure, Rehovoth, 1967.
† Work performed under the auspices of the U.S. Atomic Energy Commission.

nucleon N as a quasiparticle of the π-nucleon (P_{11}) system. The calculations extant are in conflict[4,5] on this point. Whether or not their differing conclusions owe to model-dependent assumptions it is difficult to say. The models do seem to agree that if you predict the Roper resonance ($T_\pi = 575$ MeV) with its corresponding inelasticity ($\eta = 0.35$), it is not possible to bind the nucleon; conversely, binding the nucleon excludes the possibility of obtaining the resonance.

The composite model of the production amplitude will state that the final particles are produced ($\pi - N'$) by the decay of the N(nucleon) and $N^*(1238)$ isobars; ($\pi - \pi$) by the decay of the conjectured[6] $\sigma(I = 0, J = 0)$ at mass 730 MeV, or thereabout, or of the ρ-meson. In effect the production amplitude is computed as the product of: off-shell, two-body amplitude for scattering from a composite; propagator function for a composite; and vertex function for dissociation of composite into a final pair of particles.

Evidently, we can attempt to assess the validity of the composite model description by computing continuum corrections to it. A method of doing this is proposed in this discussion. Essentially the corrections take into account the $\pi - \pi$ and $\pi - N'$ rescatterings. This amounts to computing the $\pi - \pi$ and $\pi - N'$ final state interactions in partial waves not dominated by the assumed quasiparticles of the composite description. By assumption these partial waves are small and hence capable of being described in Born approximation. In particular a particle from the decay vertex interacts with the odd, or spectator, particle to make the final state. This process is not included in the iterations of the composite-model theory. For example, the S_{11}, $\pi - N'$ interaction following the 2π decay of the ρ-isobar is now taken into account as a continuum correction by our procedure. The S_{11}-channel of the $\pi - N'$ system is troublesome[7,8] to describe and similar remarks pertain to our understanding of the $\pi - \pi$ system as Weinberg's[9] work underscores. Our model is not likely to be decisive in clarifying the outstanding difficulties here. It is however capable of describing the triangle-graph singularities, themselves continuum corrections, which arise from specific approximations[10] to the three-body, final-state wave function.

Previous calculations of the π-production process are numerous. This extensive literature is summarized in the two references, Olsson and Yodh[11] (OY) and Goebel and Schnitzer[12] (GS), which we take to be representative of our present understanding. A schematic representation of the two calculations is shown in Figure 1. The GS calculation is the more easily understood. It is based upon the static model of Chew and Low.[13] In the spirit of that model the production is described by retaining single nucleon and one meson-nucleon intermediate states. This

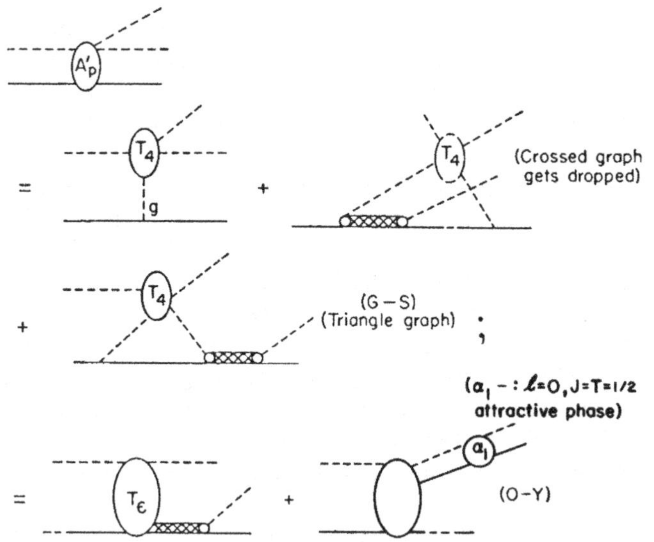

Figure 1. The π(– – –) production amplitude from a nucleon (———). Two models are shown for computation in this amplitude. In both of these $N^*(1238)$, ·XXX·. formation is accounted for. Detailed explanation of symbols appears in the text

results in the three graphs illustrated in Figure 1. The $\pi - \pi$ scattering is described in scattering length approximation for *S*-states and *P*-states and the crossed graph is dropped to get a simple integral equation. It is also assumed that $\pi - N'$ rescattering occurs in the (3, 3) state only. The absence of the crossed cut is a feature shared by the composite model. The latter handles three-particle unitarity exactly and this aspect is approximate in GS. The natural question as to the relative importance of crossing and unitarity is not answered in either calculation. This is a problem of some immediacy. The GS fit to available data is unsatisfactory on at least two counts. The fit to the invariant mass of the $\pi^+\pi^-$ system predicts suppression of high masses whereas one actually observes a considerable enhancement. (The OY model also fails in this same sense.) The GS model does not predict any real isobar production. The threshold for such production is in fact $T_\pi = 450$ MeV an energy well below that at which GS compute their cross sections.

The OY description, also depicted in Figure 1, allows for formation of the $N^*(1238)$ and its decay into final $\pi + N'$. These particles are permitted to rescatter via the attractive phase α_1 of the S_{11}-channel. It is found that this continuum contribution is essential for a quantitative fit of the production data. It appears that OY perform their evaluations with the propagator for the composite particle (N^*) taken on its mass shell. This is not done in the composite model treatment and is further inappropriate to three-body problems. There is also the omission of the nucleon isobar contribution to $\pi - N'$ states and of some interaction between final pions. The latter is required, as OY note, to account for asymmetry in the $\pi^+\pi^-$ mass spectrum. We will not discuss the detailed fits, generally quite good, of OY to the production cross sections and angular distributions. The reader is referred to their papers for these. Their succinct parametrization of the contributing partial wave cross sections is altogether valuable and, the reproduction of results following therefrom is a boundary condition for any proposed theory.

Apart from the question of the $\pi - \pi$ interaction noted above, one studies the production process to get some feeling for the dependent and perhaps larger features of the strong interactions. Among these is that having to do with whether the nucleon bootstraps in the sense of the $SU(6)$ static model or is bound by elastic forces. We can comment upon this aspect in the following sense. If our three-body representation of π-nucleon scattering in the P_{11}-channel describes both the nucleon pole and Roper resonance one has the suggestion that the nucleon is not too complicated an object. The resonance would be said to arise from mainly three-particle intermediate states, $\pi + N^*(1238)$. Its inelasticity would be ascribed to coupling to other channels open at the energy in question.[2,14] Failure to get an acceptable description of the P_{11}-channel would in all likelihood indicate the necessity for inclusion of strange particle channels. The SU_3[15] classification of hadron states would be employed for the subsequent discussion.

The Model

The formulation of a completely soluble model for three-body states has been given, especially by Amado and collaborators,[1,2] and by Lovelace.[3] The former group predicates its discussion upon the assumption of a Lee[16] model Hamiltonian or one of the many variants thereof.[17] We follow this presentation here. Lovelace starts with his version of the Faddeev[18] equations and introduces separable potentials[19] into these to achieve a soluble problem. The equivalence of the two methods in the limit of vanishing wave function renormalization $Z_3 = 0$[20] has been exhaustively discussed. In the nonrelativistic approximation implied, the conservation law is that for the mass and the symmetry implied is Galilean invariance. The static description of the Lee model is acceptable on these grounds. To go beyond this to a fully covariant, relativistic set of equations describing three-body states is not altogether straightforward. Lovelace *et al.*,[21] and Blankenbecler-Sugar[22] have given one formulation of this problem. No numerical evaluations seem to be available. A second formulation has been given by Aaron *et al.*[23] The numerical evaluations of the $\pi - N$ phases with only nucleon exchange in the model seem to be in reasonable agreement with experiment for this level of approximation. Both formulations cited go beyond Hamiltonian theories. One is forced to sum certain sets of graphs consistent with two-particle and three-particle unitarity, obtaining thereby linear, off-shell integral equations with only the unitarity cut. All of the remarks of this paper imply the existence of some model Hamiltonian system. Such is adequate to exhibit the general structure of the equations describing production and scattering.

The Hamiltonian we need to discuss the three-particle problem $\pi + N \to 2\pi + N'$ is arrived at as follows. Assume that N the initial nucleon is a quasiparticle, a nucleon, of the π-nucleon system. Then, next, the final states are dominated by two-body resonances. This says the pairs of final particles come from the virtual decays of intermediate particles N, N^* and σ, ρ. Finally, note that the production amplitude A_P, Figure 2, is given as the product of three factors: (three-particle,

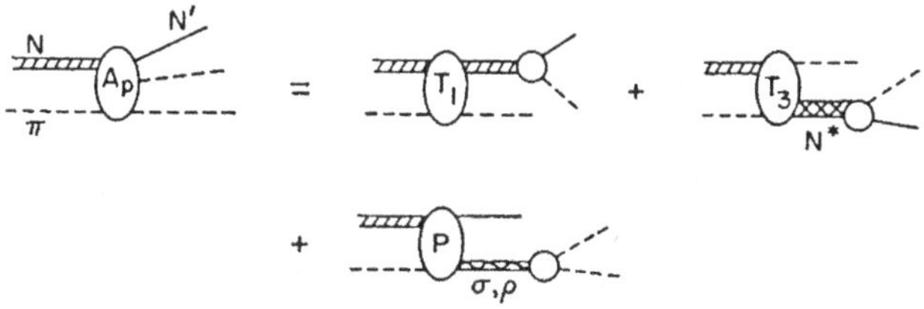

Figure 2. The composite, three-particle model for the production amplitude. The nucleon is both elementary and composite, /////, in this scheme. The resonant $\pi - \pi$ state, σ, ρ /\/\/\/\/\ appears here.

off-shell amplitude for $\pi + N \to$ extra particle + quasiparticle) × (three-body propagator for quasiparticle, and, free extra particle) × (decay vertex, or rather bound-state form factor for the quasiparticle). The Hamiltonian is selected to be of such a generality that: (1) we can compute the three off-shell amplitudes, T_1, T_3 and P of Figure 2; (2) the bound state, nucleon pole appears in both the two-body, $\pi - N'$, and three-body, $\pi - N$, representations of π-nucleon (P_{11}) scattering; and (3) the rescatterings $\pi - N'$ and $\pi - \pi$ are accounted for. To this end we introduce a Hamiltonian containing the following interactions, in the standard notation of the Lee[16] and Zachariasen[24] models.

$$
\begin{aligned}
V &\leftrightarrow N_1 + \theta \\
V' &\leftrightarrow N_1 + \theta
\end{aligned} \tag{1}
$$

Thus $V^i + \theta \rightarrow V^j + \theta$ (i, j denote V and V') is determined.

$$V + \theta \leftrightarrow U \tag{2}$$

where U is the Bronzan[25] particle representing an additional fermion degree of freedom. The direct bound state pole in $V - \theta$ scattering is provided by U. We discuss the inclusion of this pole in the Appendix to this paper.

$$B + \bar{B} \leftrightarrow A \tag{3}$$

As $V + \theta \rightarrow N + 2\theta$ is contained in Equations 1, that interaction together with Equation 3 makes it possible to compute P, namely, $V + \theta \rightarrow A + N'$. The new interaction, Equation 3, as well as that of Equation 2, does not spoil the $N - \theta$ sector of the usual Lee model. Both Equations 1 and 3 account for the desired rescatterings. To the foregoing, we add the following key:

$$
\begin{array}{llll}
V \equiv N & V' \equiv N^* & N_1 \equiv N' & U \equiv N' \\
\theta \equiv \pi & B \equiv \pi & \bar{B} \equiv \pi & A \equiv \sigma, \rho
\end{array} \tag{4}
$$

The fermion degrees of freedom V, V', N_1 and U couple through the bosons, $\theta \equiv B \equiv \bar{B}$ and A. The physical objects relevant to our computation are of course $N(938)$, $N^*(1238)$, $N'(938)$, $\pi(140)$, $\sigma(730)$, $\rho(750)$. The difference between N and N' is that the former is composite and the latter is represented by an elementary field. Our Hamiltonian is now fully specified except to note that the usual cutoff function $u(\omega)$, $\omega = \sqrt{k^2 + \mu^2}$, $\mu \equiv m_\pi$, of the Lee model is implied throughout our discussion

The three-body, off-shell amplitudes satisfy coupled equations. These are illustrated in Figures 3 and 4. The equations couple the amplitudes for $\pi - N$ scattering and the inelastic processes $\pi + N \rightarrow \pi + N^*$ and $\pi + N \rightarrow N' + \sigma$. The equations differ from those discussed by Aaron in that one additional amplitude, P, now contributes to the system of simultaneous equations. In the event that the σ-meson does not exist as a particle, its effect, owing to a reasonably large phase (at least 60° at 730 MeV in all fits to $\pi - \pi$ scattering), is taken into account even so through a scattering amplitude. This amplitude can be used to generate the corresponding composite particle propagator and form factor ($\sigma \rightarrow 2\pi$-vertex function). Evidently the ρ-meson does not present us with the sort of difficulties noted. In any event the model is no way in

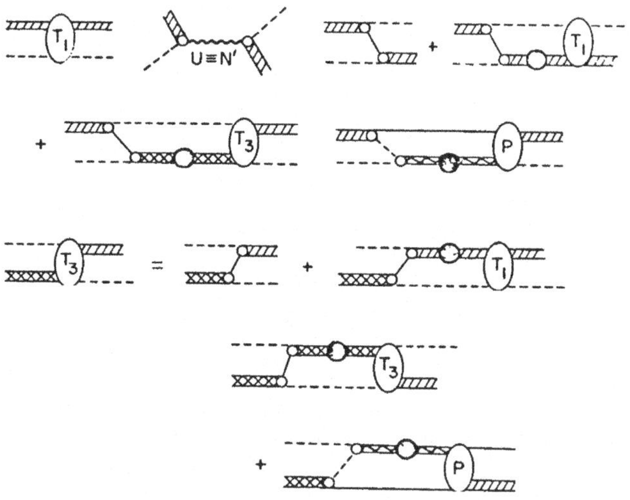

Figure 3. Integral equation for three-body off-shell amplitudes for $\pi - N$ scattering.

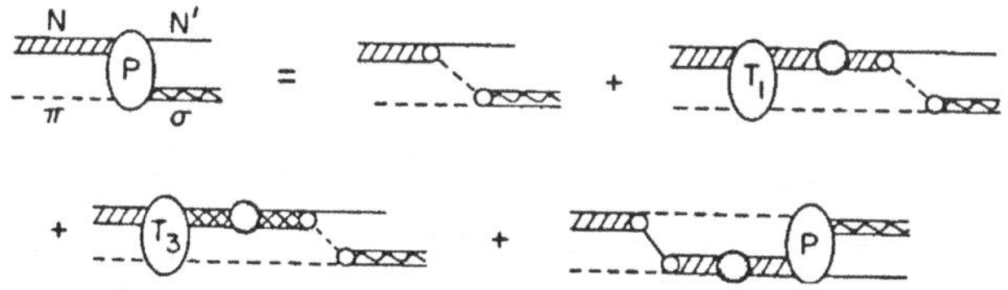

Figure 4. Integral equation for three-body off-shell amplitude introduced in addition to those of Fig. 3 as required for the discussion of the production process.

apparent conflict with the small scattering lengths preferred by Weinberg.

It is understood that the off-shell amplitudes T_1 and T_3 are coupled matrix functions with elements in the (1, 1), (1, 3), (3, 1) and (3, 3) P-wave channels. This fact is, and the corresponding details are, *e.g.*, the four crossing matrices, spelled out in References 2 and 3. In the S_{11}-channel the amplitudes become single functions and this reduces the numerical work considerably. The amplitude P is required for the coupling transitions $S_{11} \leftrightarrow P_{11}$ between the $\pi - N$ and $N' - \sigma$ channels.

The production amplitude, symmetric in the labels of the final pions, Figure 5, is labelled with three vector momenta in the static limit. The matrix element of this amplitude is written as

$$\langle \mathbf{q}_1{}^0 | A_P(s) | \mathbf{q}_1 \mathbf{p}_1 \rangle + \langle \mathbf{q}_1{}^0 | A_P(s) | \mathbf{q}_2 \mathbf{p}_2 \rangle \qquad \text{(nucleon elementary)}$$

or as

$$\langle \alpha_1{}^0 \mathbf{q}_1{}^0 | A_P(s) | \mathbf{q}_1 \mathbf{p}_1 \rangle + \langle \alpha_1{}^0 \mathbf{q}_1{}^0 | A_P(s) | \mathbf{q}_2 \mathbf{p}_2 \rangle \qquad \text{(nucleon composite)}$$

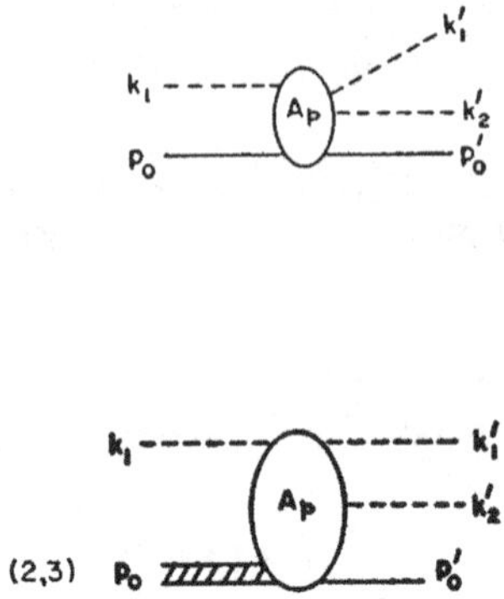

Figure 5. Momentum labels for the production amplitude: the physical amplitude and its composite model representation.

The choice of variables here is that suggested by Lovelace, but, not carrying his specific normalizations. (Take the limit $M \to \infty$ for the static model.)

$$
\begin{aligned}
\mathbf{q}_1{}^0 &= \frac{1}{M+\mu}(M\mathbf{k}_1 - \mu\mathbf{p}_0) \\
\mathbf{q}_2 &= \frac{1}{M+2\mu}[(\mu+M)\mathbf{k}_2{}' - \mu(\mathbf{k}_1{}' + \mathbf{p}_0{}')] \qquad (5) \\
\mathbf{p}_1 &= \frac{1}{M+\mu}(M\mathbf{k}_1{}' - \mu\mathbf{p}_0{}')
\end{aligned}
$$

It is clear from these relations how one constructs the complete set of coordinate pairs $(\mathbf{q}_i \mathbf{p}_i)$, $i = 1, 2, 3$. The variable s is the barycentric energy in this nonrelativistic discussion. We label the states of the composite $\pi - N'$ system by $\alpha_1{}^j$ and $\alpha_2{}^j$, where the labels 1 and 2 refer, as in Equation 5, to the extra, i.e., noninteracting, pion. Then, $j = 0$, as a superscript, denotes the nucleon bound state, $j = 1$ the isobar $N^*(1238)$; $\alpha_3{}^1$ is the composite 2π-state, σ (or ρ in the $I = J^\pi = 1^-$ state).

The off-shell amplitude for three-body states appears when we write the matrix element for production in the preliminary form

$$
\begin{aligned}
\langle \xi_i | Ap(s) | \xi_f \rangle &= \langle \alpha_1{}^0 \mathbf{q}_1{}^0 | T_1(s) | \alpha_1{}^0 \mathbf{q}_1 \rangle \langle \alpha_1{}^0 | \hat{A}_1(s) | \mathbf{p}_1 \rangle \\
&\quad + \langle \alpha_1{}^0 \mathbf{q}_1{}^0 | T_3(s) | \alpha_1{}^1 \mathbf{q}_1 \rangle \langle \alpha_1{}^1 | \hat{A}_1(s) | \mathbf{p}_1 \rangle \\
&\quad + \langle \alpha_1{}^0 \mathbf{q}_1{}^0 | P(s) | \alpha_3{}^1 \mathbf{q}_3 \rangle \langle \alpha_3{}^1 | \hat{A}_3(s) | \mathbf{p}_3 \rangle \\
&\quad + (1 \to 2, \text{ in the final states}) \qquad (6)
\end{aligned}
$$

where $\xi_i \equiv (\mathbf{k}_0, \mathbf{p}_0)$ and $\xi_f \equiv (\mathbf{k}_1', \mathbf{k}_2', \mathbf{p}_0')$ for the physical problem. The amplitudes $\hat{A}_i(s)$ $(i = 1, 2, 3)$ refer to and specify the decays in the two-body subsystems. Some discussion of these quantities is to be given subsequently. It is imperative to note that no inconsistency appears in the labels of the off-shell amplitude $\langle \alpha_1^0 \mathbf{q}_1^0 | T(s) | \alpha_1^0 \mathbf{q}_1 \rangle$ in which a nucleon label, α_1^0, stands on both sides of the matrix element. On the left, the nucleon is on its mass shell. This is not so on the right where we require an improper vertex function, or pionic form factor, to describe the ultimate decay into $\pi + N'$. This point has been discussed by Lovelace. Finally, it should be apparent from the result, Equation 6, that the three-body scattering amplitudes T_1, T_3 and P_1 are indeed half, off-shell quantities. There is no problem in computing full, off-shell amplitudes in the sort of theory considered here.

Having completed discussion of the off-shell, three-body amplitudes, we turn to the two-body amplitudes. These will be brought to a form which when substituted into Equation 6 provides an expression for the production amplitude agreeing with intuition based upon Feynman diagrams. We begin in what seems an unmotivated fashion evaluating the vertex operator, $V \to N' + \theta$, of the usual Lee model as

$$\langle N'\mathbf{k}(-)| j_v^+ |0\rangle = \rho(\omega)K(\omega)g \qquad \rho(\omega) = \frac{u(\omega)}{\sqrt{2\omega\Omega_0}} \qquad \Omega_0 = (2\pi)^3 \quad (7)$$

The renormalized coupling constant is g and K is the improper vertex function having only a right-hand cut. It is straightforward to show that

$$K(\omega) = D^{-1}(\omega) \quad (8)$$

It is moreover true that the propagator function for a stable V is given by

$$(\omega - m_v)\Delta(\omega) = 1/D(\omega) \quad (9)$$

where m_v is the observed mass. In the quasiparticle case it is sufficient to modify the D-function through the introduction of a C.D.D. pole.[26] The form factor vanishes at the real mass of the unstable particle. The residue of Im D is directly related to the decay width or inverse lifetime of the unstable particle. All of these statements can be made in the small-width approximation. This then involves second-sheet continuations of the D-function and one talks about a complex mass for the unstable particle. Knowing then how to make modifications in the theory to cover the contingency of an unstable particle we will manipulate the various formal expressions without regard to the stability of the V. We have then

$$\langle N'\mathbf{k}(-)| j_v^+ |0\rangle = (M + \omega - m_v)\langle Nk(-)| \psi_v^+ |0\rangle \qquad (\psi_v \equiv Z^{-1/2}\psi_v) \quad (10)$$

Now the quantity $\rho(\omega)$ plays the role of a bound state form factor $\langle \alpha_i{}^j | \mathbf{p}_i \rangle$ in the description of composite particle scattering[1] ($Z_V = 0$; $N' + \theta \leftrightarrow V$) in the $N' - \theta$ sector. This form factor is identified as follows. If the scattering of the pair of particles is dominated by bound states or resonances, then the corresponding scattering amplitude,

$$t_\ell(E) = \frac{-N_\ell(E)}{D_\ell(E)} \tag{11}$$

has a numerator function $N_\ell(E)$ such that

$$\langle \alpha_i{}^j | \mathbf{p}_i \rangle_\ell = \sqrt{N_\ell(E_i)} \qquad \left(E_i = \frac{1}{2\mu_i} p_i{}^2\right) \tag{12}$$

The analogy with potential scattering is easy to extract. Assume for definiteness that particles 2 and 3 scatter ($i = 1$ in Equation 12) and that this pair forms a bound state or resonance denoted by the superscript j. We take the Fourier transform of the Schroedinger equation to obtain, apart from normalization,

$$\langle \alpha_1{}^j | \mathbf{k} \rangle \equiv (E - \mathbf{k}^2)\psi_{\alpha_1}{}^j(\mathbf{k}) = \int d\mathbf{k}' \langle \mathbf{k} | v | \mathbf{k}' \rangle \psi_{\alpha_1}{}^j(\mathbf{k}') \tag{13}$$

in a self-evident notation. The partial wave information is contained in the statement

$$\langle \alpha_1{}^j | \mathbf{k} \rangle = \sum_\ell (2\ell + 1) P_\ell(z) \langle \alpha_1{}^j | \mathbf{k} \rangle_\ell \tag{14}$$

A discussion of the bound state form factors has been given by Blankenbecler[27] in another connection.

We make the choice

$$\rho(\omega_i) \equiv \langle \alpha_i{}^j | \mathbf{k}_i \rangle \tag{15}$$

for the reasons noted and then Equations 7 and 10 give

$$g\langle \alpha_i{}^j | \mathbf{k} \rangle K(\omega) = (M + \omega - m_v) \langle N'\mathbf{k}(-) | \psi_v{}^+ | 0 \rangle \tag{16}$$

where j denotes the composite state V; and

$$g\langle \alpha_i{}^j | \mathbf{k} \rangle \Delta_i{}^j(\omega) = \langle N'\mathbf{k}(-) | \psi_v{}^+ | 0 \rangle \equiv \langle \alpha_i{}^j | \hat{A}_i(\omega) | \mathbf{k} \rangle \tag{17}$$

This result then enables us to write the production amplitude matrix element of Equation 6 as

$$\begin{aligned} \langle \xi_i | A_P(s) | \xi_f \rangle = {} & \langle \alpha_1{}^0 \mathbf{q}_1{}^0 | T_1(s) | \alpha_1{}^0 \mathbf{q}_1 \rangle g \Delta_1{}^0(s) \langle \alpha_1{}^0 | \mathbf{p}_1 \rangle \\ & + \langle \alpha_1{}^0 \mathbf{q}_1{}^0 | T_3(s) | \alpha_1{}^1 \mathbf{q}_1 \rangle g \Delta_1{}^1(s) \langle \alpha_1{}^1 | \mathbf{p}_1 \rangle \\ & + \langle \alpha_1{}^0 \mathbf{q}_1 | P(s) | \alpha_3{}^1 \mathbf{q}_3 \rangle g' \Delta_3{}^1(s) \langle \alpha_3{}^1 | \mathbf{p}_3 \rangle \\ & + (1 \rightarrow 2, \text{in the final states}) \end{aligned} \tag{18}$$

This expression has been previously obtained by Lovelace[3] and his normalization of the momenta q_j is required to suppress reduced mass factors in the various propagators. The coupling constant g and g', that symbolizing σ-decay, will be absorbed into new form factors. This change and one other which improves the notation is summarized as follows:

$$g\langle\alpha_1{}^0\,|\,\mathbf{p}_1\rangle \equiv f_N(\mathbf{p}_1) \qquad g\langle\alpha_1{}^1\,|\,\mathbf{p}_1\rangle \equiv f_{N^*}(\mathbf{p}_1) \qquad g'\langle\alpha_3{}^1\,|\,\mathbf{p}_3\rangle \equiv f_\sigma(\mathbf{p}_3)$$
$$\Delta_1{}^0(s) \equiv \Delta_N(s - \sqrt{\mu^2 + q_1{}^2}) \qquad \Delta_1{}'(s) \equiv \Delta_{N^*}(s - \sqrt{\mu^2 + q_1{}^2})$$
$$\Delta_3{}'(s) \equiv \Delta_\sigma(s - q_3{}^2) \tag{19}$$

The propagators now exhibit their correct energies for a three-body problem. The redefined form factors are to be directly compared with those, Equation 13, coming from potential theory. The coupling constants g and g' are of course determined in the two-body sectors of the theory. That g' is associated with the Yukawa coupling in $A \leftrightarrow B + \bar{B}$ of the Zachariasen model.

We can specify the determination of the form factors somewhat beyond the prescription of Equation 12. In particular for the $\pi - N'$ channels, the scattering of the decay products is nicely summarized by the separable potential representation described by Bolsterli and MacKenzie.[28] In their representation the potential form factor $v(p)$, that which we have termed a bound state form factor, is specified by considerations of analyticity of the scattering amplitude as

$$\lambda_0\, v_\ell{}^2(p) = A(p)\exp\left[⨍ ds'\,\frac{\delta_\ell(s')}{s' - s}\right] \qquad (s = p^2) \tag{20}$$

The function $A(p)$ is known and depends upon the observed phase shift $\delta_\ell(s)$, this being required at all energies. The coupling constant of this representation is designated as λ_0. It is clear from these remarks that we would write

$$\langle\alpha_i{}^j\,|\,\mathbf{p}\rangle_\ell = v_\ell(p) \tag{21}$$

for the decay channels being discussed. Actually things are somewhat more complicated than we have described. The result, Equation 20, holds for a single uncoupled channel. In the case of coupling, as interests us, it is necessary to use appropriate eigenphase representations. This complication will not be explored by us.

The essential appeal of the prescription summarized by Equations 20 and 21 is its simplicity. Moreover, although semiempirical (the high-energy behavior of the phase, $\delta_\ell(\infty) = 0$, is subject to considerable ambiguity and hence adjustment), we have the feeling that one has specified the numerator function $N_\ell(s)$ as well as possible within limitations of our current understanding. We should like therefore to apply our prescription to the scattering of the $\pi - \pi$ decay products.

The numerator function for $\pi - \pi$ scattering in our simple model is[24]

$$N(s) = \lambda + \frac{g'^2}{s - m_\sigma^2} \tag{22}$$

Then with

$$t = \frac{-N}{D} \equiv \frac{1}{\rho(s)} \varepsilon^{i\delta(s)} \sin \delta(s) \qquad \rho(s) = \frac{1}{16\pi} \sqrt{\frac{s - 4\mu^2}{s}} \tag{23}$$

the D-function being modified as described earlier so as to provide an unstable σ, we determine a relation between λ and g' through[9]

$$k \operatorname{ctn} \delta(s) = \frac{1}{a} \qquad a = 0.2\,\mu^{-1} \tag{24}$$

The second relation between the coupling constants is furnished by the "known" decay width Γ of the σ. The D-function near threshold, $s = 4\mu^2$, is approximated by the functional form holding in the vicinity of $s = m_\sigma^2$, the unstable particle mass

$$D(s) \approx 1 + i \frac{m_\sigma \Gamma}{s - m_\sigma^2} \qquad \Gamma = \frac{g'^2}{16\pi} \left(\frac{m_\sigma^2 - 4\mu^2}{m_\sigma^2} \right) \tag{25}$$

This questionable approximation which ought not to be too bad for a resonance near threshold can be replaced by a more suitable one with a little thought. At any rate for the purpose of this survey we retain it and assert that Equation 23 then gives us a $\delta(s)$, $\delta(\infty) = \pi$. We insert this into a relation such as Equation 20, appropriately subtracted, to obtain the corresponding form factor. Admittedly this treatment of the $\pi - \pi$ system has been very crude and has to be rethought. The remarks of the remainder of this presentation are rather less tentative.

The propagators Δ for the quasiparticles are the remaining items of Equations 18 and 20 requiring any comment. These are specified by Equation 9. Each such propagator sums the bubble diagrams in the two-particle approximation to the self-energy consistent with elastic unitarity. The three-body propagators, not discussed by us, but implicit in the integral equations of Figures 3 and 4 are defined so as to be consistent with full, three-particle unitarity.

We conclude our presentation with a few selected comments. In the event the model described does not altogether correspond to physical reality and the three-body final states are not fully dominated by two-body composites, some corrections may be proposed. These come from the continuum interactions, *i.e.*, the nonresonant background rescattering, of final pairs of particles. Such contributions are summarized in Figure 6 and include the S_{11}, $\pi - N'$ channel treated by OY,[11] and, the

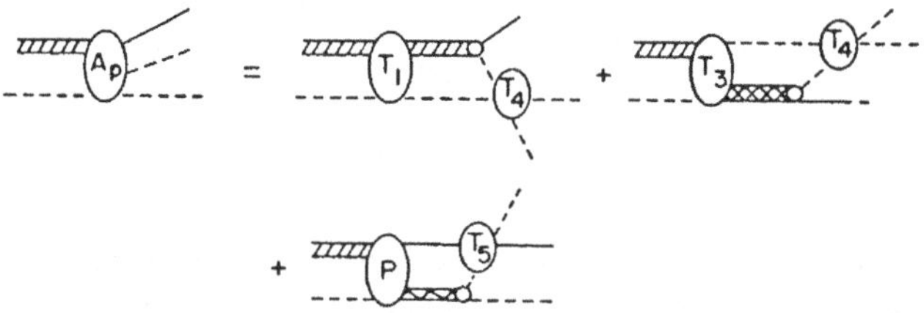

Figure 6. Continuum corrections to the composite model for production.

final state interaction of the $\pi - \pi$ system. The nonresonant contributions involve two-particle amplitudes to be approximated in the effective range or scattering length regimes. The contributions so obtained are then added to those already computed in the composite model theory, Figure 2. Finally, Amado[10] has commented upon the triangle graph singularities and their role in deciding what sorts of approximations to three-body wave functions are likely to be viable. In the theory presented here the relevant graphs are those of Figure 7. According to remarks already made we would only, here, compute these as a part of the nonresonant background contribution.

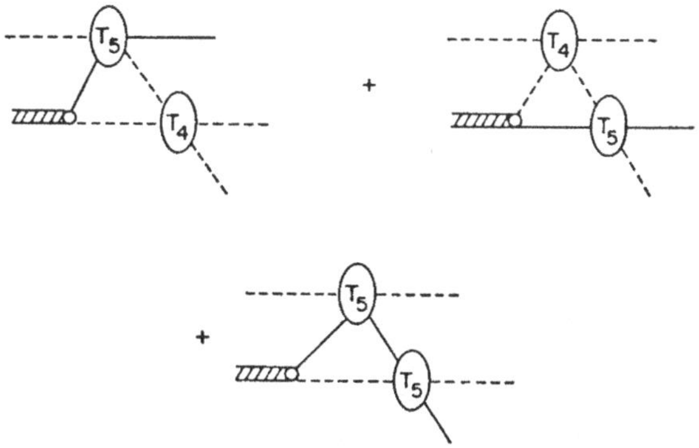

Figure 7. The triangle graph contributions to the continuum corrections to production.

We would like to acknowledge the receipt of helpful comments from R. Aaron concerning his work in progress on the class of problems discussed here. One of us (J.E.Y.) is indebted to R. D. Amado and André Martin for several discussions and continued encouragement. Both R. Aaron and R. D. Amado are thanked for reading the manuscript.

Appendix

The U-particle, or Direct Nucleon Pole

We record here the modification of the equations to be made in the presence of the direct nucleon pole. The representation of that pole as in Figure 3 is only meant to be suggestive. A correct solution of the problem is as follows. Solve the coupled equations of Figures 3 and 4 with $U = 0$, or the direct pole being suppressed. Denote the $\pi - N$ (previously called T_1) so obtained by the symbol A. The amplitude containing the nucleon pole is shown in Figure A1 and will involve an

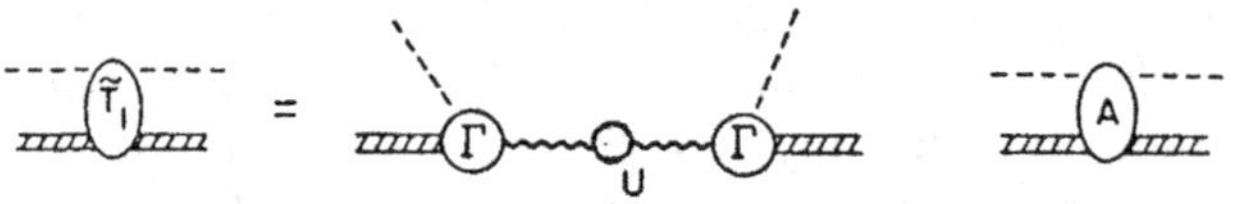

Figure A1. Amplitude for $\pi - N$ (particle scattering from a composite) scattering exhibiting the nucleon pole.

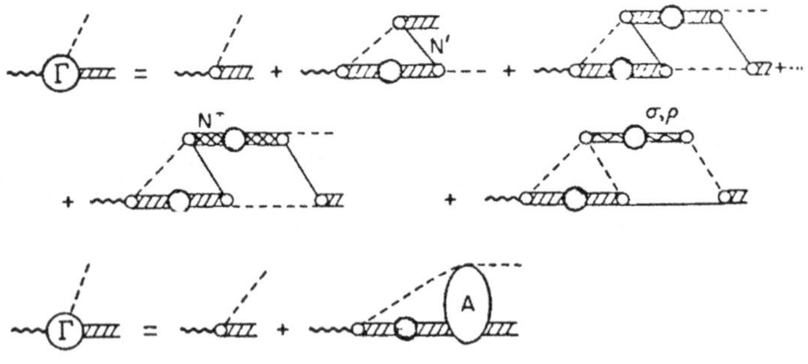

$$= g^{0}_{\pi NU} + g^{0}_{\pi NU}\,\Delta^{(r)}_{N}\,\Delta_{\pi}\,A_{N\pi}$$

Figure A2. The proper vertex function required for construction of the $\pi - N$ amplitude.

integral over *A*. This is made clear by ths diagrammatic representation of the vertex function, Figure A2. Our notation is described in Figure A3. There are some technical complications associated with our

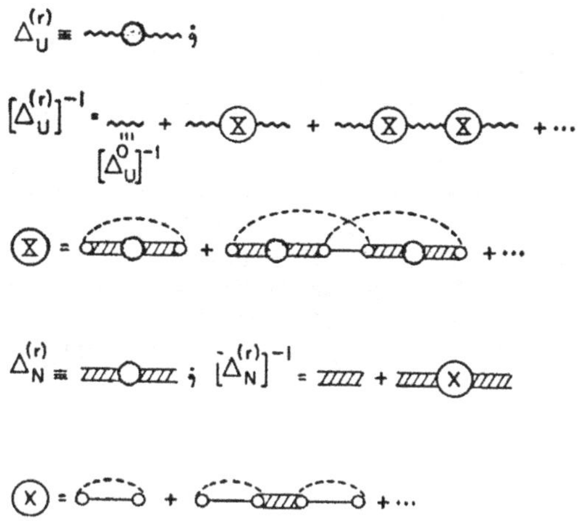

Figure A3. The renormalized *U*-particle propagator associated with the $\pi - N$ amplitude. The propagator has a pole at the physical mass of the nucleon.

procedure, indeed, an alternative one is known to Amado and Aaron, but these are not the proper subjects for further discussion here.

References

1. R. D. Amado, *Phys. Rev.*, **132**, 485 (1963).
2. R. Aaron, *Phys. Rev.*, **151**, 1293 (1966).
3. C. Lovelace, *Phys. Rev.*, **135**, B1225 (1964).
4. P. Narayanaswamy and L. K. Pande, *Phys. Rev.*, **136**, B1760 (1964).
5. P. Nath and K. V. Vasavada, *Phys. Rev.*, **152**, 1254 (1966).
6. W. D. Walker, J. Carroll, A. Garfinkel, and B. Y. Oh, *Phys. Rev. Lett.*, **18**, 630 (1967); P. E. Schlein, *Phys. Rev. Lett.*, **19**, 1052 (1967); E. Malamud and P. E. Schlein, *Phys. Rev. Lett.*, **19**, 1056 (1967).
7. H. Goldberg and E. L. Lomon, *Phys. Rev.*, **131**, 1290 (1963); ibid., **134**, B659 (1964).
8. J. Hamilton and W. S. Woolcock, *Rev. Mod. Phys.*, **35**, 737 (1963).
9. S. Weinberg, *Phys. Rev. Lett.*, **17**, 616 (1966).
10. R. D. Amado, *Phys. Rev.*, **158**, 1414 (1967).
11. M. G. Olsson and G. B. Yodh, *Phys. Rev.*, **145**, 1309 (1966); ibid., **145**, 1327 (1966).
12. C. J. Goebel and H. J. Schnitzer, *Phys. Rev.*, **123**, 1021 (1961).

13. G. F. Chew and F. E. Low, *Phys. Rev.*, **101**, 1570 (1955).
14. L. F. Cook and B. W. Lee, *Phys. Rev.*, **127**, 297 (1962).
15. M. Gell-Mann, *Phys. Rev.*, **125**, 1067 (1962).
16. T. D. Lee, *Phys. Rev.*, **95**, 1329 (1954).
17. A. I. Jaffe and A. S. Reiner, *Phys. Rev.*, **152**, 1304 (1966).
18. L. D. Faddeev, *Sov. Phys. — JETP*, **12**, 1014 (1961).
19. J. E. Young and P. R. Stein, *Nucl. Phys.*, **26**, 97 (1961).
20. L. Rosenberg, *Phys. Rev.*, **135**, B715 (1964).
21. D. Freedman, C. Lovelace, and J. Namyslowski, *Nuovo Cimento*, **43**, 258 (1966).
22. R. Blankenbecler and R. Sugar, *Phys. Rev.*, **142**, 1051 (1966).
23. R. Aaron, R. D. Amado, and J. E. Young (to be published).
24. F. Zachariasen, *Phys. Rev.*, **121**, 1851 (1961).
25. J. B. Bronzan, *Phys. Rev.*, **139**, B751 (1965).
26. M. Gell-Mann and F. Zachariasen, *Phys. Rev.*, **124**, 953 (1961).
27. R. Blankenbecler, *Nucl. Phys.*, **14**, 97 (1959).
28. M. Bolsterli and J. MacKenzie, *Physics*, **2**, 141 (1965).

F. J. Corbató
Massachusetts Institute of Technology

19 A Paging Experiment with the Multics System

Paging was first introduced in the Atlas computer[1] and promptly had great influence on system designers, for paging allows the solution of two problems in a systematic way: (1) It is possible to execute programs that are only partially loaded into primary memory or that require primary memory space larger than that available, and (2) It is possible to interrupt and remove programs from primary memory and later restore them with minimal storage allocation difficulties. These issues become especially important when one designs large-scale multiple-access systems in which multiplexing among user programs, which are competing for primary memory, is generally the case. Despite the obvious benefits that can result from paging, there is also a danger of extreme degradation of system performance whenever excessive paging activity occurs.[2,3] System degradation as a result of paging is, of course, caused by a mixture of overall system design (*e.g.*, a relatively small core memory) and the particular paging strategy employed. Since there still is not a great deal of insight into paging strategies, it is valuable to examine them so as to isolate this component of the general problem.

Introduction

Paging strategies and algorithms have been studied previously, for example, by Belady.[4] From these results and others, it is clear that there

* Work reported herein was supported (in part) by Project MAC, and M.I.T. research program sponsored by the Advanced Research Projects Agency, Department of Defense, under Office of Naval Research Contract Number Nonr-4102 (01).

is considerable statistical fluctuation in performing experiments, and more importantly, that there is a dependence on the characteristics of the program samples chosen and the computer environment in which the programs are operated. The strength of this dependence is still not well understood and is one reason for the present study. Moreover, because the Multics system[5] contains many radical departures from previous systems, it is a good candidate for investigating the invariance of results. Some of the departures that could easily affect paging behavior are: the use of recursive and pure procedures, the use of segmentation, the employment of a PL/I subset as a language for system programming, the homogeneous paging of much of the supervisor on the same basis as user programs, as well as the relatively large amount of pageable memory, and the particular page size.

With the above motivation for paging studies, it is important to recognize the goals and limitations that are being considered. It is, in principle, possible to design computing systems where all paging sequences are specified by the programmers. In this way, one would obtain more effective operation with near optimum sequencing of programs and data to and from secondary and primary memory. However, such a procedure is in g eneral impractical in the Multics sytem for a variety of reasons; one immediate obstacle is that the system is designed for the simultaneous multiplexing of multiple users and processors who in turn share a single copy of all common procedures or data. Thus a procedure may be simultaneously a member of several processes (*i.e.*, programs) each with different paging specifications. But even beyond this fundamental difficulty, there is the inconvenience to the programmer of determining and analyzing program behavior. Even granting that he were able to do it accurately and that he could specify data-dependent variations, there is still a problem that such specification represents unwanted clerical tedium. Thus, in large systems such as Multics, there is an important need for an automatic paging strategy that adapts to all situations, changes of programming habits, and even variations in the programming style or sharing patterns among the user population.

Having established that an automatic paging strategy is desirable, it also follows from the above considerations that such a strategy is constrained to using past performance and events to predict the future. It is only to the extent that there is a correlation of past and future behavior that a paging algorithm can have any effect. A key strategic question that a paging algorithm must answer whenever a new page is needed is: "Which page should be removed from core memory?" Two possible strategies of page removal are those of first-in-first-out (FIFO) and least-recently-used (LRU). In the case of the FIFO algorithm, the justification is that it is easy to implement; it is particularly valid in those cases where pages are brought in for a brief initial burst of activity and then abandoned for long periods. In the case of the LRU algorithm, it is not obvious how to implement it precisely without expensive hardware design assistance or without incurring paging overhead several orders of magnitude greater than can be tolerated; nevertheless, the LRU paging strategy is probably a good one since one would expect a high correlation of least-recently-used pages with pages unneeded by a program in the immediate future.

The Multics Paging Algorithm

In the Multics system a paging algorithm has been developed that has the implementation ease and low overhead of the FIFO strategy and is an approximation to the LRU strategy. In fact, the algorithm can be viewed as a particular member of a class of algorithms which embody for each page a shift register memory of length k. At one limit of $k = 0$, the algorithm becomes FIFO; at the other limit as $k \to \infty$, the algorithm is LRU. The current Multics system is using a value of $k = 1$, and it is the purpose of the present study to explore the effect of variations of k.

To understand more clearly the significance of k, a simplified view of the Multics algorithm is first given. The algorithm that is called upon whenever a new page is needed keeps track of all potentially removable pages with a pointer into a circular list, an example of which is shown in Figure 1, and cyclically evaluates each page for either retention or replacement by the new page. Each page has associated with it a usage bit in its page table entry and the usage bit is turned on (*i.e.*, set to 1) by the processor hardware whenever the page is referenced (*i.e.*, read or written). Whenever the algorithm examines a page entry, it extracts the associated usage bit and enters it into the high-order position of a k-bit shift register after shifting the contents of the register one bit-position lower. Then if the shift register is nonzero, the page is retained; if the shift register is zero, the page is replaced by the new page. In either case the usage bit for the page is turned off and the circular list pointer is advanced.

Some properties of this class of algorithms can be deduced from the above description. It is clear that the case of $k = 0$ is nothing but the FIFO algorithm. Further, as k increases, the pointer of the circular list must on the average revolve k times as far as it does in the case of $k = 1$

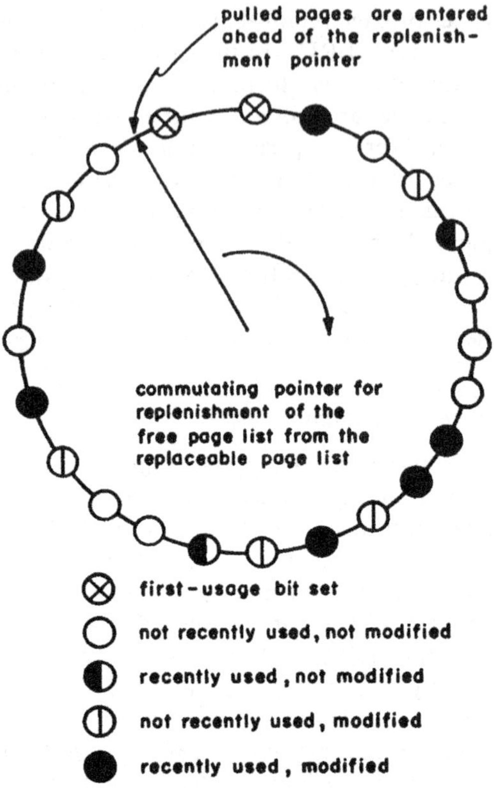

Figure 1. An example of the replaceable page list.

to yield as many page replacements. For large values of k, the mechanism takes on the character of a periodic sampled-data system where the number of leading (high-order) zeros in the shift register of a page is a direct measure of the number of cycles through the list since the page was used and hence the approximate time; in this way the LRU algorithm is reached in the limit.

An important aspect of this class of algorithm is that the overhead of operation increases with increases of k. It is therefore possible to engineer a practical algorithm by adjusting k to a value such that the reduction in page replacements due to the algorithm's improved predictive properties is just balanced by the increased overhead.

A critical aspect in the design of a paging strategy is the maintenance of stable operation under transient program behavior. Many of the algorithms studied by Belady have the property of cyclic performance variations in that periodically all pages have their usage bits reset simultaneously. Similarly, an early version of the Multics algorithm suffered from a periodic sampling of page usage bits which was only loosely coupled with page replacement frequency. To consider this issue further and to examine the stability of the present Multics algorithm, the example of $k = 1$ is considered. In this case, the usage bits are distributed on the average such that half are on and half are off. For if the circular pointer sweeps through the usage bit 1, it will leave behind it the usage bit 0. Conversely, when usage bit 0 is swept through, usage bit 1 is left behind since the replacing page brought in will in the case of demand paging certainly be used. Furthermore, with periodic page replacements, if there is an excess of zero usage bits, the pointer rotation rate will slow down and more one bits will be created; conversely, for the opposite case, the pointer rotation rate will speed up. Thus, the algorithm's decision making mechanism, because it is synchronized to the traffic of page replacements and by its design contains negative feedback, is quite stable in its performance.

In the area of algorithms used by system programs, it is easy for misunderstanding to arise and for nontrivial aspects to be overlooked. For this reason a more detailed view of the precise mechanism of the Multics algorithm is given in Figures 2 and 3. As will be noted in the following explanation, several additional refinements over what has already been discussed are needed to create a working strategy.

Figure 2 gives the flow of control which occurs when a program finds a page needed from secondary storage and a page fault occurs. The initial test in Box 1 normally takes the "no" branch. Because its purpose is easier to understand after the main logic of the algorithm is established, an explanation is given below rather than here. In Box 2 a test is made to ascertain if the pool of empty blocks of primary storage is nearing exhaustion. (A free block pool is used so as to be able to service a page fault more rapidly without waiting for a page to be removed from primary core memory.) If the free block pool is low, Box 3 is entered to replenish the pool. (Figure 3 represents an elaboration of Box 3.) Control next passes to Box 4 where a free block is designated for the new page and a pointer to this block is entered into a list of pages being read (*i.e.*, "pulled") into core memory. In Box 5, the location on secondary storage of the new page is determined from a master data base called the Active Segment Table (AST) and the secondary storage drum controller is directed to transmit the page into core memory. However, before this transmission is complete the program in Boxes 6 and 7 proceeds to do several bookkeeping steps for all

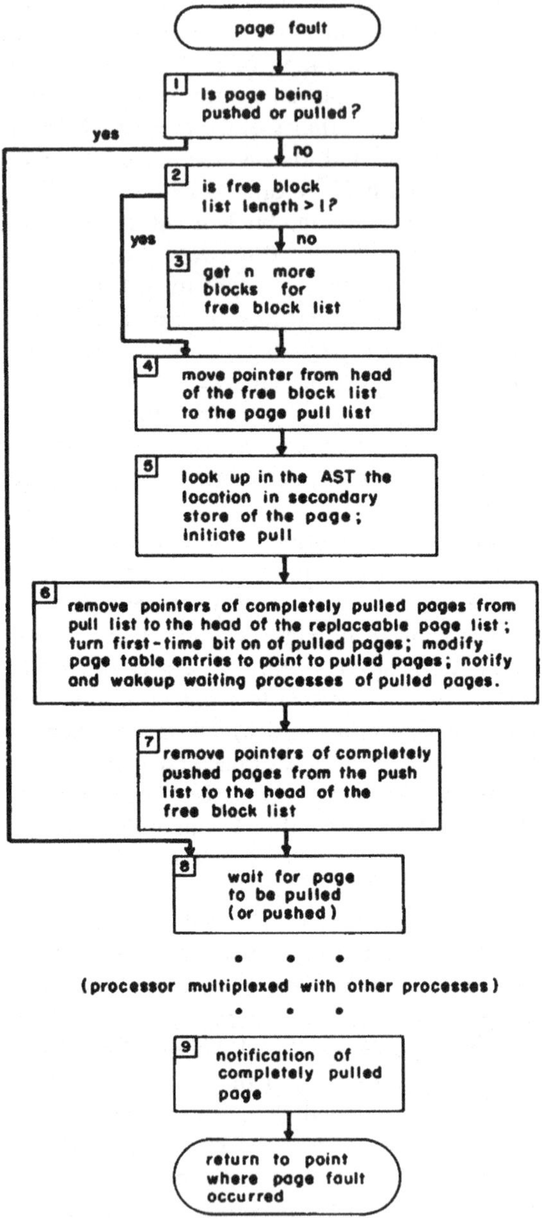

Figure 2. The Multics page fault algorithm.

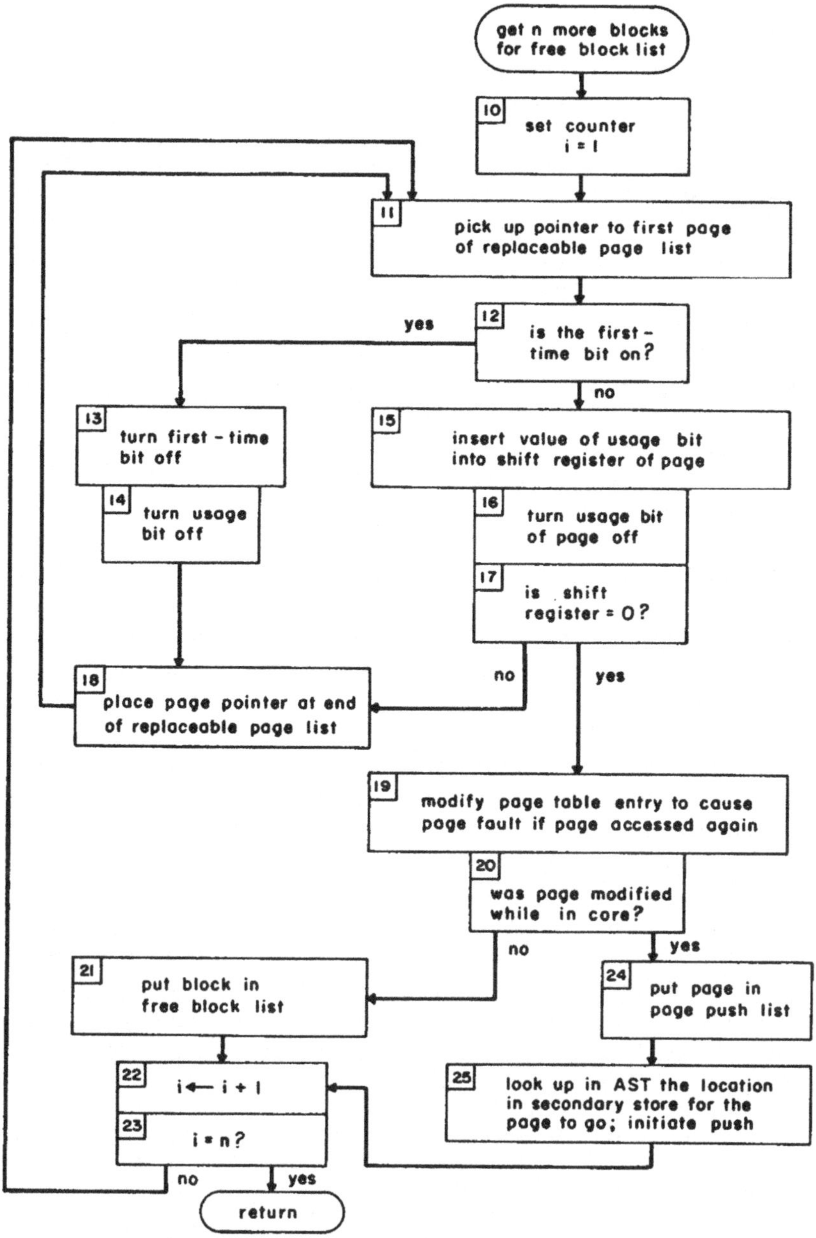

Figure 3. Replenishment of the free block list.

the drum-core transfers that have been completed since the drum status was last checked. One step is the switching of the page pointers from the pull list to the *head* of the replaceable page list. The reason for selecting the head rather than the tail of the list is that in addition a special "first-time" 1-bit switch is turned on for each of the newly pulled pages. As will be seen later in Figure 3, this switch makes it possible for the algorithm to ignore for a few page faults the initial burst of references to a page before it begins to monitor usage. (Otherwise a new page would always appear used after the first pointer rotation since the attempt to reference it is what triggered the pull.*) After setting the switches, the page table entries for the pages are adjusted so that attempts to reference the pages will succeed and the other processes which had been waiting for the arrival of pages now in primary storage are scheduled for processor service. Similarly in Box 7, the blocks corresponding to the pages which have been removed (*i.e.*, "pushed") from core memory to the drum are transferred to the free block list. At this point the process is unable to proceed until its needed page is pulled; hence in Box 8, control is transferred to the processor multiplexing section of the supervisor to await notification from another process. Finally at some later time when the page is pulled, another process schedules the original process for a processor, and in Box 9 control is returned to the point where the page fault occurred so that processing may resume as if uninterrupted.

The initial test in Box 1 only takes the "yes" branch in those relatively rare instances when a desired page is either being pulled into or pushed from primary memory. This situation can come about from two cases. In one case the needed page was in primary memory but became sufficiently inactive that a previous page fault has initiated its removal and the push is not yet complete. In the second case another process has developed a need for the page and has already initiated a pull. In either case, however, the flow of control is to Box 8 where the processor is released until the page transit is completed. At this time the process is restarted at the faulting location. If the page was being pulled, the process proceeds; if the page was being pushed, a page fault occurs and a fresh attempt is made to obtain the page.

* It should be noted that for the "first time" bit mechanism of the algorithm to be effective, the size of the free block pool should be somewhat larger than the number of processes being multiprogrammed; if this is not the case, the process which causes a page fault will probably not have an opportunity to run before the usage bit of the page in question is reset during free block pool replenishment. Under such circumstances the "first-time" bit mechanism can be removed with little effect on the overall algorithm.

Figure 3, which is largely self-explanatory, shows the flow of control required to replenish the free block list with n more blocks. (In Multics, n is currently set to a value of 3.) A few items are of note however. One is that the test in Box 12 is where the first-time switch allows the initial burst of usage of a pulled page to be ignored. Another item of note is in Box 20 where a test is made to avoid pushing pages which have not been modified while in core. (For this purpose the processor hardware assists by setting a modified bit in the page table entry whenever a page is modified.) In any case, however, control returns from the replenishment section when n blocks are selected for the free block list.

Finally it should be noted in passing that the algorithm presented in Figures 2 and 3, although basically correct, lacks many critically important features which are present in the version in the Multics system: These ignored features are needed for proper treatment of several complications. Some are

1. There are additional mechanisms needed for segment descriptor tables and for page tables. (In Multics these pages are either of two sizes: 64 or 1024 words.) In general there must be machinery for segment management.
2. Mechanisms are needed for the dynamic "wiring down" and "unwiring" of pages in core memory for purposes such as page tables and preprocess tables.
3. Data base interlock mechanisms and protocols must be established to allow multiple processors simultaneously to operate and take page faults.
4. Special attention must be paid in the details of the algorithm so as to allow simultaneous sharing of the pages of a segment by different processes.
5. Refinements are needed to allow for special situations such as pages which are created (with zeros) in growing segments.
6. In rare instances, the free block pool can become empty due to statistical fluctuations and special consideration must be made.
7. In the interest of coding efficiency certain mechanisms are not used exactly as described. For example, the push and pull lists are not formally created but are effected by the use of indicators in the elements of the replaceable page list.
8. In the case where the paging drum has completed several pulls but no page faults have occurred in other processes for an inordinate period of time, special provision must be made to reschedule for execution the processes which can proceed.

The Experiment and the Results

One of the purposes of this paper is to describe an experiment with the Multics paging strategy in which the value of k, the shift register length, was varied. To perform the experiment it was necessary to modify the system programs slightly but not in a way that seriously perturbs the results.

To serve as a test load, two cases were selected. The first case was the standard initialization computation which the system always performs to bootstrap itself in from the system tape, generate the configuration-dependent system environment, establish the paging and segmentation machinery, and then proceed to load the remainder of the system into secondary storage. Although there is a great deal of paging in this case, an objective is to establish whether or not the computation is typical.

The second case of the paging test was a short program written for the purpose which proceeded to call automatically a sequence of about ten basic, noninteractive commands. It was not felt that any more careful choice of sample was warranted since the Multics system is still undergoing rapid development and evolution. Moreover, past experience indicates that the commands chosen represent typical operations of a user population.

The environment of the experiment was a reproducible one-user, time-sharing system with a uniform page size* of 1024 36-bit words and a replaceable page pool of about 170 pages. Five distinct runs were made with different values of k and the values obtained are given in Tables 1 and 2. The page fault counts contain both those occurring explicitly as page faults and those contained implicitly within segment faults. (A segment fault occurs when there are neither pages nor segment page table in primary memory.) The Central Processor Unit (CPU) times given are those required to service the combined page and segment faults and do not include either the user terminal typing time or the delay due to the rotation time of the secondary storage drum.

It should be noted the average fault service times given are particularly large because of the present state of the Multics system at this time (July 1968). A soon-to-be-implemented strategy of reducing the number of effective segments by binding related groups together is expected to lower the average fault service time by at least a factor of four to the

* The Multics system potentially could have other effective page sizes as well as nonuniform page sizes. Future experiments may explore these directions although none are now contemplated.

neighborhood of a few milliseconds; the effect of this change will be to reduce the differences between the results for different values of k.

Conclusions

In examining the results given in Tables 1 and 2, several conclusions

Table 1. Page Faults *vs.* CPU Fault Service Time in System Initialization (total CPU time = 564.2 sec for $k = 1$)

k	Page Faults	Total CPU Fault Service Time (sec)
0	8309	184.5
1	4250	107.9
2	4098	106.5
4	4205	112.5
7	4317	120.2

Table 2. Page Faults *vs.* CPU Fault Service Time in a Sample Set of Basic Commands (total CPU time = 74.7 sec for $k = 1$)

k	Page Faults	Total CPU Fault Service Time (sec)
0	3628	72.9
1	1659	36.4
2	1635	37.5
4	1598	38.7
7	1725	44.3

can be drawn. First it is clear that the two cases give similar effects and that there is a dramatic improvement in going from $k = 0$ to $k = 1$. Second, a value of k in the range of 2 to 4 appears to give a paging strategy with the number of page faults down to a level where possible further improvements are small compared to statistical fluctuations. Third, when one examines the total fault service times, it is clear that as k increases so does the computational overhead of the paging algorithm. Since this latter basis is the pertinent one for comparison, the optimum value of k lies in the range of 1 to 2. Fourth, with so little difference between the results in the range of k from 1 to 4, a value of $k = 1$ is indicated. The reason for the latter choice is because a lower value of k

should produce an algorithm which is more stable and adaptive to transient changes in paging behavior. Fifth and finally, there should be some caution in extending the present results to other circumstances since: (1) the results are based on a small sample, (2) the Multics system may have properties which are uniquely its own, and (3) the system is still evolving and changing rapidly.

Acknowledgements

The Multics system is being developed on a cooperative basis by members of the Bell Telephone Laboratories, the General Electric Company, and Project MAC of M.I.T. The development and implementation of the paging strategy has been principally done by a team led by P. G. Neumann and R. C. Daley and including M. R. Wagner and G. F. Clancy. In addition, A. J. Goldstein made early background contributions, and F. J. Corbató assisted in developing the particular method of synchronizing page usage monitoring with page fault frequency.

The author would like to thank P. J. Denning, E. G. Coffmann, C. T. Clingen, and R. C. Daley for helpful discussions. Appreciation is given to J. W. Gintell and D. R. Vinograd who implemented the procedures used to meter the fault processing data. Finally, special and warm gratitude is extended to T. H. Van Vleck for his expert knowledge of Multics and his assistance in performing the experiments in the face of the extreme complexities which are present in the development of a large system.

References

1. T. Kilburn, " One-Level Storage System," *IRE Trans. Electronic Computers*, Vol. **EC-11**, No. 2 (April 1962).
2. P. J. Denning, " The Working Set Model for Program Behavior," *Commun. ACM*, Vol. **11**, No. 5 (May 1968).
3. P. J. Denning, " Resource Allocation in Multiprocess or Computer Systems," (Ph. D. Thesis), *Project MAC Report MAC-TR-50* (May 1968).
4. L. A. Belady, " A Study of Replacement Algorithms For a Virtual Storage Computer," *IBM Systems Journal*, Vol. **5**, No. 2 pp. 78–101 (1966).
5. F. J. Corbató and V. A. Vyssotsky, " Introduction and Overview of the Multics System," *AFIPS Conference Proceedings*, Vol. **27**, Fall Joint Computer Conference (1965).

C. S. Draper
Massachusetts Institute of Technology

20 Basic Principles of Guidance, Navigation and Control for Flight Vehicles

Introduction

Natural laws control the behavior of all things great and small. Thorough understanding of these laws, the spark of human creativity, and broad knowledge of formal reasoning processes make it possible to set up mathematical equations for describing situations that evolve from interactions among entities of the physical world. Throughout history, a few individuals have made truly impressive contributions to the power of mankind for dealing with the problems that appear from these interactions. Among these great ones, Philip M. Morse is distinguished by his mastery in the overall integration of conceptual understanding, theoretical reasoning, and practical computing to effectively predict observable results for complex systems of many kinds. His record of personal achievements, his extensive writings, and his ability to impart understanding and inspiration to students and associates have brought him unusual recognition from the world in which he lives. It is my objective in this paper to add a small stone to the edifice of appreciation to Phil Morse for his contributions in science, education, and engineering.

As a graduate student in physics during the 1930's, I had the good fortune to have Philip Morse as my registration officer, teacher, thesis advisor, and best of all, a personal friend and companion in devotion, among other things, to Chinese food. The sum of total experiences that came my way from these associations certainly had a strong influence on my career. Particularly valuable as a result of my long time emulation of an able mentor has been the development of habits leading toward independent thought in attacks on problems of all kinds. This viewpoint has for me surely been a strong factor in success with many situations involving pioneering technology and engineering on the frontiers of human endeavor.

Science, in my work for a doctorate in physics, and aeronautical engineering, as a field of simultaneous study and professional teaching, involved differences in details, but under the guidance of Phil Morse the various areas easily became complementary endeavors. The notions of fields, force, inertia, angular momentum, dynamics, and other basic concepts associated with actions inside and among atoms all came to have useful applications to everyday particles, rigid bodies, electronic devices, and real-world systems of all kinds. The considerable part of my education provided by my studies in the Physics Department has served me well during pioneering excursions into technology regarded as the area of human activity devoted to the realization of devices for changing the natural environment toward situations having improved compatibility with the needs and desires of mankind.

Navigation, Guidance, and Control

Flight vehicles, their power plants, and controls have occupied a substantial part of my attention during the last four decades, but my primary concern has been navigation and guidance of craft ranging from deep diving submersibles, surface vessels, vertical take-off and landing machines, aircraft, orbiters, and ships to travel in the space among the planets and the stars. In practice, *navigation*, the process of determining vehicle location, speed, and direction of motion is complemented by *guidance*, the process of determining *commands* for the *control* system which causes the vehicle involved to accomplish its mission by obeying these commands. The fundamental problems are those of geometry with respect to some space selected as the basic reference within which the location, orientation, and velocity of another space fixed to the vehicle may be described in quantitative terms. For operational use, the orientation and angular derivative components of vehicle space with respect to the chosen external reference space are combined with the vectors and derivatives of linear relative motion as the essential geometrical quantities for the purposes of control, navigation, and guidance. The situation is not one of science with observations directed only toward the determination of geometrical quantities. Rather it involves applications of capabilities to be developed by technology for actively changing these quantities so that the vehicle is forced toward fulfillment of some specified plan. The overall process is too complicated for a complete discussion here, but certain aspects of the basic problems in geometry may be of interest as a demonstration of how a sound and inspiring education may provide the fundamental knowledge and a mental attitude capable of producing significant results in widely divergent fields.

Reference Space, Vehicle Space, and Computation

Navigation, guidance, and control for manned or unmanned vehicles moving in environments of all kinds always depend upon a selected *external reference space* in which the desired path of the controlled vehicle is specified, and to which vehicle orientation and location are referred. These elements are defined in terms of geometrical relationships between the external reference space and a *vehicle space* chosen for convenience and fixed to structure. Coordinates representing this space have six degrees of freedom with respect to coordinates associated with the external reference space. The problems of engineering and technology are those of measuring and controlling the geometrical quantities that are essential for describing and forcing vehicle motions into successful patterns of action. Under the circumstances of modern technology, any commands determined from adequate information about vehicle orientation, location, and associated derivatives may be made effective by control equipments that have been generally available for some time. The frontiers of navigation and guidance have been, and still remain, in the region of sensing geometrical quantities, generating signals to represent these quantities, and processing these signals to realize the essential location and velocity information. With recent spectacular developments of computer science and technology, it is now feasible for navigational equipment to accept input signals of adequate quality, and to produce output indications good to a few feet in location, a few thousandths of a foot per second in velocity and fractional minutes of arc in orientation. All these results depend upon clock signals accurate within ranges including ten and more orders of magnitudes. Computing and time keeping are not limiting factors in applications of generalized geometrical theory that from the standpoint of navigation and guidance is effectively complete and ideal. It follows that results from properly designed and constructed subsystems to process information do not determine the overall inaccuracy levels of practical operations. The limiting sources of imperfection for navigational indications remain in the generation of signals representing physical quantities from which proper computing systems can produce satisfactory indications.

Radiation Contact Navigation

The technology of radiation signals based on contacts between vehicle space and known locations in external reference space now has a level of development that allows distance to be determined within a few feet and makes angular indications possible to fractional milliradians. When optical, ultra violet, infra-red, radar, radio, or laser contacts are continuously available between cooperative stations having fixed or moving known positions and orientations with respect to an external reference space, location and angle signals fed to proper computers result in good location and velocity indications of vehicle coordinates with respect to external reference space. The quality of these outputs is now adequate to meet the general requirements of navigation but do not usually provide high accuracy information on vehicle orientation and its derivatives.

Navigation, based on radiation contacts, may be satisfactory when such contacts are not effectively eliminated by clouds, fog, land, water, distance, or other obscuring effects, but in all situations depend upon inherently noisy signals that require increasingly careful filtering and correlation treatments as they become weaker. These treatments require time that results in averaging delays and uncertainties in magnitude that are troublesome when instantaneously available indications of high quality are desired. Cooperative stations in the external reference space are very helpful for improving the quality of navigational information and are now generally in operational use. Situations in which stations of this kind do not exist because of lack of equipment, earth curvature, hostile ground environments, undersea conditions, and great distances associated with solar system and interplanetary spaces may be met by equipment which complement radiation contact devices by self-contained vehicle-carried systems.

Self-Contained Vehicle-Carried Systems

Satisfactory systems, self-contained and vehicle-carried, must provide acceptable results for substantial time periods without external contacts after initial settings or up-dating information from external reference space. Submarines, operating well below the surface, aircraft flying over enemy territory, commercial aircraft over broad ocean areas, and space vehicles operating in regions from which lines of sight are obscured, exemplify situations in which good navigation has great importance under circumstances in which radiation contacts with cooperative stations are not dependable or do not exist. To overcome the difficulties associated with situations of this kind by continuously generating low inaccuracy indications of location and velocity and providing signals representing orientation and angular velocity of vehicle space with respect to external reference space, it is necessary to realize on-board systems capable of operation during significant intervals of complete isolation from the extra-vehicular environment. Among the subsystem functions essential in any navigation system, those of vehicle space, computation, and indication with adequate performance and acceptable size are well within the capabilities of equipment available from existing technology. The remaining functional requirement, that of sensing physical quantities from which vehicle location and velocity with respect to external reference coordinates may be derived has been and is more difficult. Means for mechanizing this function with high accuracy and great reliability did not exist two decades ago at the beginning of modern developments in navigation, and they still remain on the frontiers of technology.

Instrumented On-Board Reference Coordinates

This situation makes it apparent that the essential pioneering region of technology associated with self-contained vehicle-carried navigation systems is that of providing directly and continuously available on-board reference coordinates related in a known and useful way to external reference coordinates, and that of sensing components in these coordinates of physical quantities from which navigational information may be calculated. To be effective the internally instrumented reference coordinates must be accurately adjustable to selected known orientations with respect to external reference coordinates by the employment of radiation and mechanical contacts. Once adjusted, the arrangement must either maintain a physically present *geometrical reference member* or provide a pattern of computer-memory-stored information from which components with adequately low inaccuracies may be related to selected directions in external reference space.

Background for Performance Requirements

My own involvement with self-contained navigational systems started from personal experiences in piloting aircraft through bad weather conditions. Observations and thoughts during these sometimes obviously dangerous experiences forced the conclusion that the basic difficulty involved had its genesis in a lack of accurate information with respect to external reference coordinates and a complete lack of knowledge about any quantities from which location and velocity with respect to external reference space could be estimated. In the 1930's gyroscopic instruments giving visual indications of orientation, the vertical and north were generally available with inaccuracy levels in the range of a few degrees. Even rudimentary instruments designed to sense quantities leading to aircraft location and velocity did not exist. There was no significant thought directed toward attempts to realize self-contained navigation systems, and as a matter of fact, the authoritative NACA reports on aircraft instruments plainly stated that navigation by on-board equipment was not possible because of the performance levels that would be required. The necessary accuracies depended then and, of course, also today on the fact that on the earth's surface an angle equal to one minute of arc between verticals corresponds to one nautical mile. It was recognized that changes in location with respect to the earth could be related to double integrations of inertia reaction force associated with acceleration, but that difficulties were introduced by the almost spherical shape of the earth, and the fact that any instrument sensitive to acceleration would respond without discrimination to gravitational force. In Germany, Shuler, working during the 1920's on gyroscopic compasses for marine purposes, showed that effects from gravity could be separated from those of inertia reactions associated with accelerations by proper tuning of the system used to support the geometrical reference member. In effect, the Shuler period of some 84 minutes causes inertial reaction forces components in the horizontal plane to tip the reference member with an angular acceleration equal to that of the vertical observed on the accelerated vehicle caused by the linear acceleration divided by the earth's radius. The correctness of Shuler's theory has long been accepted but the technology of his day and for many years afterward was not able to produce mechanization to provide adequate performance from practical systems. One hindrance to the development of proper equipment was an almost universal tendency of engineers to regard instruments as individual gadgets with little attention to basic functions and to combinations of components, subsystems, and systems for forming overall systems having desired characteristics.

Performance Goals

This state of affairs existed when my own contributions to self-contained navigation systems started from a simple analysis of performance requirements for geometrical reference members and receivers to sense *specific force* (the resultant of gravity and inertia reaction force) components along convenient coordinate directions fixed in these members. In order to establish general performance goals, navigation over the earth's surface with inaccuracies of one mile or less for each hour of flight was chosen for consideration. With this in mind, and remembering that one minute of arc error in the direction of the indicated vertical corresponds to an inaccuracy of one nautical mile, it was obvious that the geometrical reference member had to limit increases of its angular errors to an average rate not greater than about one minute of arc per hour. For convenience, this *drift rate* may be considered as approximately one meru, (milli earth rate) defined as one thousandth of the earth's rate of 15 degrees (900 minutes) per hour. These considerations established an order of magnitude specification for geometrical reference member performance as about one meru in angular velocity. In 1945 (and also today) drift rates associated with aircraft instruments were about one earth rate, one thousand times greater than the minimum required for the geometrical reference member of a one-mile-per-hour inaccuracy navigation system.

From the standpoint of setting approximate performance goals for the specific force receiving instruments to provide adequate signals in one-mile-per-hour systems, it is effective, if not exact, to assume that the 6000 feet error builds up uniformly in 3600 seconds, an order of magnitude rate of about 1.7 feet per second. Under the working hypothesis that an effectively constant acceleration input error is responsible for this buildup of error, the basic relationship of distance equal to one half acceleration multiplied by the square of elapsed time shows that the resultant inaccuracy-causing acceleration corresponds to about one thousandth of a foot per second per second, which is approximately 3.1×10^{-5} earth's gravity. Sensors capable of handling acceleration inputs with this level of inaccuracy are able to provide signals allowing one mile per hour navigational errors.

These considerations establish rough specification levels of about one meru for geometrical reference member drift and capacity to sense specific force magnitudes of approximately 10^{-5} g for instruments to be used in one-mile-per-hour-error, self-contained navigation systems. It is also pertinent to note that applications for both civilian and military technology exist in which the assumed inaccuracy of 6000 feet after one hour of operation must be reduced by two or more orders of magnitude. This means that drift rate levels of one millionth of earth's rate and specific force input inaccuracies in the region of 10^{-8} g both have more than academic interest for the technology of self-contained navigation equipment. These objectives coupled with the requirement of reliable operation for many thousands of hours from components producible by industry were the challenges that originally caught my attention and attracted my best efforts for over three decades.

Applications of Inertial Principles

The fundamental problem of instrumenting a reference space able to maintain itself with accurately known geometrical relationships to another arbitrarily selected external reference space without mechanical or radiation contacts between the two spaces does not appear to be solvable by any means except those based on the principles of inertial reaction as described by the laws of Newton. These principles state that the time rate of change of vector velocity of any particle with respect to inertial space is proportional in magnitude to the magnitude of the applied force and has the direction of this force; and second, that an equal and opposite effect comes into existence as the instantaneous response to any applied force. It follows that any mass moving under no applied force will have no linear motion with respect to inertial space and could be used to identify an origin of coordinates nonmoving with respect to inertial space. Conversely, any mass constrained to move with a vehicle-carried geometrical reference must receive applied force components equal in magnitude to the magnitude of the linear acceleration of the vehicle and opposite in direction to this imposed acceleration. Leaving relativistic effects out of consideration, certainly a valid assumption for the velocities achieved by controllable vehicles built by men, this acceleration—inertial reaction force relationship—is quantitative and linear without either hysteresis at low accelerations or saturation for high accelerations.

Gyroscopic Principles and Angular Deviation Sensors

Geometrical reference coordinates present problems based on the continuous maintenance of orientation related to external reference coordinates. This becomes possible if means are available for detecting angular deviations with respect to inertial space from arbitrarily chosen and adjusted reference orientations of instrument units. Newtonian reaction forces acting on the particles of a rotating rigid body provide the basic effects needed for sensing angular deviations. The overall action of such a spinning wheel is conveniently described in terms of angular momentum which for a rotor with cylindrical symmetry is equal in magnitude to the moment of inertia about the axis of symmetry and the angular velocity of spin about this axis. Angular momentum is a vector quantity which is usually represented by a vector along the spin axis related to direction of spin by the right-hand rule convention of vector analysis.

For the special but generally typical situations that are associated with symmetrical rotors spinning with high angular velocities and subjected to only very low angular rates of the spin axis about directions at right angles to itself, it is valid to consider that the total angular momentum involved is identical with that generated by the spin of the rotor. Rotors of this kind, carefully maintained at a constant angular velocity of spin, determine the essential actions associated with gyroscopic instruments, and for this reason are called *gyroscopic elements.* Elements of this kind obey a law based directly on Newton's principles. The resultant action is simply that the time rate of change for the angular momentum vector of a gyroscopic element with respect to inertial space is along the vector (defined by the right-hand rule) of the applied torque. This means that the angular momentum vector turns directly toward parallelism and like sense with the applied torque at an angular velocity magnitude proportional to the magnitude of the applied torque and inversely proportional to the magnitude of the angular momentum. In situations where vibratory motion is not important, this action is called *precession* and is quantitative without hysteresis or saturation for substantially unlimited ranges of practical operation. This statement is, of course, based on the assumption of effectively constant spin angular velocity and rotors with substantially perfect rigidity to give unchanging moments of inertia. In practice, good mechanical design and excellent materials provide substantially perfect rotors while synchronous rotor drive motors supplied from power sources frequency controlled by modern techniques are able to provide angular momentum magnitudes that are constant for practical operation within a part in some one hundred million.

Sensor Calibration

Mechanization of geometrical reference members depends upon the sensing of angular deviations from adjustable reference orientations associated with the physical structure of instrument units which generate output signals that quantitatively represent the input deviations. Similarly, the function of computing navigational indications must be provided with signals based on the sensing of specific force components in terms of coordinates associated with the geometrical reference member.

Because the performance requirements of satisfactory angular deviation sensors and specific force sensors are especially severe, the factors that determine the nature and quality of the *calibrations* which quantitatively describe the essential input-to-output relationships involved have basic importance for design and performance in operation. Calibrations depend upon interaction of functions in two subsystems that appear in all sensing instruments; first, the transformation of input changes into physical effects under exactly known laws, and second, the generation of counterbalancing reaction effects that are accompanied by events capable of serving as output signals. As a matter of general practice, output signals are either continuous or pulsed changes in electrical quantities. The special problems of calibration for *inertial quantity* sensors stem from the low thresholds that must be reliably handled with certainty and the wide range of input levels over which calibrations must be effective and accurate. For angular deviations, lower limits are in the vicinity of one hundredth to one thousandth of a second of arc applied at angular velocities ranging from less than one second of arc per day up to many thousands of arc seconds per second of time. For specific force sensors the threshold lies in the vicinity of 10^{-8} or less of earth's gravity and has high limits of hundreds of earth gravities.

Inertial reaction forces associated with imposed linear accelerations and inertial reaction torques accompanying forced changes in orientation of angular momentum vectors are inherently linear over the required wide spectrum of input levels. This means that massive elements and gyroscopic elements are inherently ideal as input receptors. From the standpoint of calibration this means that the accurate determination of calibration for a single point is sufficient to provide calibration information unquestionably covering full ranges of operation.

Signal generation for small changes of a continuously varying input, which is usually the problem associated with angular deviation receivers, is easily provided by electro-magnetic or electro-static arrangements especially suitable for sharp indications of zero deviation as a null between small angular excursions. On the other hand, specific force inputs are generally integrated once to produce velocity information (after correction for the effects of gravity) and then integrated to compute changes in location with time. Because considerable periods are common with acceleration acting at various levels, output signals representing time integrals of specific force inputs after initial conditions are selected, often cover many orders of magnitude with the necessity of always low inaccuracies maintained throughout all levels. Signals of this kind are generally very difficult to realize in terms of analogue variations in any physical quantity. On the other hand, pulses adapted to describe information by means of digital formats permit the representation of arbitrarily great output ranges in numerical terms. Signals of this kind are particularly adapted for use as computer inputs especially from the standpoint of additional integrations that can be reduced to processes of counting.

Angular deviation sensor output signals for null indications are usually analogue in nature and do not present special problems in calibration. However, the design and calibration of these receivers for changes of their individual reference orientations in response to command signals is a more troublesome matter outside the scope of the present discussion.

Specific force receivers providing outputs that represent one time integration of input levels are well served by signal generators based on relative rotation of internal parts with an electrical pulse produced for each angular increment of known size. With relative angular velocity accurately proportional to input quantity level, no restriction on the number of revolutions allowed and an accurate calibration of the output angle content for each pulse, the requirements for performance and calibration over an indefinitely extended range can be satisfied for *specific force integrating sensors*.

Practical Inertial Quantity Sensors

For the reasons already discussed, inertial reaction forces and torques acting on masses and gyroscopic elements restrained by output-signal-generating arrangements are inherently capable of the wide range of operation required of instruments to sense angular deviations and specific force in mechanizations of self-contained geometrical reference members and specific force integrating receivers needed for self-contained on-board navigation systems. The fundamentals are unquestionably adequate from the standpoint of accuracy and range. The difficult problems are associated with the realization of practical systems which allow basic principles to operate without unacceptable performance imperfections being introduced by interfering effects. These effects appear because of the necessity of maintaining equipment in operation on moving bases attached to vehicles with sometimes erratically moved vehicles which must often sustain relatively high accelerations for considerable periods of time. Troubles appear in two general regions; the protection of basic behavior of specific force receiving masses and gyroscopic elements from disturbing forces and torques and the support of larger components and subsystems with effectively frictionless angular freedom.

My own contributions have been largely in these two regions through the conception and design of arrangements to allow Newton's laws to operate in effectively ideal environments. Two general methods of attack have been applied, the larger and heavier equipment parts have been suspended by conventional bearing and fluid pressure arrangements with servo drives controlled by nulling signals from angular deviation receivers to overcome the frictional, unbalance and other forces associated with the rotation-allowing supporting elements. This approach to the control of interference effects in navigation with three degrees of angular freedom devices was first applied in my Instrumentation Laboratory at M.I.T. over twenty years ago and is universal practice today for both military and civilian applications.

Interference Effect Control for Input Quantity Receiving Elements

It has been noted that a gyroscopic element responds to an applied torque by precessing about an axis at right angles to the torque axis and the angular momentum vector, the angular velocity magnitude being proportional to the magnitude of the applied torque, and that this action is suitable for the reception of angular deviation and for providing calibrated elements for specific force sensors. This gyroscopic behavior is the fundamental action for the realization of pracitical instruments to provide high performance in the presence of relatively great interfering effects.

A feeling for the difficulties involved may be based on the level of interference that corresponds to the torque from a threshold level angular deviation rate. As an example, with numbers having typical orders of magnitude, a gyroscopic element of one million gram centimeters squared per second may be considered. An imposed input angular velocity of 0.726×10^{-4} radians per second, which corresponds to one earth rate, would require a reaction torque of 72 dyne centimeters. Thus, if a threshold angular deviation rate of one millionth of earth's rate is considered, the corresponding torque would be 0.72

$\times 10^{-4}$ dyne centimeter. From this fact it appears that for practical instruments to reliably deal with one millionth of earth's rate, interfering torques must have levels not greater than a fraction of one ten thousandth of one dyne centimeter. The actual situation involved is, of course, much more complicated than can be illustrated by the simple example chosen, but the order of magnitude of allowable interference torque is generally correct. Before my concern for gyroscopic devices began some thirty years ago to introduce new viewpoints, instruments based on spinning rotors were generally characterized by the tendency of their spin axes to hold directions in space. Action of this kind is merely the special case that exists when torque applied to a gyroscopic element is kept at zero. No torque means no precession and no change in spin axis direction with respect to inertial case. Difficulties of a practical nature appear when spinning rotors are suspended with a two-degree-of-freedom (actually spin represents a third-degree-of-freedom as far as the rotor is concerned) gimbaled system commonly called a Cardan suspension. This arrangement, which dates back to the days of Foucault, is effective for illustrating gyroscopic action but involves so much friction that quantitative results are of low order. Attempts to improve the situation by using larger rotors and higher spin velocities have been universally self-defeating because interfering effects have increased as fast or even faster than controlling torques. The addition of servo drives has helped, but disturbances acting directly on the gyroscopic rotor have distorted the nulling signals to an extent that has prevented the realization of high performance.

Theoretical discussions of gyroscopic phenomena appear in many texts and have formed the subject matter for books of substantial size. The assumption of three degrees of angular freedom with relatively low friction as in the conventional top or electro-statically supported spinning spheres leads to extensive exercises in mathematics describing complex motions determined by initial conditions, weight unbalances, and mass distributions. These motions involve not only precession but also oscillations called *nutation* and conical paths for axes of spin under the forces associated with gravity and acceleration. Spinning bodies carried by inviscid field suspensions that serve for both spin and angular motions of the spin axis about directions at right angles to itself, have been studied for many years as the essential elements in gyroscopic instruments but have never reached the stage of routine use for operational systems.

My own approach to the use of angular momentum for the purposes of sensing angular deviations and receiving specific force inputs with one integration for output signals, has been derived from a knowledge of basic physics and experience with advanced mathematics. Many engineers of the M.I.T. Instrumentation Laboratory have worked with me on the problems of gyroscopic instruments and have made many design contributions, but the basic concept of restraining effective gyroscopic action to a single degree of freedom so far as input angular velocity and output torque are concerned has been an approach to practical instruments conceived, defended and developed by myself as the principal protagonist involved. Today the ideas conceived thirty years ago are working in many operational systems.

The basic notion involved is that if precession is restrained to a single *output axis* rigidly maintained at right angles to the spin axis while a preferred direction is established for the spin axis itself by a null-type signal generator acting between the rotor-carrying gimbal and the instrument case, it is possible to achieve output signals that depend on rotations of the case about an *input axis* at right angles to the *spin reference axis* (the direction of the spin axis for the signal output at its null point) and to the output axis. With such an arrangement, an output signal appears only if the instrument case is rotated about the input axis. Rotations about any other direction with respect to the case have no effect on an instrument using the configuration as described so that it is truly a single-degree-of-freedom angular deviation receiver sensitive only to rotations about its input axis.

The basic principles for designing and building inertial quantity sensors were clear many years ago, but techniques of mechanization for supporting spinning rotors with interfering torque levels less than one ten thousandth of a dyne centimeter in the presence of environmental vibration and acceleration of widely varying frequencies and amplitudes in ten and hundreds of earth gravities remained to be worked out. Obviously such low torque levels could not be expected from parts with surface contacts even in the presence of good lubrication. Flotation could be used to separate solid members, but certainly introduced erratic torques from eddy currents and convective motion. The key to the basic problem was supplied by the recognition that a clean, very viscous fluid used to float a spinning rotor contained in a hermetically sealed can could be made substantially free from erratic torques from a thin layer of fluid but unavoidably introduced strong viscous shear effects. These effects resulted in drag torques proportional to the relative angular velocity between inner and outer cylindrical shells, with the fluid contained in the clearance volume. With a gyroscopic element output torque proportional to the input axis angular velocity and the viscous shear drag reaction torque proportional to relative angular velocity, it was feasible to use viscosity effects for realizing a single integration of input angular velocity. Using this action it became possible to realize angular deviation receivers in which viscous shear was not an interference effect but actually provided an essential function in the chain of sensor component actions. This concept became clear some twenty-five years ago and provided the starting point for a long sequence of ever improving gyroscopic angular deviation receivers.

In practice, it is never possible to realize perfect floatation of a gyro rotor hermetically sealed within a cylinder. However, the volume coefficient of temperature expansion of a fluid is never exactly the same as the same coefficient for an immersed float. This makes it feasible to use controlled temperature variations for adjusting buoyancy toward the ideal of zero linear force tending to translate a float within its supporting fluid. Even with excellent thermal conditions, residual effects remain that must be reduced to zero by changeable electro-static or magnetic support units that need to exert only relatively weak forces because of near-perfect floatation. Substantially ideal results are being achieved from single-degree-of-freedom angular deviation sensor designed to apply the principles described.

Specific force integrating receivers may be designed to use angular deviation receivers with the features discussed above as reaction torque generators for an unbalanced mass located on the rotor spin axis to serve as the sensing element for specific force inputs along a direction parallel to the gyro unit input axis. By mounting this unit so that its input axis is parallel to the axis of rotation of a single-degree-of-freedom servo driven member, specific force acting on the unbalanced mass along this servo axis causes the torque to disturb the spin axis from its reference position so that a gyro unit output signal is sent to the servo system electronics. This signal causes the servo motor to drive its shaft in the proper direction for the gyro rotor output torque to oppose the unbalanced mass torque due to input acceleration. The resultant of these various actions is that the servo driven member rotates with an angular velocity proportional to the magnitude of the specific force input and a sense uniquely related to the direction of the input. Digital signals, with each pulse having a constant and known content of integrated specific force, provide an output signal capable of being started at any desired point and increasing without limit as a sequence of numbers.

Specific force integrating sensors in which the principles discussed above are applied now exist by the thousands and are giving excellent long-term performance in many applications.

Summary

Education bears many fruits often in strange places and extraordinary ways, but its effectiveness can surely be judged in terms of the creativity it has engendered. Phil Morse was my teacher in quantum mechanics and boundary value problems many years ago. The ideas and methods of attack he cultivated in this one student have effectively revolutionized the ways gyroscopic actions are utilized in measurements of geometrical quantities. Incidentally, the possible improvements in the performance of devices to sense angular deviations and specific force inputs started an era of great advances in the technology of systems for control, navigation and guidance. Today this technology has produced thousands of operational systems for civilian and military applications. Even with tangible results the surface has scarcely been scratched by comparison with new developments that will come during the years of progress that surely lie ahead.

Phil Morse has taught me to think in fundamental principles, leading me to truly realize the meaning of the statement that "the time rate of vector change of angular momentum with respect to inertial space is equal to the applied torque." Application of this general principle instead of "rigidity in space" or treatment of a rotating body as the structure with which a complex set of differential equations may be associated led me toward engineering and technology. Forty years spent in these fields has been very interesting and rewarding. I am eternally grateful to a great teacher for the knowledge and inspiration I received from him so long ago.

Herbert P. Galliher
University of Michigan

21 Exponential-Poisson Stress Processes, Linear Aging, and the Gompertz Law

It is a matter of common experience that the reliability law $\bar{R}(t)$ of a system — the probability that the system operates successfully for a length of time greater than t — will depend upon the kind of deployment to which the system is being subjected and upon the kind of environment in which the deployment occurs. A system that is tested in the laboratory may fail differently in the field. We may hear it said "a change in environment will do you good."

Thus there is in principle no *absolute* reliability law $\bar{R}(t)$ for a given system. This suggests trying to analyze a system's failure in terms of two factors, namely (1) the *stress* that is applied on the one hand, and (2) the inherent *strength* of the system on the other. Failure occurs when stress exceeds strength. Under stress is to be put all of the potentially harmful influences that are peculiar to the external situations of the system. Strength, on the other hand, is an idealized ability to withstand stress that is carried by the system into each situation. This strength thus becomes the basis for predicting the law $\bar{R}(t)$ for any specified exposure to stress.

The *age* of a system, though often measured in terms of the amount t of time during which the system has been exposed to stress, may be better represented by the system's current strength, and we sometimes use the term age in this way. But we shall not here.

In order to gain a complete accounting of the system's reliability, it is obviously necessary to include under stress not merely the intended stresses, but also the unexpected ones, the *noise*. Because failures so often appear to occur randomly during time, one is actually concerned with time patterns of stress, each of which may be well represented as a sample $A(t)$ of a stochastic process. Likewise the strength of the system is described in terms of a strength function of time. A common objective of observing the system's performance is to estimate its strength function and its current strength.

Consequently, interest develops in discovering the useful, common and parameterizable kinds of stochastic stress processes, and strength functions, as well as in determining the kinds of reliability laws that are entailed by specified combinations of stress processes and strength functions. This is potentially a large subject. In this paper we shall merely consider two rather interesting problems of this sort that relate to some recent work in biological research and that bring out some further potentials of the statistical theory of extremes. In any investigation of reliability, the role of a system's components quickly assumes importance. In connection with some medical questions, we were led to examine the oft-observed Gompertz law for human mortality. As we shall see, a system whose reliability follows this law under a certain simple family of compound-Poisson stress processes can be hypothesized to be a series system composed of a very large number of components whose strength functions decrease linearly or asymptotically linearly. This result is basically just a formal one in the statistical theory of extremes, which has been emphasized by Gumbel.[1] In passing, we shall also show that a certain series-parallel structure, with random stress and stationary component strength functions, that has been proposed by L. Szilard[2] for the length of human life, characteristically does not increase rapidly enough in failure rate, with increasing age, to correspond to a Gompertz reliability law.

1. Compound-Poisson Stress Processes

All the stress processes that we consider here are based on an underlying Poisson process of *stress times*. These latter are discrete events, or instants, whose average rate of occurrence we term the *stressing-rate* b. The probability $a_n(t)$ that during a time interval of length t there occur n of these stress times is thus given by the formula $(bt)^n \exp(-bt)/n!$. Obviously b might be nonstationary, but we confine ourselves to the stationary case here.

At a stress time there is a *stress magnitude* M. Typically, M is a random magnitude, with probability distribution $G(x) = \text{prob}[M \leqslant x]$. We shall use $\bar{G}$ for $1 - G$. In this paper, we shall be mostly interested in the case when $\bar{G}(y) = \exp(-my)$, which we refer to as the *exponential-Poisson.* It is the continuous-magnitude version of the *stuttering-Poisson.*[3,4]

Examples of compound stress processes will come readily to mind. While the tool wears steadily, nevertheless the final failure is typically triggered by some excessive momentary stress magnitude. The electrical circuit fails under an excessive voltage fluctuation or sudden physical damage. The rubber band, while it hardens with exposure to air, is nevertheless ultimately broken only at an occasion out of a process of relatively discrete occasions of use. The human body often succumbs during an acute medical episode; the same is true of large operational systems, as when the airport-acceptance system breaks down under too large a current traffic load. As to whether a compound stress pattern is compound Poisson in form or not depends primarily upon conditions of statistical independence.

In order to discuss some examples, it will be helpful if at this point we record some important principles, and specific terminology, from reliability theory. The function $h(t) = -(d/dt)\ln \bar{R}(t)$ is variously termed the hazard-rate, hazard-function, or age-specific failure or mortality rate or function of the system whose reliability is described by $\bar{R}(t)$. With $f(t) = dR(t)/dt$, the quantity $f(x; t) = f(x + t)/\bar{R}(t)$ is the age-specific probability density for the length of life that remains for a system of age t. Then, $h(t) = f(0; t)$. It is a useful fact that any probability density function $f(L)$ can be written in the form of the canonical age-specific exponential $f(L) = h(L)\exp[-H(L)]$, where $H(L)$ is the cumulative hazard $H(L) = \int_0^L h(t)\,dt$.

In Reference 5, a number of reliability principles are shown to be related to the type of monotonicity, if any, characterizing $h(t)$. Morse[6] had noted such principles for exponentially phased lifetimes, *e.g.*, it does not pay to maintain a parallel system of exponential phases. Manufacturing or genetic defects can make $h(t)$ decreasing for small values of t. Wearout goes with an increasing function $h(t)$. Failure from pure chance corresponds to $h(t)$ being a constant a in which case $H(t) = at$, proportional to the duration of exposure to hazard, and $\bar{R}(t) = \exp(-at)$, the (stationary) exponential. If the hazard rate is itself proportional to time, say $h(t) = ct$, then lifetime has the Weibull distribution. However, this distribution can also be regarded as a distribution of extreme values, thereby suggesting a mechanism by which a linear hazard rate is accomplished. Finally, for a system which is a complex of components or subsystems, the important relationship is that between the hazard rates of the components and the hazard rate of the system.

2. Two Basic Examples

The simplest example is that in which strength is independent of stress history. That is, $S(t)$ is independent of $A(T)$ for all t and T. Thus, $S(t)$ is simply a fixed function of time, and the system is nonplastic. In a biological system, perhaps such a strength function could only be genetically programmed. Suppose now that $A(t)$ is a compound-Poisson stress process. Then it is evident that

$$k(t) = \frac{h(t)}{b} = \begin{cases} \bar{G}[S(t)] & \text{for} \quad S(t) > 0 \\ \infty & \text{for} \quad S(t) \leqslant 0 \end{cases} \tag{1}$$

Clearly $k(t)$ increases if $S(t)$ decreases, and conversely. Since $\bar{G}(y) \leqslant 1$ for all y, the value of $h(t)$ can never exceed the value of the stressing rate b. This puts limitations upon the types of reliability laws that can hold for all values of the time t. Specifically, suppose that $h(t)$ is to be the Weibull ct. Then t will have to be kept less than b/c in value, a maximum life that may not be consistent with the range of observed values of t. If, in particular, $A(t)$ is the exponential Poisson, then for $t \leqslant b/c$, it follows that $S(t) = (-1/m)\ln(ct/b)$. The maximum life will be the same for all such stress processes, which again may not be consistent with observed facts.

As a second example, suppose that each increment of stress decreases strength by an equal amount, strength not changing between stress times. The system is then a mechanism for accumulating the experienced stress. Its strength at time t is now just $S(0) - A(t)$, and $A(t)$ is the compound-Poisson variate. The lifetime-to-failure is now the first-passage time for the compound-Poisson process, whose probability density function is

$$f(t) = b \sum_{n=1}^{\infty} a_n(t) g_n[S(0)] \tag{2}$$

where $g_1(x) = \bar{G}(x)$ and $g_n(x) = g(x) * g_{n-1}(x)$, the $*$ denoting convolution. It is readily verifiable from the defining expression for $h(t)$ that $h(t)$ cannot exceed the stressing rate b. On reflection, this is clear from the compound nature of the stress process. Again, therefore, limitations are placed upon the types of reliability laws that could characterize a system of this simple accumulating nature. For higher values of the hazard rate to be observable, it thus begins to appear that the system must have some series-like substructure.

Incidentally, whether or not $h(t)$ is monotone can depend upon the nature of the distribution $\bar{G}$. If M were always a constant, with zero variance, then as Morse[6] showed $h(t)$ would be increasing; L then has the Erlang distribution. However see Reference 5, pp. 36, 37 for the possibility that $h(t)$ might not be increasing.

In this example the value of the *strength* of the system at time t is a random variable which depends upon the cumulative stress to date. Thus the terminology *strength function* has to be replaced by *strength process*. This, we see, will in general be true of strength that is dependent upon the history of random stress.

The accumulating of stress can be accomplished by many alternatives of system structure. For example, if a parallel system is so designed that the current stress is borne by one component at a time, then the lifetime of the system is essentially a sum of random variables.

3. *Multicomponent Structures and Human Mortality*

We turn now to two examples of multicomponent structures that might serve as models for human mortality. The first of these is a proposal of L. Szilard. He suggests[2] that a somatic cell of an adult would be nonfunctional if both of any pair of the homologous chromosomes in the cell were defective. He hypothesizes a stationary Poisson process of *aging hits, i.e.*, stress times. An aging hit on a chromosome renders it defective. A number r of the chromosomes may have been defective at birth due to faulty genes, and this number is assumed to be random with a Poisson distribution. Of the total number N of somatic cells, when the number that are functional declines to a certain critical level k, the organism dies. In the terminology of multicomponent structures, Szilard's model of the cell is one of homologous chromosomes in parallel, with 23 pairs of chromosomes in series. The whole organism is a "k-out-of-N" (see Reference 5, p. 216) structure of cells. Using the 1950 U.S. Census Life Tables for white females aged 70 to 90, Szilard estimates that the average number of inherited faulty genes is between 2 and 4, and that the ratio k/N should be between 1/4 and 1/12.

This proposed model has been criticized (see Reference 7, p. 92ff.) in that no attempt is made to compare the consequences with the Gompertz law that the observed mortality function for humans increases exponentially with age (see Figure 1). We shall now show that in fact the

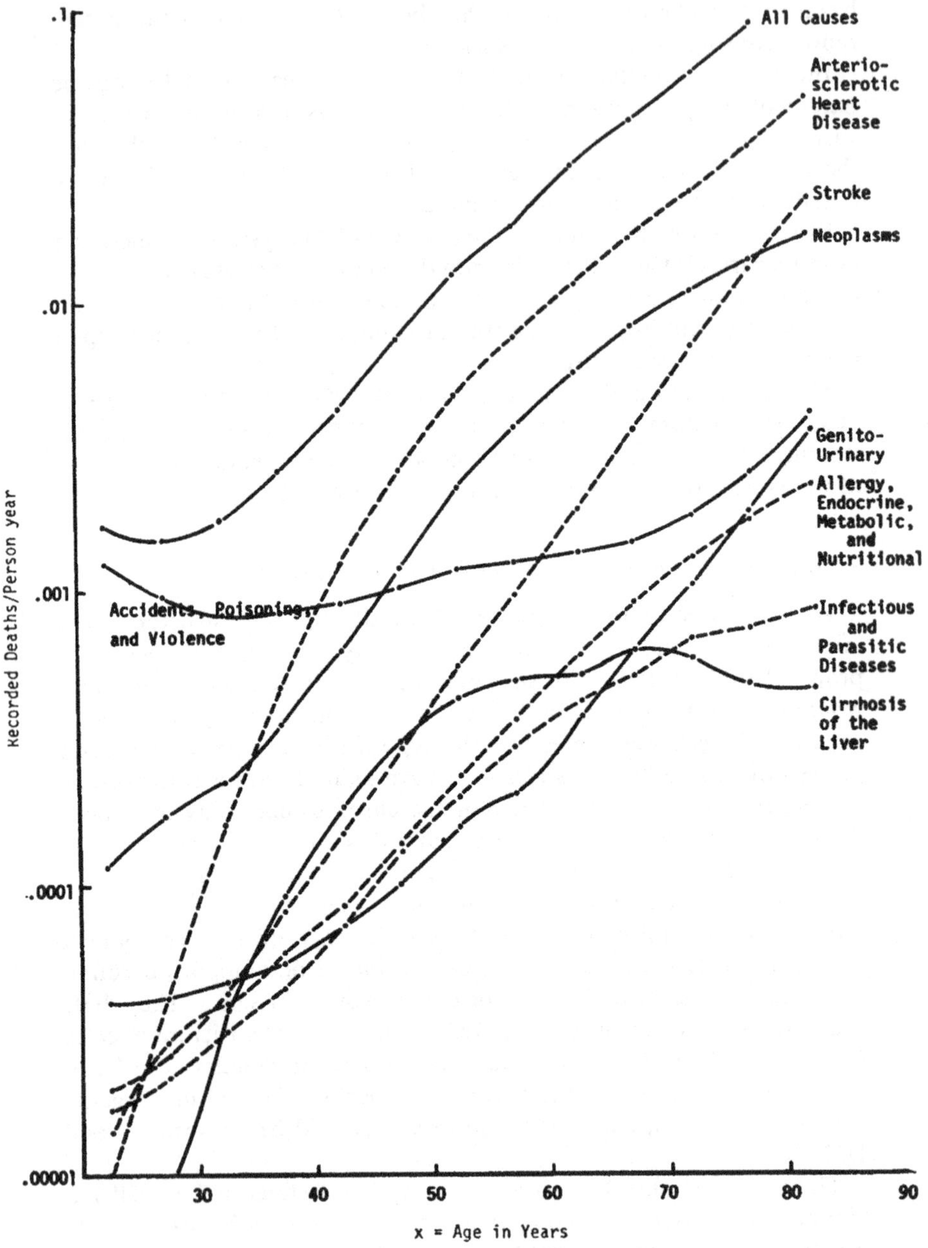

Figure 1. Death rates from various recorded causes. White males (U.S. 1960).

hazard function for this particular k-out-of-N model does not increase in a way characteristic of the exponent function.

There are $m = 23$ chromosome pairs in the human. Under Szilard's physical assumptions (we work directly from these, avoiding his approximations), the survival (reliability) law for the cell is $\bar{C}(t) = [1 - (1 - \exp(-bt)^2]^{m-r}[\exp(-bt)]^r$, from which the hazard function for the cell is readily found to be

$$h_C(t) = mb[1 - (1 - r/m)/(2\exp(bt) - 1)].$$

For the cell, $h_C'(t) > 0$, $h_C(0) = rb$, and $h_C(\infty) = mb$. The reliability law for the organism is, by assumption,

$$\bar{F}(t) = \sum_{i=k}^{N} m_i(t) \qquad \text{With } m_i(t) = \binom{N}{i}\bar{C}^i C^{N-i} \tag{3}$$

From this one obtains the k-out-of-N hazard function as

$$h(t) = kh_C(t)m_k(t)/\bar{F}(t) \tag{4}$$

For $0 < k < N$, $h(0) = 0$ and $h(\infty) = kmb$. Moreover, it is not difficult to show that $h'(t)$ is positive. Since k is large, it might therefore be argued that $h(t)$ could serve as a crude approximation to a function of the form $A\exp(Bt)$. However, it follows from Equation 4 that

$$u(t) = \frac{h'(t)}{h(t)} = \frac{h_C'(t)}{h_C(t)} + \frac{h_C(t)[N\bar{C}(t) - k]}{C(t)} + h(t) \tag{5}$$

From this it can be determined that $u(0) = \infty$ while $u(\infty) = 0$, since $h_C'(\infty) = 0$. A simple proof that $u'(t)$ is negative for all t does not seem possible. Nevertheless, were $h(t)$ of the form $A\exp(Bt)$, $u(t)$ would be the constant B, a requirement qualitatively different from the nature of $u(t)$ here. While admittedly r is also a random variable, nevertheless its range is deemed too small, in significant frequency, to provide any sufficient offsetting effect. Thus, the k-out-of-N model will require a more rapidly declining strength process for the chromosome in order to match a Gompertz-like mortality law for the organism.

The oft-observed Gompertz law is again rather evident in the plot in Figure 1 of human mortality from all causes. The rates in the figure are for white males, computed using the 1960 U.S. Census count as the population at risk, and the age-specific (5-year groupings) counts of deaths reported in Vital Statistics for 1960.[8] The approximate linearity with age of the logarithm is also evident for some groups of causes by which the deaths were classified, but not for all. Before inferring too much from such data, a number of cautions should be heeded. Classification of the cause of death may reflect inaccuracies, inadequacies, fashions, arbitrariness, and even falseness of diagnosis: Often there are at the time of death multiple medical conditions in the individual any one of which could be imminently fatal; the cause of death and the physiological route of death may be confused in the classifying; the classifications themselves are groups of potentially heterogeneous diseases; white males can constitute a statistical mixture of biological populations; recorded deaths are surely a process that exhibits random fluctuations in magnitudes.

From a purely formal standpoint, any one of three hypotheses suffices to yield a Gompertz-like law:

I. Exponential increase with age, of the stressing-rate. Suppose that a system were to grow in number of components, during time, by component *mitosis*, and that no successor components fail unless one of them acquires a certain condition. Under a stationary stress process, the hazard rate will be asymptotically exponential in the age of the system. No consideration of component strength function need arise.

II. The Strehler-Mildvan model. This model (Reference 7, pp. 103ff.) is a form of the first example, in Section 2 above. In it, $S(t) = S(0)[1 - at]$ and $\bar{G}(y) = \exp(-my)$, *i.e.*, a linearly declining strength function and an exponential-Poisson stress process. Accordingly, for $t \leqslant a^{-1}$,

$$h(t) = b \exp[-mS(0)(1 - at)]. \qquad (6)$$

Strehler argues that the stresses are fluctuations in survival-energy requirements, so that these fluctuations should have the Maxwell-Boltzmann exponential distribution of the kinetic energy of gas molecules. Strehler and Mildvan[9] provide data showing that various measurable body functions do decline approximately linearly with age, at rates whose average suggests a maximum life of the order of 150 years. They also point out that if the quantities b, m, and a are constants for all humans, then there should be an inverse linear correlation between $\ln \bar{R}(0)$ and $amS(0)$. Such correlation is in fact rather observable in data from various countries (Reference 9, p. 17).

It can be asked whether there might be "components" of the body for which this model holds. For example, the body might be a series system in such components. If each had a linear decline in strength function and if for the ith component the product $m_i a_i S_i(0)$ were the same common constant, the same Gompertz law would hold for each component and for the whole system. Strehler and Mildvan do not consider this question in detail in relation to the body functions which they examine; nor for each of those functions is the quantity $a_i m_i$ the same in the data they exhibit showing linear declines in strength.

One might hope to find additional evidence in the other curves in Figure 1. While the individual curves do rise approximately parallel to their sum, nevertheless each represents possibly a gross mixture of physiological functions and of kinds of stress. Moreover, if coronary deaths, which account for 40% of all deaths in each year of age from about 50 on, are mostly caused by the atherosclerotic plaque, possibly a model such as I above describes a growth in the number of potential sites for thrombus formation or for occlusion. Again, accidental deaths should probably first be removed from the data. And the curve for death from neoplasms may correspond better to the model which Szilard proposes. The subject cannot be pursued farther here. An extensive inquiry into cellular longevity will be found in Reference 7.

III. Smallest Value Statistics. We turn now to a purely formal, and elegant, type of model that results in a Gompertz reliability law, and that suggests further interesting inquiries into kinds of structures of systems composed of large numbers of components.

Suppose that we have a random variable with a distribution function $F(t)$. For samples of size N, define c_N by $F(c_N) = N^{-1}$. c_N is called the characteristic smallest value. Now let s_N be the smallest value in a sample of size N and let $\bar{F}_N(t) = \text{prob}[s_N > t] = \bar{F}(t)^N$. For example, suppose that $-\infty < t < D_N$ and that $F(t) = \exp[p(t - D_N)]$. Then $c_N = D_N - p^{-1}(\ln N)$. And, as $N \to \infty$, the quantity $y = s_N - c_N$ is distributed as

$$\bar{F}_\infty(y) = \exp[-\exp(py)] \tag{7}$$

which has a Gompertz hazard function. In Gumbel[1] will be found proofs that this result holds for any choice of the distribution $F(t)$ if the value of $F(t)$ decays at least exponentially fast for small t. There is a corresponding simple theory for the distribution of largest values. Moreover, the Gompertz distribution, such as Equation 7, is stable in the sense that the distribution of the smallest values from it has the same distribution. Thus, says Gumbel, "at each age death takes away the weakest members" (Reference 1, p. 247).

Extreme value statistics are important in many contexts, and Gumbel has collected numerous illustrations. There are only 3 types of asymptotic distributions of extreme values. One of these is the Weibull, which occurs when the *a priori* variate is limited in range. In Reference 10 the Weibull distribution is employed to represent the time to occurrence of carcinoma in tissue. In Reference 11 the Weibull distribution is found as a model of the duration of conflict situations, by a mental health researcher.

Now suppose that as $N \to \infty$, the value of D_N is kept fixed at the value $p^{-1} \ln N$. Then $F(0) = N^{-1} \to 0$, and the variate s_N approaches a Gompertz distribution. Thus if a system has a Gompertz failure law, and if it can be divided in finer and finer fashion into components which are statistically identical and independent, with linear strength functions, and if the system is a series one in the components, then it would be quite consistent that the length of component life should increase without limit at the rate of $\ln N$. A qualitative consequence of this sort is of course not surprising considering the physical construction of animate life and inanimate matter. It is consistent with the assertion of Weismann (Reference 7, p. 34) that "natural death occurs only among multicellular beings; it is not found among unicellular organisms." For our point this last claim need be only comparatively true.

Appreciation is expressed to Dr. W. J. Horvath for drawing the author's attention to some of the references.

References

1. E. J. Gumbel, *Statistics of Extremes*, Columbia University Press, New York (1958).
2. L. Szilard, " On the Nature of the Aging Process," *Proc. Natl. Acad. Sci.*, **45**, 30–45 (1959).
3. P. M. Morse, " Solutions of a Class of Discrete Time Inventory Problems," *Operations Research*, **7**, (1959) 1 (Jan.-Feb.) 67–78.
4. A. Kaufmann and R. Cruon, " Le Processus de Galliher," *Revue Francaise de Recherche Operationelle*, **3**, 137–144 (1959).
5. R. E. Barlow and F. Proschan, *Mathematical Theory of Reliability*, John Wiley & Sons, New York (1965).
6. P. M. Morse, *Queues, Inventories and Maintenance*, John Wiley & Sons, New York (1958).
7. B. L. Strehler, *Time, Cells and Aging*, Academic Press, Inc., New York (1962).
8. *Vital Statistics of the United States*, 1960, Volume *II*, Part A, Table 5-9. U.S. Dept. of Health, Education and Welfare, Washington, D.C., 1963.
9. B. L. Strehler and A. S. Mildvan, " General Theory of Mortality and Aging," *Science*, **132**, (1960) 3418 (July 1) 14–20.
10. M. C. Pike, " A Method of Analysis of a Certain Class of Experiments in Carcinogenesis," *Biometrics*, pp. 142–161 (March 1966).
11. W. J. Horvath, " A Statistical Model for the Duration of Wars and Strikes," *Behavioral Science*, **13**, pp. 1, 18–28 (1968).

Walter Helly
Polytechnic Institute of Brooklyn

22 A Residential Land Use Prediction Model

Knowledge of the future spatial distribution of an urban population is useful when planning municipal services such as roads, rapid transit, schools, fire stations, and public utilities. This paper exhibits a prediction model which may suffice for some of these purposes.

1. General Discussion

We assume that the overall population growth, if any, and the overall housing construction rate are estimated exogenously. A detailed model may also require a breakdown of the population into several socio-economic classes, a land use inventory, and a zoning survey. None of these are considered here; the present investigation is restricted to the question of where a given number of houses will be erected to house a given population.

In the aggregate, an urban population is not very mobile. Because it lives in substantially constructed houses that last for many years, the spatial distribution tends to change only as the result of the combined effects of new construction and of variation in the size of the overall population. This does not mean that there are no rapid redeployments of extant social classes. For example, a middle class may rapidly abandon the central old housing of a city and move to new housing in the suburbs. But the housing thus abandoned usually is claimed instantly by poorer persons (the *trickle down* process) who either are becoming more numerous through births and migration or are increasing their housing standards by occupying more rooms per capita. It is true that some central housing is abandoned because of great age or because land is taken over by commerce or highways, but such net demolition has rarely occurred on a very grand scale as compared to the overall city size. Thus the effect on spatial distribution of a middle class out-migration is a simple spreading of the city, coupled with a decline in room occupancy only if housing construction is large compared to population growth. When there is a decline in room occupancy, we may observe a net decline in population in the centrally located zones.

The above discussion is intended to emphasize the near obvious point that people live only where there are bedrooms. They redistribute themselves substantially only after additional bedrooms are constructed to receive them. (These comments apply only in the absence of war or other sudden upheaval.) Thus the problem of predicting population distribution is really one of predicting where new bedrooms will be built.

An additional bedroom can be built only on a site not yet occupied by a bedroom; replacement housing does not affect the population distribution. Thus the first requirement in urban growth prediction is to locate and enumerate all vacant sites suitable and legal for housing. Because new housing will survive a long time compared to occupancy by a particular tenant, the room — rather than the house or apartment — is a proper unit for a housing site census. The maximum number of rooms that may be built in a unit land area is restricted by zoning or custom. For any given zone, the available number of room sites is the maximum number minus the number already built on that zone.

At any given time, a prospective house builder — individual, commercial, or government — will have a choice among a number of vacant sites or *opportunities* from among which he selects one or more as required by his project. We hypothesize that he will tend to favor the sites most accessible to the main activities of the urban area. If the area is spatially symmetric, the centroid of such activities will be in the center of the area and probably will exhibit some central business district characteristics. If the area is not symmetric, it still is possible to establish one or more centroid points. Accessibility to a centroid usually is measured in terms of travel time (sometimes coupled with cost) by the predominant available transportation. One should note that a site's accessibility may change in time as a result of increased congestion or the construction of new facilities.

The available site opportunities are ranked in order of decreasing accessibility.

The underlying hypothesis of this paper's *intervening opportunity* approach is that a builder first considers the most accessible vacant site. There is some probability that he accepts it. If not, he considers the next most accessible vacant site. If this is not acceptable, he proceeds onward until he does accept a site. Thus the likelihood that a site is accepted is a monotonically decreasing function of the number of more accessible sites or *intervening opportunities*. This viewpoint has been used before with some success as documented in Section 2 of this paper.

The hierarchy of site opportunities is likely to differ for different socioeconomic classes. Thus a good model may have to be worked out for each of several such hierarchies within the same urban area. It should be pointed out that in the United States today almost all new site construction is for the higher income classes. Though there is some new housing for the poor, this is largely in the form of redevelopment of already occupied sites.

Naturally, the distribution of available site opportunities changes as time progresses. Areas near the centroid, or centroids, will tend towards saturation with few remaining opportunities. The suburbs will become more and more *attractive* simply because uncommitted sites will exist there. The process is likely to be accelerated as transportation improves, changing the order of accessibility, or as central regions are rezoned, reducing the number of remaining opportunities therein.

A builder usually plans his projects several years before they are first occupied by residents. Thus his estimate of the number of intervening opportunities, between a centroid and his project, is exaggerated compared to the situation faced by the ultimate householders. Such a time lag will distort the process and may be a cause of localized boom-bust cycles in housing construction. At the very least, time lags will result in the saturation of a central region somewhat earlier than intended by society.

The author believes that the relatively mechanical processes described above can explain much of the change in population distribution usually ascribed to relatively transitory social attitudes. Whether or not this belief is justified will require some detailed historical studies, here viewed as a subsequent model-validating project. In any event, the model of Section 3 below provides for all the processes enumerated thus far.

2. *Previous Formulations*

To the present author's knowledge, extant formulations of the intervening opportunity model are time independent; they represent the housing site selection process as occurring at a single moment in time.

The first proposal of such a model was by S. A. Stouffer.[1] Somewhat more elegant versions have been used by M. Schneider[2] for transportation planning and by G. G. Sherratt[3] for gas and sewer system planning. Rather than starting with an intervening opportunity viewpoint, Sherratt immediately postulates Schneider's result and proceeds to elaborate it to application on somewhat idealized asymmetric cities. Sherratt's model is discussed further by C. A. Wilkins.[4]

The general approach of these authors is as follows: Let there be one centroid so that sites nearest to it are considered first. Let $M(r)$ be the number of sites within a circle of radius r about the centroid. Because sites are very numerous in a real city, it is reasonable to treat integer-valued $M(r)$ as a continuous variable.

Let dM be the number of sites between M and $M + dM$. The intervening opportunity argument used is that there is a probability pdM that a given house's site is selected in the infinitesimal range dM, given that the site was not selected nearer to the centroid than M. Here p is a constant.

Let $\Pr(M)$ be the probability that any given house will be built further from the centroid than the M most centrally located sites. Then

$$\Pr(M + dM) = \Pr(M)(1 - pdM)$$

The solution to this equation is

$$\Pr(M) = e^{-pM} \tag{1}$$

If a total of N_T houses are built, the number built on the M most central sites is $N(M) = N_T[1 - \Pr(M)]$ and the density $n(M)$ of houses is

$$n(M) = \frac{dN(M)}{dM} = pN_T e^{-pM}$$

Note that the model does not ensure that $N(M) < M$. Thus the constants must be chosen to satisfy this practical requirement.

Suppose that sites are available at a uniform density d throughout the area. Then $M(r)$ is proportional to r^2 and

$$n(r) = n_c \exp(-ar^2) \tag{2}$$

$n(r)$ is the density of houses as a function of r, and $n_c = pdN_T$ is the peak house density at the centroid, and a is a redefined constant. Equation 2 is the formulation used by Schneider and by Sherratt.

Other models have been proposed by W. G. Hansen,[5] who offers a gravitational analogue, and by C. Clark,[6] whose model is based on a best possible fit of a negative exponential curve to data from several cities. H. Weiss[7] offers improvements on Clark's model.

3. The Basic Model

Suppose that all housing sites in an area, or all housing which appeals to a particular socioeconomic class in that area, are ordered so that the most accessible or desirable one is number one, the second is number two, and so forth. Thus the available sites at time zero can be described by the set of integers $1, 2, \ldots, m, \ldots$. No last integer is specified for a least accessible site because the total number of possible sites on earth is clearly a very large number compared to the number actually used by one urban area. In principle, all sites on earth might be used by that area.

Let $M(m, t)$ be the number of sites, no less desirable than site m, still vacant at time t. Then $M(m, 0) = m$. Because sites are numerous, little is lost in treating m and M as continuous variables. Further, for the reasons specified in Section 1, the room — rather than the house — will be viewed as the basic unit for m and M.

The site selection process for each new room is viewed as follows: The most accessible, still available site is considered first. There is some probability that it will be accepted. If not, the next most accessible site is considered, and so forth until a site is accepted. This process can be described by defining pdM as the probability that a given room's site will be in the range dM, given the prior rejection of all vacant sites more accessible than the range dM.

In the following argument, it is assumed that p is a constant independent of m, of time, or of the number of rooms already built. This is of course a simplification. At an early stage of a city's development, the number of sites at the immediate periphery of development is small. Thus the magnitude of p, which can be viewed as the probability that a given site is acceptable, is likely to be large compared to its value for an old large city with very many equivalent sites. Thus one way to improve the present model would be to let $p = p(m)$ and have it decrease monotonically with increasing m.

Let $\Pr(M) = \Pr[M(m, t)]$ be the probability that a new room's selected site will be less accessible than the M most accessible vacant sites. Then

$$\Pr(M + dM) = \Pr(M)(1 - pdM)$$

so

$$\Pr(M) = e^{-pM} \tag{3}$$

Equation 3 is the same as Equation 1, except that now $M = M(m, t)$.

Let $N(t)$ be the total number of sites occupied at time t. This is to be supplied exogenously so that $M(m, t) = M(m, N)$, as will now be shown.

In an infinitesimal time dt, some dN rooms will be built. Of these, a fraction $[1 - \Pr(M)]$ are expected to be built on the M most accessible, still available sites. The number of rooms built in dt on these M best sites is equal to the number of these sites that become occupied and thus removed from the vacant site stock in dt. Thus

$$\frac{dM(m, t)}{dt} = \frac{dM(m, N)}{dt} = -\frac{dN}{dt}[1 - \Pr(M)]$$

and explicit time may be cancelled from the equation. If we rearrange and substitute from Equation 3,

$$\frac{dM}{dN} = -(1 - e^{-pM}) \tag{4}$$

The solution to Equation 4 is found by separation of variables and by integration. The constant of integration is evaluated by setting $M = m$ at $N = 0$. One obtains

$$M(m, N) = \frac{1}{p}\log[(e^{pm} - 1)e^{-pN} + 1] \tag{5}$$

Let $F(m, N)$ be the probability that site m is occupied when N sites are occupied. In other words, $F(m, N)$ is the fraction of sites occupied in the range $(m, m + dm)$. Then

$$F(m, N) = \frac{d[\text{number of sites occupied in the range } (0, m)]}{d[\text{total number of sites in the range } (0, m)]}$$

$$= \frac{d[m - M(m, N)]}{dm}$$

From Equation 5,

$$F(m, N) = \frac{e^{pN} - 1}{e^{pN} + e^{pm} - 1} \tag{6}$$

Figure 1 exhibits the form of Equation 6 for various values of pN and of pm. The abscissa is in units of pm. Large values of the abscissa variable correspond to the less accessible, less desirable sites with large m. The exhibited family of curves shows the progression of the city from its beginning with few houses, pN small, to its later fully developed stages, pN large, when most of the more desirable sites have been built upon.

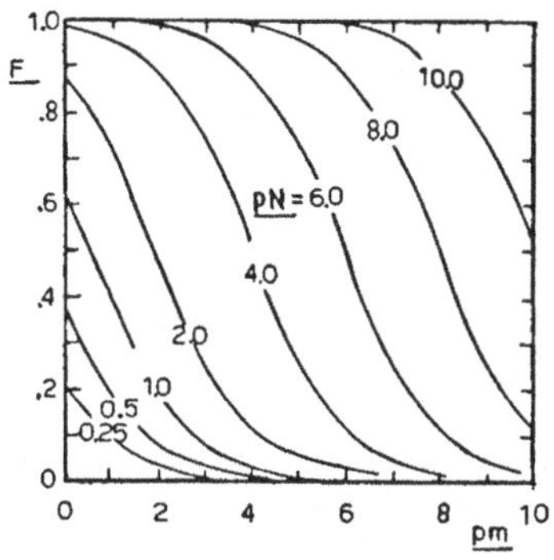

Figure 1. Housing site occupancy: F = probability that a site is occupied, m = number of that site in decreasing order of accessibility, N = total number of occupied sites, p = probability that a *considered* site is accepted.

Because of saturation of the inner (small m) sites, later building occurs almost entirely on the outer fringes (large m). It should be recalled that here we deal only with new site development not rebuilding of existing structures.

4. *Illustrations and Extensions*

Symmetric City with Constant Site Density. It is not normal for United States cities that the density of sites remains constant as distance from the centroid is increased. However in some European cities, such as Paris, a tradition of residence in apartments tends towards fairly uniformly high-density zoning throughout the area.

In a circularly symmetric city with constant site density d, the number of sites within a radius r of the centroid is $m = \pi r^2 d$. From Equation 6, the density, in number of sites per unit area, of developed sites is

$$n(r) = F(\pi r^2 d, N) \cdot d = \frac{(e^{pN} - 1)d}{e^{pN} + \exp(pd\pi r^2) - 1} \tag{7}$$

For small N and large r, Equation 7 approaches

$$n(r) \doteq pNd \exp(-pd\pi r^2) \tag{8}$$

This has the same form as the time independent formulation, Equation 2.

Symmetric City: Site Density Declines with Increasing Distance from Centroid. In most United States cities, zoning and custom dictate a declining housing site density with increasing distance from the centroid.

To illustrate the effect of this policy, we postulate a site density $d(r)$, in rooms per square mile, consisting of two terms. The first declines exponentially with r; the other is constant and represents the maximum site density in open country. Thus

$$d(r) = c_1 e^{-ar} + c_2$$

where a, c_1, and c_2 are constants. The number of sites between the centroid and r is given by

$$m(r) = 2\pi \int_0^r (c_1 e^{-ax} + c_2)\, x dx$$

$$= 2\pi \left\{ \frac{c_2 r^2}{2} + \frac{c^1}{a^2} [1 - (ar + 1)e^{-ar}] \right\} \tag{9}$$

The fraction of sites that are occupied, $F[m(r), N]$, is given by Equation 6 with $m(r)$ from Equation 9. Then the density of occupied sites is $n(r) = F[m(r), N] \cdot d$.

For a numerical example, suppose that $c_1 = 100{,}000$ rooms per square mile, $c_2 = 5000$ rooms per square mile, $a = 0.2$, and $p = 10^{-6}$. If the average housing unit consists of five rooms, $p = 10^{-6}$ implies a chance of one in 200,000 that a particular *considered* site is accepted. As remarked on before, this probability is an overall average, being too small for the city in its earliest stages and too large when it has a large population.

Figure 2 shows $n(r)$, in rooms per square mile, calculated for the given parameters and for a total number of rooms ranging from $N = 250{,}000$ to $N = 10{,}000{,}000$. At an average occupancy of 0.5 persons per room, this shows a hypothetical city development up to a population of 5,000,000.

Effect of a Time Lag. Suppose that there is a time lag between a site's selection and sufficient activity on that site to influence subsequent site selections. One may represent this situation by assuming that site selection, at the development stage of N occupied sites, is based on the earlier stage when $(N - \Delta N)$ sites were occupied. If so, Equation 4 becomes

$$\frac{dM(m, N)}{dN} = -[1 - e^{-pM(m, N - \Delta N)}] \tag{10}$$

An approximate solution can be obtained, to first order and for constant ΔN, by setting

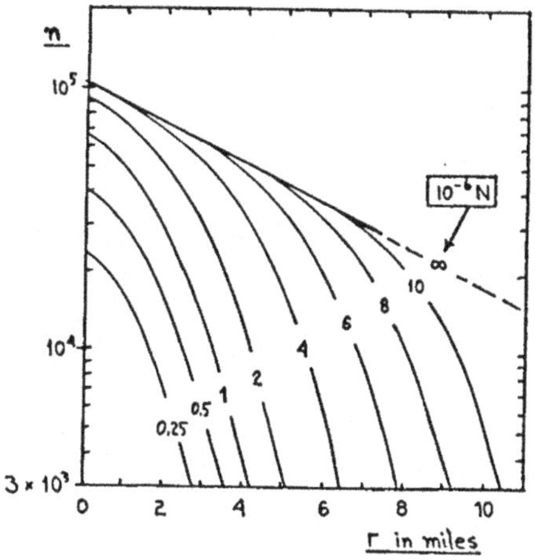

Figure 2. Housing density $n(r)$ in a circularly symmetric city where site density declines exponentially with radius r; N = total number of occupied sites.

$$M(m, N - \Delta N) \doteq M(m, N) - \Delta N \frac{dM(m, N)}{dN}$$

The solution is

$$M(m, N, \Delta N) \doteq \frac{1}{p} \log\{[1 + e^{-pm}(e^{pN} - 1)]^{-p\,\Delta N}(e^{pm} - 1)e^{-pN} + 1\} \quad (11)$$

The fraction of sites occupied in the range $(m, m + dm)$ is

$$F(m, N, \Delta N) = \frac{d[m - M(m, N, \Delta N)]}{dm}$$

$$= F_0\left\{1 + \frac{p\,\Delta N}{1 + \beta}\left[\frac{\log(1 + \alpha)}{\alpha} - \beta\right]\right\} \quad (12)$$

where

$$\alpha = \frac{e^{pN} - 1}{e^{pm}} \qquad \beta = \frac{e^{pm} - 1}{e^{pN}}$$

and F_0 is the value of F when $\Delta N = 0$, as given by Equation 6,

$$F_0 = \frac{e^{pN} - 1}{e^{pN} + e^{pm} - 1} = \frac{\alpha}{\alpha + 1}$$

The behavior of the rather complicated Equation 12 can be visualized by considering its behavior near the centroid and at the outer edges of the city. Near the centroid, where m approaches zero,

$$\lim_{m \to 0} F(m, N, \Delta N) \doteq F_0 + p^2 N\,\Delta N e^{-pN}$$

Since here $F > F_0$, the time lag results in a greater construction volume at the center than would occur without a time lag.

At the outer edge of the city, m is large and

$$\lim_{m \to \infty} F(m, N, \Delta N) \doteq F_0(1 - p\,\Delta N)$$

Here $F < F_0$ and the development is sparser than without a time lag.

A quantity of interest is the time of peak *overdevelopment* at the city center. Since $N = N(t)$, supplied exogenously, the value of N at this peak is equivalent to the time. The maximum overdevelopment at the centroid occurs at

$$\frac{d}{dN}[F(m = 0, N, \Delta N) - F_0(m = 0, N)] = 0$$

which yields

$$N \doteq \frac{1}{p} \tag{13}$$

One might speculate on the possibility of a setback in the housing market immediately after the number of rooms exceeds $(1/p)$.

Change in Occupancy. The relation between overall population and the housing supply changes rather slowly. For the United States, there has been a decline from about one person per room in 1900 to 0.7 persons per room in 1960. The decline in the most crowded central *slum* areas of large older cities has been more drastic. In order to relate future population distribution to the distribution of the housing supply in such older city centers, it will be necessary to extrapolate past local occupancy trends. The Census Tract Series of the decennial United States Census may provide some of the pertinent data.

Demolition. In most of the urban area, land zoned for and devoted to residences remains thus dedicated for an indefinitely long time. However, at the city center, where most housing is old and where competition for land by commercial and civic users is intense, there is some demolition and consequent permanent reduction in both the housing supply and the number of remaining vacant sites. Thus, for old central business districts it might be appropriate to use an intervening opportunity model, as described here, to predict the growth of commercial and governmental structures in areas previously devoted to housing. The sites claimed by these nonresidential users should be subtracted from the housing site inventory.

5. Model Validation

The present paper proposes a land use prediction model without a parallel presentation of objective evidence regarding its validity. It thus remains to calibrate the model for several urban areas and to use the calibrated model to *predict* growth in these areas over a reasonably prolonged period in the past. The model can be considered useful only if one thus obtains reasonable parallels to actual developments. An effort to this end will be instituted.

References

1. S. A. Stouffer, " Intervening Opportunities," *Am. Soc. Rev.*, Vol. **5**, No. 6 (1940).
2. M. Schneider, in *Chicago Area Transportation Study — Final Report*, Vol. **2**, City of Chicago, Chicago, Ill. (1960).
3. G. G. Sherratt, " A Model for General Urban Growth." *Management Sciences, Models and Techniques*, Vol. **2**, 147–159, Pergamon Press, New York (1960).
4. C. A. Wilkins, " Methodology of Urban Population Distributions," *Operations Res.*, **16**, 1–9 (1968).
5. W. G. Hansen, " Land Use Forecasting for Transportation Planning," *Highway Research Bulletin*, **253**, 145–151, Highway Research Board, Washington, D.C. (1960).
6. C. Clark, " Urban Population Densities," *J. Roy. Stat. Soc.*, **114**, Pt. IV, 490–496 (1951).
7. H. Weiss, " The Distribution of Urban Population and an Application to a Servicing Problem," *Operations Res.*, **9**, 860–874 (1961).

Ralph L. Miller
General Motors Corporation

23 Analysis of Variance Techniques: An Expository Review

Introduction

There are a large number of statistical techniques which fall under the heading of analysis of variance. Workers in most fields are familiar with contingency tables and F-tests, correlation, and linear regression. Less widely used are discriminant analysis, components analysis, factor analysis, and canonical correlation. In this paper we shall take a look at all of these techniques and attempt to provide some perspective as to what each can and cannot do, as well as how they relate to one another.

This paper is basically expository. While some of the findings are original, most are well known. However, the literature on this topic is difficult for analytic-minded people — or at least it was for us. Much of it is from the field of psychology, and the mathematical simplicity and authenticity are shrouded in metaphysical discussions of *real causes, underlying structure*, etc. Most of the remaining literature is from theoretical statistics, and while mathematical rigor is preserved, it is difficult to maintain a sense of application.

One convenient use of analysis of variance techniques is in projects involving large amounts of data. They can be applied in a *preanalysis* phase to reduce the data and get a feel for it, or they may be used in final models. The purpose of this paper is to briefly explicate them in terminology meaningful to analytic workers.

Correlation

The concept of correlation will undoubtedly be familiar to most readers. It will be discussed here only so that it can serve as a base point, relative to which the other techniques can be positioned.

The correlation between two variables u and v is defined to be

$$r_{uv} = \frac{\operatorname{cov}(u, v)}{\sigma_u \sigma_v} \tag{1}$$

If we seek a linear relationship between u and v

$$u = \alpha v + \beta \tag{2}$$

it is reasonable to seek to minimize the expected squared deviation. Since σ_u^2 is fixed, we can minimize

$$\phi = \frac{E[(u - \alpha v - \beta)^2]}{\sigma_u^2} \tag{3}$$

Taking partial derivatives we find that the optimal solution is given by

$$\alpha = \frac{r_{uv}\sigma_u}{\sigma_v} \tag{4}$$

$$\beta = \bar{u} - \alpha \bar{v} \tag{5}$$

$$\phi = 1 - r_{uv}^2 \tag{6}$$

Thus, we find a good fit whenever r_{uv} is near plus or minus one.

Note that this derivation differs in an important theoretical respect from classical derivations. Normally, v is treated as given, and α and β become random variables conditional on v. Either minimum variance or maximum likelihood estimates are then made for α and β, again conditional on v. The results are not necessarily optimal (in any sense) for any other values of v. Yet it is common practice to use them for predictive purposes with new values of v.

The approach taken here is quite different; v is treated as a random variable and α and β as constants. Least-squares estimates are then made for α and β. These estimates involve parameters of the distributions of u and v. If these parameters are known, there is a theoretical nicety about this approach. However, more commonly these parameters must be estimated from the data. If the usual estimates for means, variances, and covariances are used, the numerical results are identical to those under the other derivation.

No claim is being made for the superiority of this approach. The intent is merely to highlight some of the often overlooked implications of this type of analysis and to place those implications into some sort of perspective. It should be clear that for large samples the two viewpoints converge theoretically as well as numerically — the conditional sample space becomes a replica of the total (unconditional) sample space, just as the parameter estimates become exact.

Multiple Linear Regression

Multiple linear regression is a very simple generalization of the above. Here v is a random variable, and may be a function of other random variables.

Suppose, in particular, that

$$v = \mathbf{b}'\mathbf{X} \tag{7}$$

where $\mathbf{X}$ is a vector of random variables (independent variables), predictors, etc., and $\mathbf{b}$ is a vector of coefficients. We can write

$$u = \alpha v + \beta \tag{8}$$

$$= \alpha \mathbf{b}'\mathbf{X} + \beta \tag{9}$$

$$= \mathbf{c}'\mathbf{X} + \beta \tag{10}$$

with $\mathbf{c} = \alpha\mathbf{b}$. We can optimize in two steps. Equation 6 gives the minimal values of ϕ for fixed r_{uv}. But r_{uv} depends on the coefficient vector $\mathbf{b}$, and we can further improve the objective function by choosing $\mathbf{b}$ so as to maximize r_{uv}. Of course, α must still be chosen according to Equation 4 in order to reach the minimum given in Equation 6.

Since a scalar multiple does not affect the correlation between two variables, it is no restriction to require that

$$\sigma_v^2 = q^2 \tag{11}$$

where q is some prechosen value (q will cancel out of later computations). We also introduce the matrix $M_{\mathbf{XX}}$ (which contains the variance and covariance of the elements of X) and the column vector $M_{u\mathbf{X}}$ (which contains the covariances of u with the elements of X). Our job is then to maximize

$$r_{uv} = \frac{1}{\sigma_u \sigma_v} \operatorname{cov}(u, v) \tag{12}$$

$$= \frac{1}{\sigma_u \sigma_v} M'_{u\mathbf{X}} \mathbf{b} \tag{13}$$

Subject to

$$\sigma_v^2 = q^2 \tag{14}$$

We form the Lagrangian

$$\phi = \frac{1}{\sigma_u \sigma_v} M'_{u\mathbf{X}} \mathbf{b} - \kappa(\mathbf{b}' M_{\mathbf{XX}} \mathbf{b} - q^2) \tag{15}$$

Partial differentiation and substitution yield

$$\frac{1}{\sigma_u q} M'_{u\mathbf{X}} - 2\kappa \mathbf{b}' M_{\mathbf{XX}} \tag{16}$$

$$\mathbf{b} = \frac{\sigma_v}{r_{uv} \sigma_u} M_{XX}^{-1} M_{uX} \tag{17}$$

Substituting for α from Equation 4, the optimal choice for the coefficient vector **c** is seen to be

$$\mathbf{c} = M_{\mathbf{XX}}^{-1} M_{u\mathbf{X}} \tag{18}$$

When the usual estimates for these parameters are used, this is identical to the traditional multiple linear regression result. Thus, we see that multiple linear regression is intimately related to simple correlation results.

Canonical Correlation

Canonical correlation generalizes the concept of simple correlation just one step further. Now both u and v are considered to be linear combinations of other variables, and the coefficients of both are to be chosen so as to maximize the correlation.

We let

$$u = \mathbf{a}'\mathbf{Y} \tag{19}$$

$$v = \mathbf{b}'\mathbf{X} \tag{20}$$

where **Y** and **X** are both random vectors (not necessarily of the same dimensionality). In addition to $M_{\mathbf{XX}}$, we now have $M_{\mathbf{YY}}$ (the variance-covariance matrix of the elements of **Y**), and $M_{\mathbf{XY}} = M'_{\mathbf{XY}}$ (the matrix of covariances between the elements of **X** and those of **Y**). We are to pick both a and b so as to maximize

$$r_{uv} = \frac{1}{\sigma_u \sigma_v} \operatorname{cov}(u, v) \tag{21}$$

$$= \frac{1}{\sigma_u \sigma_v} \mathbf{a}' M_{\mathbf{YX}} \mathbf{b} \tag{22}$$

Since a scalar multiple does not affect correlation, it is no restriction to normalize both **a** and **b** so that

$$\sigma_u = p \tag{23}$$

$$\sigma_v = q \tag{24}$$

where p and q are some prechosen constants (which will eventually cancel from our analysis)

We form the Lagrangian

$$\phi = \frac{1}{\sigma_u \sigma_v} \mathbf{a}' M_{\mathbf{YX}} \mathbf{b} - \gamma(\mathbf{a}' M_{\mathbf{YY}} \mathbf{a} - p^2) - \kappa(\mathbf{b}' M_{\mathbf{XX}} \mathbf{b} - q^2) \tag{25}$$

Taking partials and substituting for σ_u and σ_v, we get

$$\frac{1}{pq} M_{\mathbf{YX}} \mathbf{b} - 2\gamma M_{\mathbf{YY}} \mathbf{a} = 0 \tag{26}$$

$$\frac{1}{pq} M_{\mathbf{XY}} \mathbf{a} - 2\kappa M_{\mathbf{XX}} \mathbf{b} = 0 \tag{27}$$

It is easy to verify that the solution to these equations is

$$(M_{\mathbf{YY}}^{-1} M_{\mathbf{YX}} M_{\mathbf{XX}}^{-1} M_{\mathbf{XY}} - \lambda_i I)\mathbf{a} = 0 \tag{28}$$

$$\mathbf{b} = \frac{q}{2p^3\gamma} M_{\mathbf{XX}}^{-1} M_{\mathbf{XY}} \mathbf{a} \tag{29}$$

$$2\kappa q^2 = 2\gamma p^2 = \sqrt{\lambda_i} \tag{30}$$

Thus, the problem becomes one of extracting eigenvalues λ_i and eigenvectors **a**.* But, unlike the case for multiple linear regression, there may be more than one solution. In fact, it can be shown that the number of different solutions will be the minimum of the number of dimensions to **Y** and **X**. Each solution is orthogonal to all others. Moreover, it may be readily verified that

$$r_{uv} = \sqrt{\lambda_i} \tag{31}$$

* The eigenvalues of a matrix A are those values of λ for which a vector **a** exists such that $A\mathbf{a} = \lambda\mathbf{a}$ .The vector **a** is called the eigenvector associated with λ. In general, there are as many distinct solutions as the rank of A. Techniques exist for finding the eigenvalues of any matrix.

Thus, maximal correlation is obtained by using the eigenvector corresponding to the largest eigenvalue. But additional orthogonal correlations can be constructed using the other values.

One of the useful applications of canonical correlations is as an exploratory tool. It provides the ability to seek relationships between a number of variables without having to select one as the *dependent variable*. Moreover, it provides more than one relationship, which is useful in the initial stages of a study. If one correlation appears spurious, others may be selected for which theoretical justification can be found.

Discriminant Analysis

Discriminant analysis is primarily concerned with problems of taxonomy. However, as we shall later show, it bears a surprisingly close theoretical resemblance to the techniques discussed so far, both in manner of derivation and a more direct relationship discussed at the end of the section. The theory of discriminant analysis is very simple; it can be an extraordinarily powerful tool, yet it is rarely employed in operation research studies.

The point of discriminant analysis is to predict, from a set of attribute ratings (or scores), to which of several groups an individual belongs. It is used widely by psychologists to predict career choice from a battery of tests scores. We have recently employed it to predict preference levels for automobiles from ratings of their various attributes (styling, performance, cost, etc.). It is characteristic of discriminant analysis that the *dependent variable* (that to be predicted) is not a numeric quantity but a statement about group membership.

Mathematically, we have a random vector of scores $\mathbf{X}$ with some distribution which is presumed to be different for every group. Let us call $\mathbf{U}_i$ the mean of $\mathbf{X}$ in group i, and the variance-covariance matrix of $\mathbf{X}$ in group i M_i. We shall denote the fraction of the total population in group i by p_i.

In general, $\mathbf{X}$ will have many components (we used 36 automotive attributes to predict preference), and this makes group patterns difficult to detect. Discriminant analysis proceeds by making a linear transformation

$$z = \mathbf{a}'\mathbf{X} \tag{32}$$

mapping the attribute vector into a scalar. The coefficient vector $\mathbf{a}$ is to be chosen in such a way as to separate the groups as widely as possible.

Let us define the matrix T to be the unconditional variance-covariance matrix of $\mathbf{X}$ (expected deviations around the grand population mean). We also define

$$
\begin{aligned}
W &= \sum_i p_i M_i \\
\bar{z}_i &= \text{mean of } z \text{ in group } i \\
\sigma_i^2 &= \text{variance of } z \text{ in group } i \\
\bar{\bar{z}} &= \text{grand population mean of } z.
\end{aligned}
\tag{33}
$$

The groups i and j will be effectively separated if $|\bar{z}_i - \bar{z}_j|$ is large relative to the variance of z in each group. But we must construct a criterion which makes this separation for *all* pairs of groups. In this aggregate objective function, it is reasonable to weight the degree of separation of group i by the frequency with which group i occurs so that commonly occurring groups are most distinguishable. To accomplish this, discriminant analysis maximizes

$$\lambda = \frac{\sum_i p_i(\bar{z}_i - \bar{\bar{z}})^2}{\sum_i p_i \sigma_i^2} \tag{34}$$

If z is normal (which is likely by the central limit theorem since it is a sum of random variables) with equal variance in each group, λ coincides with the most powerful test for equality of group means. But in any case, λ provides a reasonable criterion of separation.

It is easy to show* that

$$\frac{1}{1+\lambda} = \frac{\mathbf{a}'W\mathbf{a}}{\mathbf{a}'T\mathbf{a}} \tag{35}$$

This ratio is minimized when λ is maximized. However, Equation 35 also admits direct interpretation. The numerator is simply the average variance of z within a single group. If the groups cluster tightly relative to the separation between groups, this numerator should be small relative to the total variance (denominator). The object is now to choose the coefficient vector $\mathbf{a}$ so as to minimize the quantity in Equation 35.

$$
\begin{aligned}
{}^{*}\ \mathbf{a}'T\mathbf{a} &= E[z - \bar{\bar{z}})^2] \\
&= \sum_i p_i E[(z - \bar{z}_i)^2 \mid z \text{ from group } i) + \sum_i p_i(\bar{z}_i - \bar{\bar{z}})^2 \\
&= \mathbf{a}'W\mathbf{a} + \sum_i p_i(\bar{z}_i - \bar{\bar{z}})^2
\end{aligned}
$$

From this the result follows directly.

To do this we can employ a trick. We define a new matrix

$$B = T - W \tag{36}$$

so that the objective function becomes

$$\frac{\mathbf{a}'W\mathbf{a}}{\mathbf{a}'B\mathbf{a} + \mathbf{a}'W\mathbf{a}} \tag{37}$$

We maximize this subject to the restriction that

$$\mathbf{a}'W\mathbf{a} = q \tag{38}$$

where q is some prechosen number (which, as before, will later cancel out of the calculations). Subject to this restriction it is sufficient to maximize $\mathbf{a}'B\mathbf{a}$. Accordingly, we form the Lagrangian

$$\phi = \mathbf{a}'B\mathbf{a} - \lambda(\mathbf{a}'W\mathbf{a} - q) \tag{39}$$

Partial differential leads to the result that $\mathbf{a}$ must satisfy

$$(W^{-1}B - \lambda I)\mathbf{a} = 0 \tag{40}$$

Substitution of Equation 40 into Equation 35 shows that the Lagrange multiplier λ of Equation 40 is identical to the criterion variable defined in Equation 34, and that this is the eigenvalue of the matrix $W^{-1}B$. The value q cancels in Equation 35 and is irrelevant to the remainder of the analysis.

Once again, there may be more than one eigenvalue, and hence more than one coefficient vector $\mathbf{a}$. In general, there will be one fewer eigenvalue than there are groups. The eigenvector corresponding to the largest eigenvalue gives the best separation, from Equation 34. However, as with canonical correlation, use can be made of all the eigenvectors.

Another frequently used measure of separation (the Wilks *lambda criterion*) is given by

$$C = \frac{|W|}{|T|} \tag{41}$$

However, Wilks also shows that this measure is related to discriminant analysis in that

$$C = \frac{1}{\Pi(1 + \lambda_i)} \tag{42}$$

where the λ_i are the set of eigenvalues obtained from Equation 40.

Relationship of Discriminant Analysis to Regression Analysis

In discriminant analysis the dependent variable is typically non-numeric. However, it is often the practice to assign numeric values to such variables in order to perform a regression. For example, survey respondents may be divided into three groups on the basis of their answer to the question:

"I buy brand X (often, sometimes, never)."

The manufacturer may wish to *explain* this behavior on the basis of certain demographic characteristics. One approach would be to assign numerical equivalents to each answer, *e.g.*,

$$\text{Often: } y = 10$$
$$\text{Sometimes: } y = 6$$
$$\text{Never: } y = 1$$

and then to run a regression between y and a vector **X**. The problem is that the choice of a scale for y will affect the results. If the *optimal* scale is chosen, the regression procedure is equivalent to discriminant analysis.

To see this, we let y_i be the value assigned to group i, and let p_i be the *a priori* probability of group i membership. Then the criterion function for regression analysis, Equation 3, can be written

$$\phi = \frac{E[(y - \mathbf{a}'\mathbf{X})^2]}{E[(y - \bar{y})^2]} = \frac{\sum_i p_i E[(y_i - \mathbf{a}'\mathbf{X})^2 \mid \text{group } i]}{E[(y - \bar{y})^2]} \tag{44}$$

(Note that once the group is known, y is no longer a random variable, and can be replaced by its value y_i.)

If the y_i are fixed, we can take partials with respect to **a** alone, and normal regression results are obtained. But, if the y_i can be selected, we must take partials with respect to them as well. It is clear that a minimum is attained when

$$y_i = E[\mathbf{a}'\mathbf{X} \mid \text{group } i], \tag{45}$$

so that

$$\phi = \frac{\mathbf{a}'W\mathbf{a}}{\mathbf{a}'T\mathbf{a}} \tag{46}$$

This is the same objective function as for discriminant analysis. Thus, in a sense, a discriminant analysis identical to the *optimal* regression analysis in that it determines the scale for the dependent variable which minimizes ϕ. Equation 6 and 35 relate the criterion parameters of the two approaches:

$$r_{uv}^2 = \frac{\lambda}{1+\lambda} \tag{47}$$

Bayesian Classification

Another technique for taxonomy is Bayesian classification. Once again, we consider a vector **X**, having a different distribution in each group. Let $f_i(\mathbf{X})$ denote its distribution in group i, and p_i denote the *a priori* probability of membership in group i. After we observe **X**, we have more information about group membership, and the posterior probabilities are given by

$$P(\text{group } i \mid \mathbf{X}) = \frac{p_i f_i(\mathbf{X})}{\sum_i p_i f_i(\mathbf{X})} \tag{48}$$

A reasonable classification rule is to classify each individual in that group for which his posterior membership probability is highest. This will result in a minimal number of misclassifications. Since the denominator is the same for each group, the rule is to assign an individual to that group i for which $p_i f_i(\mathbf{X})$ is maximal, or briefly

$$\max_i p_i f_i(\mathbf{X}) \tag{49}$$

An important special case of this technique arises when **X** has a multinormal distribution with mean $\mathbf{U}_i$ and variance-covariance matrix M_i in group i. In this case, Equation 49 becomes

$$\max_i p_i |M_i|^{-1/2} \exp\left[-\frac{1}{2}(\mathbf{X}-\mathbf{U}_i)' M_i^{-1}(\mathbf{X}-\mathbf{U}_i)\right] \tag{50}$$

or, more simply,

$$\min_i (\mathbf{X}-\mathbf{U}_i)' M_i^{-1}(\mathbf{X}-\mathbf{U}_i) + \log|M_i| - 2\log p_i \tag{51}$$

The first term of Equation 51 is simply the Euclidean distance from **X** to $\mathbf{U}_i$ measured in standard deviation units. Even if **X** is not normal, this is a reasonable classification criterion. The *form* of the next two terms depends on the normal assumption, but even in a nonnormal case they have at least the right sort of properties. If **X** has a great deal of dispersion around its mean, in group i, then **X** becomes correspondingly less compelling evidence for classification in that group. On the other hand, if group i is very likely *a priori*, its posterior likelihood is increased. In short, reasonable classification results can be expected from Equation 51 even if **X** does not have a multinormal distribution.

A final special case occurs when M_i is equal in all groups. In this case, the second term of Equation 51 can be ignored since it appears for all i. Expanding the first term of Equation 51 we also find an $\mathbf{X}'M_i^{-1}\mathbf{X}$ which is the same for all i. Thus, Equation 51 reduces in this case to

$$\max_i \mathbf{U}_i'M_i^{-1}\mathbf{X} + 2 \log p_i - \mathbf{U}_i'M_i^{-1}\mathbf{U}_i \qquad (52)$$

which is linear in **X**.

Equation 52 is often confused in the literature with discriminant analysis. The reason for this confusion is not too hard to understand. Equation 52 gives a linear transformation of **X** which is *best* for purposes of group distinction. But discriminant analysis does the same thing. Despite this similarity of intent, however, the two techniques are very different because the mathematical statements of the objectives are not the same. Equation 52 gives a linear transformation of **X** which maximizes the posterior probability of correct classification based on certain normality assumptions. Equation 40 (discriminant analysis) gives a different linear transformation which maximizes the ratio of within-group variance to total variance under less restrictive assumptions.

As the names of the techniques imply, discriminant analysis is useful when the problem is to understand the extent and nature of group differences. Bayesian classification is more naturally related to the objective of predicting the group membership of new individuals.

There is one very common practice against which the reader should be warned. Bayesian classification is computationally cumbersome if the dimensionality of **X** is large. Discriminant analysis makes a set of linear transformations which reduce the number of dimensions to one less than the number of groups. Because the discriminant functions (loosely speaking) *preserve the separation between groups*, it is an appealing idea to perform a discriminant analysis first, and then to do the Bayesian classification in the reduced discriminant space.* However, this is another case of being misled into thinking similarity of intent implies

* Cooley and Lohnes suggest this practice.

similarity of mathematical results. Indeed, it is easy to construct examples to show that Bayesian classification in the discriminant space is *not* equivalent to Bayesian classification in the original space, and in fact is significantly less powerful.

However, in the special case where the variance-covariance matrix is the same in every group, there *is* a different linear transformation of **X** into a space with the same dimensionality as the discriminant space in which Bayesian classification can be performed. To see this, note that Equation 51 implies that we must look at differences of the form:

$$(\mathbf{X} - \mathbf{U}_i)'M^{-1}(\mathbf{X} - \mathbf{U}_i) - (\mathbf{X} - \mathbf{U}_j)'M^{-1}(\mathbf{X} - \mathbf{U}_j) + 2\log\left(\frac{p_j}{p_i}\right) \quad (53)$$

Except for the constant term involving p_j and p_i, this can be rewritten:

$$2\left[\mathbf{X} - \frac{1}{2}(\mathbf{U}_i + \mathbf{U}_j)\right]' M^{-1}(\mathbf{U}_j - \mathbf{U}_i) \quad (54)$$

If **X** is p-dimensional and there are m groups, we shall show that there is a transformation

$$\mathbf{Y} = B_1\mathbf{X} \quad (55)$$

such that **Y** has only $m - 1$ dimensions, but such that the reduced space is isomorphic to the original space with respect to the classification statistic; that is, such that Equation 54 is invariant under the transformation.

Let the row vectors $\mathbf{b}_1, \ldots, \mathbf{b}_{m-1}$ be an orthogonal basis for the $m - 1$ dimensional space spanned by the differences

$$\mathbf{U}_m - \mathbf{U}_i \qquad i = 1, 2, \ldots, m - 1 \quad (56)$$

Let the vectors $\mathbf{b}_m, \mathbf{b}_{m+1}, \ldots, \mathbf{b}_p$ complete an orthogonal basis for E_p. Finally define

$$B_1 = \begin{pmatrix} \mathbf{b}_1 \\ \vdots \\ \mathbf{b}_{m-1} \end{pmatrix} M^{-1} \quad (57)$$

$$B_2 = \begin{pmatrix} \mathbf{b}_m \\ \vdots \\ \mathbf{b}_{m-1} \end{pmatrix} \quad (58)$$

$$\mathbf{Y} = B_1\mathbf{X} \quad (59)$$

$$\mathbf{Z} = B_2\mathbf{X} \quad (60)$$

Then for any $\mathbf{X}_0$, $\mathbf{Y}_0$, $\mathbf{Z}_0$

$$\mathbf{X}_0'M^{-1}(\mathbf{U}_j - \mathbf{U}_i) = (\mathbf{Y}_0'\mathbf{Z}_0')\left[\begin{pmatrix} B_1 \\ B_2 \end{pmatrix} M(B_1'B_2')\right]^{-1}\begin{pmatrix} \overline{\mathbf{Y}}_j - \overline{\mathbf{Y}}_i \\ \overline{\mathbf{Z}}_j - \overline{\mathbf{Z}}_i \end{pmatrix} \quad (61)$$

We now make several observations about Equation 61,

$$B_1MB_2' = \begin{pmatrix} \mathbf{b}_1 \\ \vdots \\ \mathbf{b}_{m-1} \end{pmatrix}(\mathbf{b}_m \ldots \mathbf{b}_p) \quad (62)$$

$$= \mathbf{0}$$

$$\mathbf{Z}_j - \mathbf{Z}_i = B_2(\mathbf{U}_j - \mathbf{U}_i) \quad (63)$$

$$= \mathbf{0}$$

since the rows of B_2 are orthogonal to the vectors spanning the space of $\mathbf{U}_j - \mathbf{U}_i$.

If we let

$$Q = B_1MB_1' \quad (64)$$

then Equation 61 simplifies to

$$\mathbf{X}_0'M^{-1}(\mathbf{U}_j - \mathbf{U}_i) = \mathbf{Y}_0'Q^{-1}(\overline{\mathbf{Y}}_j - \overline{\mathbf{Y}}_i) \quad (65)$$

In particular, if we let

$$\mathbf{X}_0 = 2\left[\mathbf{X} - \frac{1}{2}(\mathbf{U}_j - \mathbf{U}_i)\right] \quad (66)$$

then

$$\mathbf{Y}_0 = 2\left[\mathbf{Y} - \frac{1}{2}(\overline{\mathbf{Y}}_j - \overline{\mathbf{Y}}_i)\right] \quad (67)$$

and in this form Equation 65 tells us that the statistic in Equation 54 is indeed invariant under the transformation B_1. Thus, classification in **Y**-space is equivalent to (although computationally easier than) classification in **X**-space.

It is not necessary to compute the vectors $\mathbf{b}_i$ explicitly. The vectors of B_2 are introduced only to show that they drop out of the analysis. The only critical aspect of the vectors in B_1 is that they be orthogonal to the vectors of B_2. This is true of the differences $\mathbf{U}_m - \mathbf{U}_i$ as well as of the $\mathbf{b}_i$. Thus, we can more simply define the elements of **Y** by

$$y_i = (\mathbf{U}_m - \mathbf{U}_i)'M^{-1}\mathbf{X} \qquad i = 1, 2, \ldots, m-1 \quad (68)$$

$$Q = \text{cov}(\mathbf{Y}) \quad (69)$$

One case of special interest is that of two groups. In this case, **Y** is a scalar, and the mapping is

$$y = (\mathbf{U}_2 - \mathbf{U}_1)' M^{-1} \mathbf{X} \tag{70}$$

It is not difficult to verify that this is identical to the discriminant analysis mapping given by Equation 40. That is, in the very special case of two groups with equal variance-covariance matrices the discriminant space is isomorphic to the original space with respect to Bayesian classification.

Errors of Classification

While the methods developed thus far guarantee the best classification results, they do not tell us how good that is. We now develop a measure of classification power. The measure we shall use is the probability of correct classification.

Equations 53 and 54 tell us to assign to group i whenever

$$\left[\mathbf{X} - \frac{1}{2}(\mathbf{U}_i + \mathbf{U}_j)\right]' M^{-1}(\mathbf{U}_j - \mathbf{U}_i) \leq \log\left(\frac{p_i}{p_j}\right) \text{ for all } j \tag{71}$$

Let

$$s_{ij} = \left[\mathbf{X} - \frac{1}{2}(\mathbf{U}_i + \mathbf{U}_j)\right]' M^{-1}(\mathbf{U}_j - \mathbf{U}_i) \tag{72}$$

Then s_{ij} is normally distributed. On the assumption that **X** was in fact drawn from group i, we get

$$E(s_{ij} \mid \text{group } i) = \frac{1}{2}(\mathbf{U}_i - \mathbf{U}_j)' M^{-1}(\mathbf{U}_j - \mathbf{U}_i) \tag{73}$$

$$\operatorname{cov}(s_{ij} s_{ik} \mid \text{group } i) = (\mathbf{U}_j - \mathbf{U}_i)' M^{-1}(\mathbf{U}_k - \mathbf{U}_i) \tag{74}$$

Thus, the probability of correctly classifying an individual from group i is simply the probability that a joint normal vector with distribution given by Equations 73 and 74 lies in the region defined by Equation 71.

Principal Components Analysis

Principal components analysis is similar to the techniques discussed so far in that it is a data reduction technique designed to give the investigator better insight than the raw data provides.

In many types of studies the variables of greatest interest are those with high variance. On the one hand, we may need to single them out for further study so that precise estimates can be made. On the other hand, if we are trying to distinguish between individuals or groups, variables which remain roughly constant are simply uninteresting. Principle components analysis simply selects an orthogonal basis for the observation space such that the variance along the individual axes is as great as possible.

To be specific, suppose

$$\mathbf{Z} = \mathbf{a}'\mathbf{X} \tag{75}$$

Then

$$\text{var}(\mathbf{Z}) = \mathbf{a}'M\mathbf{a} \tag{76}$$

Once again, we maximize this subject to an arbitrary constraint, say $\mathbf{a}'\mathbf{a} = 1$. (This affects only the length, not the orientation, of the basis vectors.) Partial derivatives of the Lagrangian yield

$$(M - \lambda I)\mathbf{a} = 0 \tag{77}$$

Thus, the eigenvectors of M are all solutions, and they form a basis for the test space (since M is nonsingular, it will have as many eigenvectors as $\mathbf{X}$ has components). Equation 77 also implies

$$\text{var}(\mathbf{Z}) = \lambda, \tag{78}$$

so the variance is greatest along the axis corresponding to the largest eigenvalue. Those axes with very small eigenvalues may be considered (for some problems) uninteresting variables and not studied further.

Examination of Equation 77 indicates that the results are sensitive to the units in which each component of $\mathbf{X}$ is measured. If we change one component from inches to feet, we change its variance and the orientation of the resultant basis. For this reason it is common to consider the $\mathbf{X}$ vector normalized so that each component has unit variance. This need not be done explicitly, but rather M can be replaced by the matrix of correlations R in Equation 77. We shall adopt this practice in the following discussion.

Factor Analysis

Factor analysis is closely related to principal components analysis. The basic model underlying factor analysis asserts that the observation vector $\mathbf{X}$ (consisting of p components) can be *understood* in terms of a smaller vector $\mathbf{Z}$ (consisting of $m < p$ components) except for a small error. The model is linear, and asserts

$$DX = AZ + E \tag{79}$$

where E is a vector of errors with independent components, and D is a diagonal matrix with entries $1/\sigma_x$ (so that, as above, each component of X is normalized by its standard deviation). In factor analysis we presume that we know the variance of the components of E. As in principal components analysis, the objective is to maximize the variance of the first component of Z, then find a second orthogonal component with maximal variance, etc.

Let R be the variance-covariance matrix of DX (the correlation matrix of X), S be the variance-covariance matrix of E (diagonal), and Q the variance-covariance matrix for Z. We define

$$R^* = R - S \tag{80}$$

which is equal to R everywhere but on the diagonal. Then

$$R^* = AQA' \tag{81}$$

As above, the eigenvalues of R^* maximize the variance of the components of Z, subject to a constraint such as $A'A = I$, and we have

$$R^*A = A\Lambda \tag{82}$$

$$Q = \Lambda \tag{83}$$

where Λ is the diagonal matrix of eigenvalues of R^*.

To this point we have said nothing about how the dimensionality of Z (*i.e.*, m) is to be chosen. However, Equation 83 makes it clear that m must be chosen equal to the rank of R^*. Note that, if E has no variance, $R^* = R$ and principle components analysis results.

A great deal of literature on factor analysis has developed, and most of it concerns two problems: *interpretation of factors* and *determination of communalities*. These problems both relate to the appropriateness of certain applications of factor analysis, and are quite divorced from its mathematical theory.

Interpretation of factors involves giving a subjective physical interpretation to the basis vectors (columns of A). An example will serve to illustrate the problem. Say that the vector X consists of scores on three tests: a general IQ test, a reading comprehension test, and a spacial apperception test. The resultant Z vector is two-dimensional, and the matrix A is given by

$$A = \begin{pmatrix} 0.141 & 0.554 \\ 0.990 & -0.055 \\ -0.028 & 0.831 \end{pmatrix} \tag{84}$$

A reasonable interpretation might be that the first component of **Z** *represents* verbal ability, and that the second component *represents* mathematical ability.

The score on the general IQ test depends on both abilities to some extent, the score on the reading comprehension depends primarily on the first, and the score on the spacial apperception test depends primarily on the second. This is an example of *interpretation of factors.*

There are no hard and fast rules for this process, although it can give the researcher deeper insight into the structure of his data. One device which is often employed is to rotate the basis axes until a position is found in which each component is either very large or very small (this feature is present in Equation 84). Interpretation is often made easier by such a process. One common rotational scheme is the *varimax* rotation which attempt to maximize the variance of the elements in each column.

However, there is a drawback to rotational schemes. Suppose R^* has rank 20, but that 12 eigenvalues are so small as to be considered negligible. In such a case, the investigator might choose to work in an 8-dimensional space. However, if he finds the results difficult to interpret and applies some orthogonal rotations (*e.g.*, varimax) to the 20 basis vectors, he can no longer restrict attention to 8 dimensions, nor will the factors be independent in the rotated space. It is a common error in using factor analysis to assume that, because the first 8 factors were significant in the factor analysis solution, the first 8 will be significant in the rotated space as well. To see the fallacy of such reasoning, realize that the interchange of the most and least significant factors is itself an orthogonal rotation. On the other hand, it would be permissible to decide, on the basis of the factor analysis solution, that 8 factors were sufficient, and then to rotate only those 8 factors (not the whole space) in order to better understand that 8-dimensional space. Of course, here too the relative importance of one axis to another is lost in the rotation.

Determination of communalities involves the assignment of variance to the elements of **E**. If factor analysis is being used as a purely exploratory tool, the investigator will not have much knowledge about the structure in Equation 79. How then is he to know the variance of the components of **E**? In fact, this question is the most serious drawback to using factor analysis without prior hypotheses about structure.

A number of *ad hoc* rules have been developed to handle this problem. The most common is to derive R^* from R by replacing each diagonal element by the maximal element in its row or column. Our own feeling is that all such rules are simply so much whistling in the dark, and that unless one has a model which allows assignment of these variances, factor analysis is not the appropriate tool. However, in such cases, principal components analysis provides a reasonable alternative. It is true that **Z** will contain as many components as **X**, but in general, several of these will involve negligible variance. *Interpretation* of the most significant factors can be done, just as in factor analysis. Armed with such data and insight the investigator can probably construct a better model than with a poorly understood *ad hoc* derivation of R^*.

Statistical Tests of Significance

In all of the techniques discussed, the coefficient vectors are given in terms of the parameters of the distribution. The usual case will be that these parameters are not known, and must be estimated from the data. This means that we have only estimates of the coefficients. A large body of literature exists for measuring the precision of these estimates, and many of the commercially available computer routines give confidence intervals or statistical tests of significance as part of their output. However, the standard statistical measures and tests can often be misleading.

As one example, consider a multiple linear regression run with 10 variables and 1000 observations. Suppose the resultant r_{uv}^2 (Equation 6) is 0.05. A standard F-test shows the value to be *statistically significant* at almost any level. Yet it is doubtful that such a poor fit would be *operationally significant* for most applications. However, care must be taken in drawing even such a conclusion as this last.

For example, suppose a regression is run with a dependent variable which assumes only two possible values, and yields an $r_{uv}^2 = 0.20$. First glance would suggest that, even though statistically significant, this value is not operationally significant. We shall show that this is not necessarily the case. The argument leading to Equation 45 implies that, if the dependent variable can assume only two values, a regression is equivalent to a discriminant analysis.† Equation 47 asserts that $\lambda = 0.25$. Finally, this can be related to Bayesian classification. For two groups, Equations 73 and 74 become

$$E(s_{12} \mid \text{group } 1) = -\frac{1}{2}\frac{\lambda}{p_1 p_2} \tag{85}$$

† The form of the regression Equation 1 shows that a linear transformation of the dependent variable does not affect the regression. But, with only two values, any scale may be achieved by a linear transformation. Hence, any regression is equivalent to the *optimal* regression, or discriminant analysis.

$$\mathrm{var}(s_{12} \mid \text{group 1}) = \frac{\lambda}{p_1 p_2} \tag{86}$$

Thus, the probability of correctly classifying a group 1 member is simply

$$\Phi\left(\frac{\log(p_1/p_2) + \frac{1}{2}\lambda/(p_1 p_2)}{\sqrt{\lambda/(p_1 p_2)}}\right) \tag{87}$$

where Φ denotes the cumulative distribution of the unit normal. If $p_1 = p_2$, the probability of correct classification is 70%, a 40% improvement over guessing. The magnitude of this improvement is not obvious from the value of r_{uv}^2.

These examples are given only to suggest that the *standard* statistical tests cannot be relied upon without question. A measure must be developed which is appropriate to the particular needs of the application.

Conclusion

The techniques discussed in this paper have been known for some time, and computer program packages for carrying out most of the computations are widely available. The obstacle to their wider use has probably been incomplete knowledge on the part of workers and a (justified) reluctance to blindly rely on the output of "canned" programs. In fact, effective use of these techniques requires an insight into their objectives, capabilities, and limitations which goes beyond that provided by an F-test of significance. It is our hope that this paper can provide the basis for such working insight.

General References

T. W. Anderson, *An Introduction to Multivariate Statistical Analysis*, John Wiley & Sons, Inc., New York (1958).

W. W. Cooley and P. R. Lohnes, *Mltivariate Procedures for the Behavioral Sciences*, John Wiley & Sons, Inc., New York (1962).

P. R. Lohnes, "Test Space and Discriminant Space Classification Models and Related Significance Tests," *Educational and Psychological Measurement*, Volume 21, 1961.

C. R. Rao, *Advanced Statistical Methods in Biometric Research*, John Wiley & Sons, Inc., New York (1952).

S. S. Wilks, *Mathematical Statistics*, John Wiley & Sons, Inc., New York (1962).

Index

Made in the USA
Monee, IL
05 November 2026

67814340R00174